Polyester-Based Biocomposites

Polyester-Based Biocomposites highlights the performance of polyester-based biocomposites reinforced with various natural fibers extracted from leaf, stem, fruit bunch, grass and wood material. It also addresses the characteristics of polyester-based biocomposites reinforced with rice husk fillers and various nanoparticles.

This book explores the widespread applications of fiber-reinforced polymer composites in the aerospace sector, automotive parts, construction and building materials, sports equipment and household appliances. Investigating the advantages of natural fibers, such as superior damping characteristics, low density, biodegradability, abundant availability at low cost and non-abrasive to tooling, this book discusses what makes them a cost-effective alternative reinforcement material for composites in certain applications.

This book serves as a useful reference for researchers, graduate students and engineers in the field of polymer composites.

Polyester-Based Biocomposites

Edited by
Senthilkumar Krishnasamy, Chandrasekar
Muthukumar, Senthil Muthu Kumar Thiagamani,
and Suchart Siengchin

CRC Press
Taylor & Francis Group
Boca Raton London New York

CRC Press is an imprint of the
Taylor & Francis Group, an informa business

Designed cover image: © Shutterstock

First edition published 2024
by CRC Press
4 Park Square, Milton Park, Abingdon, Oxon, OX14 4RN

and by CRC Press
6000 Broken Sound Parkway NW, Suite 300, Boca Raton, FL 33487-2742

CRC Press is an imprint of Taylor & Francis Group, LLC

© 2024 The right of Senthilkumar Krishnasamy, Chandrasekar Muthukumar, Senthil Muthu Kumar Thiagamani and Suchart Siengchin to be identified as the authors of the editorial material, and of the authors for their individual chapters, has been asserted in accordance with sections 77 and 78 of the Copyright, Designs and Patents Act 1988.

British Library Cataloguing-in-Publication Data
A catalogue record for this book is available from the British Library

Library of Congress Cataloging-in-Publication Data
Names: Krishnasamy, Senthilkumar, editor. | Muthukumar, Chandrasekar, editor. |
Thiagamani, Senthil Muthu Kumar, editor. | Siengchin, Suchart, editor.
Title: Polyester-based biocomposites / edited by Senthilkumar Krishnasamy,
Chandrasekar Muthukumar, Senthil Muthu Kumar Thiagamani and Suchart Siengchin.
Description: First edition. | Abingdon, Oxon; Boca Raton, FL : CRC Press, [2023] |
Includes bibliographical references and index. |
Identifiers: LCCN 2022061059 | ISBN 9781032220468 (hbk) | ISBN 9781032220475 (pbk) |
ISBN 9781003270980 (ebk)
Subjects: LCSH: Polymeric composites. | Fibrous composites—Materials. |
Polyesters. | Thermoplastic composites.
Classification: LCC TA455.P58 P65 2023 | DDC 620.1/923—dc23/eng/20230118
LC record available at https://lccn.loc.gov/2022061059

ISBN: 978-1-032-22046-8 (hbk)
ISBN: 978-1-032-22047-5 (pbk)
ISBN: 978-1-003-27098-0 (ebk)

DOI: 10.1201/9781003270980

Typeset in Times
by codeMantra

Dedication

Dedicated to my family members

S. Krishnasamy

K. Malliga

K. Rajesh

Contents

Preface

Fiber-reinforced polymer composites with thermoset matrices such as epoxy, polyester, vinyl ester and bismaleimide resins have been used in widespread applications in the aerospace sector, automotive parts, construction and building materials, sports equipment and household appliances. Natural fiber-based polyester composites have inferior thermal, physico-chemical and mechanical properties over the conventional synthetic fiber-reinforced polyester composites and is considered a potential substitute in applications requiring moderate strength and stiffness. Natural fibers have advantages such as superior damping characteristics, low density, biodegradability, abundant availability at low cost and non-abrasive to tooling, which makes them a cost-effective alternative reinforcement material for composites in certain applications.

This book highlights the performance of polyester-based biocomposites reinforced with various natural fibers and is organized in the following ways: Chapters 1–8 focus on the characterization of polyester-based biocomposites reinforced with natural fibers extracted from leaf, stem, fruit bunch, grass and wood material. Chapters 9–12 address the characteristics of polyester-based biocomposites reinforced with rice husk fillers, various nanoparticles and two or more natural fibers, respectively. Chapters 13–16 highlight the suitability of the polyester-based biocomposites in various applications.

Each chapter of this book has been written by experts with publications. It is our pleasure to have worked with authors who are established researchers in the field of biocomposites and we express our gratitude to the publisher and their staff members associated with this book support. The content in this book could be of help to undergraduate and postgraduate students, research scholars, academic researchers, professionals and scientists looking for fundamental knowledge on the characterization of polyester-based biocomposites, latest research trends and the suitability of such composites in various applications.

Editors

Dr. Senthilkumar Krishnasamy is an Associate Professor at the PSG Institute of Technology and Applied Research, Tamil Nadu, India. He graduated with a Bachelor's in Mechanical Engineering from Anna University, Chennai, India, in 2005. He then chose to continue his Master's in CAD/CAM from Anna University, Tirunelveli, India, in 2009. He obtained his PhD from the Department of Mechanical Engineering, Kalasalingam University in 2016. He then worked in the Department of Mechanical Engineering, Kalasalingam Academy of Research and Education (KARE), India, from 2010 (January) to 2018 (October). He completed his post-doctoral fellowship at Universiti Putra Malaysia, Serdang, Selangor, Malaysia, and KMUTNB under the research topics of "Experimental investigations on mechanical, morphological, thermal and structural properties of kenaf fibre/mat epoxy composites" and "Sisal composites and fabrication of eco-friendly hybrid green composites on tribological properties in a medium-scale application," respectively. His area of research interests includes the modification and treatment of natural fibers, nano-composites, 3D printing and hybrid-reinforced polymer composites. He has published research papers in international journals, book chapters, and conferences in the field of natural fiber composites

Dr. Chandrasekar Muthukumar is presently working as an Associate Professor at the Department of Aeronautical Engineering, Hindustan Institute of Technology & Science, Chennai, India. He graduated with a Bachelor's in Aeronautical Engineering from Kumaraguru College of Technology, Coimbatore, India. He obtained his Master's in Aerospace Engineering from Nanyang Technological University-TUM ASIA, Singapore. He earned his PhD in Aerospace Engineering from Universiti Putra Malaysia (UPM), Malaysia. His PhD was funded through a research grant from the Ministry of Education, Malaysia. During his association with the UPM, he obtained internal research fund of 16,000 and 20,000 MYR from the University. He has five years of teaching and academic research experience. His field of expertise includes fibre metal laminate (FML), natural fibers, biocomposites, aging and their characterization. His publications are based on the fabrication and characterization techniques of bio-composites, aging studies in biocomposites and creep analysis of biocomposites. He has authored and co-authored 32 research articles in SCI Journals, 24 book chapters and 5 articles in the conference proceedings. He is currently co-editing six books which are to be published by CRC Press, Wiley, Springer and Elsevier. *Natural Fiber-Reinforced Composites: Thermal Properties and Applications* has been submitted to Wiley and is under production. He is a peer reviewer for *Journal of Composite Materials, Polymer Composites, Materials Research Express* and *Journal of Natural Fibers.*

Dr. Senthil Muthu Kumar Thiagamani is working as an Associate Professor in the Department of Mechanical Engineering at Kalasalingam Academy of Research and Education (KARE), Tamil Nadu, India. He received his Diploma in Mechanical

Engineering from the Directorate of Technical Education, Tamil Nadu, India, in 2004; obtained his B E in Mechanical Engineering from Anna University, Chennai, India, in 2007 and MTech in Automotive Engineering from Vellore Institute of Technology, Vellore, India, in 2009. He earned his PhD in Mechanical Engineering (specialized in Biocomposites) from KARE in 2018. He has also completed his post-doctoral research from the Materials and Production Engineering Department at The Sirindhorn International Thai-German Graduate School of Engineering (TGGS), KMUTNB, Thailand. He started his academic career as an Assistant Professor in Mechanical Engineering at KARE in 2010. He has 11 years of teaching and research experience. He is also a visiting researcher at KMUTNB, Thailand. He is a member of international societies such as the Society of Automotive Engineers and the International Association of Advanced Materials. His research interests include biodegradable polymer composites and characterization. He has authored several articles in peer-reviewed international journals, book chapters and conference proceedings. He has also published edited books in the theme of biocomposites. He is also serving as a reviewer for various journals such as *Journal of Industrial Textiles*, *Journal of Polymers and the Environment*, *SN Applied Sciences*, *Mechanics of Composite Materials* and *International Journal of Polymer Science*.

Prof. Dr.-Ing. habil. Suchart Siengchin is President of King Mongkut's University of Technology North Bangkok (KMUTNB), Thailand. He received his Dipl.-Ing. in Mechanical Engineering from the University of Applied Sciences Giessen/Friedberg, Hessen, Germany, in 1999; MSc in Polymer Technology from the University of Applied Sciences Aalen, Baden-Wuerttemberg, Germany, in 2002; MSc in Material Science at the Erlangen-Nürnberg University, Bayern, Germany, in 2004; Doctor of Philosophy in Engineering (Dr.-Ing.) from Institute for Composite Materials, University of Kaiserslautern, Rheinland-Pfalz, Germany, in 2008 and postdoctoral research from the Kaiserslautern University and School of Materials Engineering, Purdue University, USA. In 2016 he received his habilitation degree at the Chemnitz University in Sachen, Germany. He worked as a Lecturer for the Production and Material Engineering Department at The Sirindhorn International Thai-German Graduate School of Engineering (TGGS), KMUTNB. He has been a full Professor at KMUTNB and became the President of KMUTNB. He won the Outstanding Researcher Award in 2010, 2012 and 2013 at KMUTNB. His research interests include polymer processing and composite material. He is the editor-in-chief of *KMUTNB International Journal of Applied Science and Technology* and the author of more than 150 peer-reviewed journal articles. He has presented his research in more than 39 international and national conferences in materials science and engineering topics.

Contributors

Shiji Mathew Abraham
School of Biosciences
Mahatma Gandhi University
Kottayam, India

MRM Asyraf
Engineering Design Research Group
(EDRG), School of Mechanical
Engineering, Faculty of Engineering
Universiti Teknologi Malaysia
Johor, Malaysia
And
Centre for Advanced Composite
Materials (CACM)
Universiti Teknologi Malaysia
Johor, Malaysia

D. Balaji
Department of Mechanical Engineering
KPR Institute of Engineering and
Technology
Coimbatore, Tamil Nadu, India

Viswanathan Balaji
Department of Biotechnology (DDE)
Madurai Kamaraj University
Madurai, India

Abhishek Biswal R
Department of Food Process
Engineering, Postharvest
Technology Research Lab, School of
Bioengineering
SRM Institute of Science and
Technology
Chennai, India

Govindaraju Boopalakrishnan
Department of Biotechnology (DDE)
Madurai Kamaraj University
Madurai, India

Aniruddha Chatterjee
Centre for Advanced Materials
Research and Technology,
Plastic and Polymer Engineering
Department
Maharashtra Institute of Technology
Aurangabad, India

Jorge Alberto Vieira Costa
College of Chemistry and Food
Engineering
Federal University of Rio Grande
Rio Grande, Brazil

Bruna da Silva Vaz
College of Chemistry and Food
Engineering
Federal University of Rio Grande
Rio Grande, Brazil

Ana Claudia Araujo de Almeida
College of Chemistry and Food
Engineering
Federal University of Rio Grande
Rio Grande, Brazil

Michele Greque de Morais
College of Chemistry and Food
Engineering
Federal University of Rio Grande
Rio Grande, Brazil

Aravind Dhandapani
University Science Instrumentation
Centre
Madurai Kamaraj University
Madurai, India

Dharini V
Department of Food Process
 Engineering, Postharvest
 Technology Research Lab, School of
 Bioengineering
SRM Institute of Science and
 Technology
Chennai, India

Divyashree J S
Department of Food Process
 Engineering, Postharvest
 Technology Research Lab, School of
 Bioengineering
SRM Institute of Science and
 Technology
Chennai, India

Anthony Chidi Ezika
Department of Chemical, Metallurgical
 and Materials Engineering,
 Faculty of Engineering and the
 Built Environment, Institute of
 NanoEngineering Research (INER),
Tshwane University of Technology
Pretoria, South Africa
And
Department of Polymer and Textile
 Engineering, Faculty of Engineering
Nnamdi Azikiwe University
Awka, Nigeria
And
DST-CSIR National Center for
 Nanostructured Industrial Research
Pretoria, South Africa

Nadeem Faisal
Mechanical Engineering
Central Institute of Petrochemicals
 Engineering & Technology (CIPET),
 CIPET: CSTS – Balasore
Balasore, India

Martin Emeka Ibenta
Department of Polymer and Textile
 Engineering, Faculty of Engineering
Nnamdi Azikiwe University
Awka, Nigeria

R A Ilyas
School of Chemical and Energy
 Engineering, Faculty of Engineering
Universiti Teknologi Malaysia
Johor, Malaysia

Siby Isac
Department of Food Process
 Engineering, Postharvest Research
 Lab, School of Bioengineering
SRM Institute of Science and
 Technology
Chennai, India

Ishwariya. A
Department of Food Process
 Engineering, Postharvest Research
 Lab, School of Bioengineering
SRM Institute of Science and
 Technology
Chennai, India

Nurjahirah Janudin
Research Centre for Chemical Defence
Universiti Pertahanan Nasional
 Malaysia
Kuala Lumpur, Malaysia

Aswathy Jayakumar
Materials and Production Engineering,
 The Sirindhorn International
 Thai-German Graduate School of
 Engineering (TGGS)
King Mongkut's University of
 Technology
Bangkok, Thailand
And
Department of Food and Nutrition
BioNanocomposite Research Center
Kyung Hee University
Seoul, Republic of Korea

Mohd Azwan Jenol
Department of Bioprocess Technology,
 Faculty of Biotechnology and
 Biomolecular Sciences
Universiti Putra Malaysia
Selangor, Malaysia

Subramanian Jeyanthi
School of Mechanical Engineering
Vellore Institute of Technology
Chennai, India

Shravanti Joshi
Functional Materials Laboratory,
    Department of Mechanical
    Engineering
Marathwada Institute of Technology
Aurangabad, India

Jasila Karayil
Department of Applied Science
Government Engineering College,
    West Hill
Kerala, India

Dheeraj Kumar
National Institute of Technology
    Durgapur
West Bengal, India

Selvaraj Vinoth Kumar
School of Mechanical Engineering
Vellore Institute of Technology
Chennai, India

Apurba Layek
National Institute of Technology
    Durgapur
West Bengal, India

Jaewoo Lee
Department of Polymer-Nano Science
    and Technology
Jeonbuk National University
Jeonju-si, Korea
And
Department of Bionanotechnology and
    Bioconvergence Engineering
Jeonbuk National University
Jeonju-si, Korea

Amol Manoj
School of Mechanical Engineering
Vellore Institute of Technology
Chennai, India

**Muhammad Syukri Mohamad
Misenan**
Department of Chemistry, College of
    Arts and Science
Yildiz Technical University
Istanbul, Turkey

Ranjan Kumar Mitra
National Institute of Technology
    Durgapur
West Bengal, India

Juliana Botelho Moreira
College of Chemistry and Food
    Engineering
Federal University of Rio Grande
Rio Grande, Brazil

Rohani Mustapha
School of Ocean Engineering
    Technology and Informatics
Universiti Malaysia
Kuala Terengganu, Malaysia

Siti Noor Hidayah Mustapha
Industrial Sciences and Technology
Universiti Malaysia Pahang
Lebuhraya Tun Razak
Gambang, Malaysia

Syed Umar Faruq Syed Najmuddin
Faculty of Science and Natural
    Resources
Universiti Malaysia Sabah
Sabah, Malaysia

Durgam Muralidharan Nivedhitha
School of Mechanical Engineering
Vellore Institute of Technology
Chennai, India

Mohd Nor Faiz Norrrahim
Research Centre for Chemical Defence
Universiti Pertahanan Nasional
    Malaysia
Kuala Lumpur, Malaysia

Norizan Mohd Nurazzi
Department of Chemistry and Biology,
  Centre for Defence Foundation
  Studies
Universiti Sains Malaysia
Kuala Lumpur, Malaysia

Victor Ugochukwu Okpechi
Department of Polymer and Textile
  Engineering, Faculty of Engineering
Nnamdi Azikiwe University
Awka, Nigeria

Henry Chukwuka Oyeoka
Department of Polymer and Textile
  Engineering, Faculty of Engineering
Nnamdi Azikiwe University
Awka, Nigeria

Jyotishkumar Parameswaranpillai
Department of Science
Alliance University
Bengaluru, India

Rajkumar Praveen
Department of Biotechnology (DDE),
Madurai Kamaraj University
Madurai, India

Sabarish Radoor
Materials and Production Engineering,
  The Sirindhorn International
  Thai-German Graduate School of
  Engineering (TGGS)
King Mongkut's University of
  Technology
Bangkok, Thailand
And
Department of Polymer-Nano Science
  and Technology
Jeonbuk National University
Baekje-daero
Jeonju, Republic of Korea

L. Rajeshkumar
Department of Mechanical Engineering
KPR Institute of Engineering and
  Technology
Coimbatore, Tamil Nadu, India

M. Ramesh
Department of Mechanical Engineering
KIT-Kalaignarkarunanidhi Institute of
  Technology
Coimbatore, India

Suprakas Sinha Ray
DST-CSIR National Center for
  Nanostructured Industrial Research
Pretoria, South Africa
And
Department of Applied Chemistry
University of Johannesburg
Johannesburg, South Africa

Michael Johni Rexliene
Department of Biotechnology (DDE)
Madurai Kamaraj University
Madurai, India

Emmanuel Rotimi Sadiku
Department of Chemical, Metallurgical
  and Materials Engineering,
  Faculty of Engineering and the
  Built Environment, Institute of
  NanoEngineering Research (INER)
Tshwane University of Technology
Pretoria, South Africa

Carlo Santulli
Geology Division, School of Science
  and Technology
Università di Camerino
Camerino, Italy

Ajinkya Satdive
Centre for Advanced Materials
  Research and Technology,
  Plastic and Polymer Engineering
  Department
Maharashtra Institute of Technology
Aurangabad, India

Periyar Selvam Sellamuthu
Department of Food Process
  Engineering, Postharvest
  Technology Research Lab, School of
  Bioengineering
SRM Institute of Science and
  Technology
Chennai, India

Nur Sharmila Sharip
Research and Development Department
Nextgreen Pulp & Paper Sdn Bhd
Kuala Lumpur, Malaysia

Siti Shazra Shazleen
Department of Bioprocess Technology,
  Faculty of Biotechnology and
  Biomolecular Sciences
Universiti Putra Malaysia
Selangor, Malaysia

Jyothi Mannekote Shivanna
Department of Chemistry
AMC Engineering College
Bengaluru, India

Suchart Siengchin
Materials and Production Engineering,
  The Sirindhorn International
  Thai-German Graduate School of
  Engineering (TGGS)
King Mongkut's University of
  Technology
Bangkok, Thailand
And
Institute of Plant and Wood Chemistry
Technische Universität Dresden
Tharandt, Germany

Jayavel Sridhar
Department of Biotechnology (DDE)
Madurai Kamaraj University
Madurai, India

Vishnupriya Subramaniyan
Department of Biotechnology, School of
  Bioengineering
SRM Institute of Science and
  Technology
Chennai, India

Saurabh Tayde
Centre for Advanced Materials
  Research and Technology,
  Plastic and Polymer Engineering
  Department
Maharashtra Institute of Technology
Aurangabad, India

Ana Luiza Machado Terra
College of Chemistry and Food
  Engineering
Federal University of Rio Grande
Rio Grande, Brazil

Bhagwan Toksha
Centre for Advanced Materials
  Research and Technology,
  Plastic and Polymer Engineering
  Department
Maharashtra Institute of Technology
Aurangabad, India

1 Polyester Resins and Their Use as Matrix Material in Polymer Composites
An Overview

Saurabh Tayde, Ajinkya Satdive, and Bhagwan Toksha
Maharashtra Institute of Technology

Aniruddha Chatterjee
Maharashtra Institute of Technology

CONTENTS

DOI: 10.1201/9781003270980-1

1.1 INTRODUCTION

The material used in the construction of any object has been one of the most important aspects of human civilization. The various construction materials that human beings have used include metals, ceramics, composites, polymers and other materials. The usage and choice of material vary depending on various factors and have changed over time. A drastic change in the materials used in building artefacts occurred in the twentieth century with the era of polymers. Towards the end of the nineteenth century, synthetic polymers came up with their actual mass-scale usage started in the 1940s with the use of polymer-based composites involved in radar technology banking upon glass fibres/polyesters (Edwards, 1998; Skrifvars et al., 1998; Khan et al., 2002). The properties of products such as being strong, inexpensive, lightweight, versatile and having good barrier properties to carbon dioxide and oxygen make them comfortable, convenient and safe for our everyday lives (Senthilkumar et al., 2015; Alothman et al., 2020; Shahroze et al., 2021). The current worldwide plastic consumption is about 370 million tons (PlasticsEurope, 2020), with annual growth in consumption of about 5% (Gupta et al., 2022). Polyolefins such as low- and high-density polyethylene and polypropylene (PP), along with polyesters such as polyethylene terephthalate (PET) are the most consumed thermoplastics (Rabnawaz et al., 2017; Yeung et al., 2021). The class of materials under resin encompasses polystyrene, polyethylene, polyvinylchloride, PP, expandable polystyrene, etc. There are reports evaluating the market size of polymer resin as high as 100 billion dollars in 2018 and expected to register a Compound Annual Growth Rate (CAGR) of over 7% by the year 2030 (Polymer Resin Market Size, Share, Growth | Report, 2030).

The resin matrix is an important composite with two classes such as thermoplastics and thermosets. Further classification could be made such as saturated polyester and unsaturated polyester. The larger class of these materials is the unsaturated polyester which is about 75% of all polyester resins consumed. The class of resin which retains its solid state at ambient temperature and loses its solid state at elevated temperature is called thermoplastic. The inability of long-chain polymers not to cross-link chemically makes them unable to cure permanently, which rules out them from the structural application (Senthilkumar et al., 2021; Nasimudeen et al., 2021; Thomas et al., 2021; Senthilkumar et al., 2022). On the other side, a thermosetting resin will cure permanently by irreversible cross-linking at higher temperatures, making them very promising candidates for structural applications. The thermoset unsaturated polyester possesses a brittleness property which reduces its usability in many applications. An appropriate additive that disrupts the cross-linked chain structure is used to improve the toughness of this material (Abral et al., 2020). The processing temperature and the amount of the catalyst concentration are critical to controlling the rate of polymerization kinetics. The temperature or the concentration of the catalyst varies with the rate of reaction. The reaction leading to the release of gases and causing voids in products is undesirable (Sheng et al., 2020). Due to the post-curing operation by heat treatment, the cross-linking increases subsequently elevating the glass transition temperature of the material. Post-cure temperature is an important factor governing the extent of cross-linking. The optimized temperatures for such processes will ultimately impact the economic and environmental aspects (Silva et al., 2020).

There are several routes explored for manufacturing reinforced polyester matrix materials including hand lay-up, filament winding, sheet moulding, prepreg moulding, resin transfer moulding, vacuum-assisted moulding and pultrusion (Kuppusamy et al., 2020). The preparation of polyester resin using zinc as a catalyst and phosphorous as a stabilizer is available in the literature (Lee et al., 2016). The type and amount of reactant and catalyst, monomers and curing temperature all are important in deciding the properties as well as the recovery of the polyester resin (An et al., 2022; Zhang et al., 2021). Polyester resin materials are the topic of this chapter. Polymer matrix composites (PMCs) are introduced first and their manufacturing and reinforcing technologies are described in detail. Finally, the properties of such composites are discussed, along with their future developments.

The overall growth in the usage of polymer materials in the automotive, electric vehicles, packaging, PVC bottles and containers, food and beverage industry is mainly driving the demand for the polymer market on account of the replacement of various metal and alloy parts (Thiagamani et al., 2022; Chandrasekar et al., 2022). The material is lightweight with mechanical strength and cost-effectiveness acts as the main cause behind such a huge demand in the market. The higher growth in lightweight electric vehicle manufacturing all over the world and favourable government initiatives are going to further flourish the growth of the market. The basic required characteristics of polymer resins are optical, thermal and mechanical properties. These properties are commonly improved by curing and cross-linking with reinforcement material. There is a need for designing the polyester matrix in such a way that it carries the load satisfactorily and avoids fatigue failure (He et al., 2017; Towo and Ansell, 2008).

1.2 POLYMER MATRIX COMPOSITES

PMCs are preferred nowadays for various advanced and performance applications because of their excellent short-term and long-term mechanical and other properties as compared to the neat polymer at no additional cost (Miskolczi, 2013). PMCs find various advantageous applications in aviation, marine, automobile, structural, etc. In the 1940s, the first PMC based on polyester and glass fibre was used for radar application (Khan et al., 2002; Tabatabai et al., 2018). PMC is a synthetic combination of resin and reinforcement present physically at a microscopic level. The polymer matrix is a continuous phase which binds the reinforcement, transfers the load to the reinforcement and protects the reinforcement from the environment, whereas the reinforcement is a discontinuous phase and imparts extra strength to the composite and enhances mechanical properties. Reinforcement distributes the transfer load provided by the matrix (Chandrasekar et al., 2020). PMCs possess a matrix of either thermoplastic or thermosetting polymer reinforced by a reinforcing agent or fibre. They exhibit excellent strength and toughness with corrosion resistance properties. They have low cost, good mechanical properties and ease of fabrication. Also, they have good strength to weight ratio and specific properties because of the low density of its constituents (Askeland and Fulay 2003; Matthews and Rawlings, 1999).

PMCs exhibit higher specific strength, short-term and long-term mechanical properties such as tensile, elongation, flexural, impact, stiffness, creep, stress relaxation and fatigue characteristics, which enable the structural design to be more versatile. The polymer matrix can be thermoplastic or thermosetting. Typical polymer matrices include unsaturated polyester, epoxy, phenolics, polyurethane, polyamide, acrylonitrile butadiene styrene, PP, etc. Due to the low density of the constituents, polymer composites often show excellent strength to weight ratio. Thermoplastic resins are chain polymers and soften on increasing the temperature and degrade thermally due to the breaking of the C–C bond at elevated temperatures. Thermosetting resins are cross-linked when heated. They get hardened when heated, and at the critical temperature, they start degrading. One of the major disadvantages of the thermosetting polymer matrix is the non-recyclability mechanically or chemically (Soutis, 2005; Manfredi et al., 2006).

1.3 POLYESTER MATRIX

1.3.1 Classification of Unsaturated Polyester Resin (UPR)

Unsaturated polyester is classified into ortho, iso, bisphenol-A fumarates and vinyl ester resins.

  i. Ortho-UPRs: These are prepared using ortho-phthalic acids such as phthalic anhydride, maleic anhydride and glycerol. It has restricted thermal and chemical resistance properties. Ortho-UPR based on 1,2-propylene glycol lowers its crystallinity and improves its compatibility.
 ii. Iso-UPRs: These are prepared using isophthalic acid, maleic anhydride and glycol. Due to its high viscosity, it requires a larger quantity of solvent. It has good chemical and thermal resistance with physical properties.
iii. Bisphenol-A fumarates UPR: These are prepared using bisphenol-A and fumaric acid. Due to the high raw material cost, this UPR is costly, but they possess excellent chemical, corrosion and thermal resistance properties as compared to iso- and ortho-UPR. Bisphenol-A is the main backbone chain rigidity to the final structure.
 iv. Vinyl ester resins: These are prepared using acrylic acid or methacrylic acid and epoxy resin. It has high viscosity and requires a larger quantity of styrene as a diluent to reduce the viscosity. It has good performance ability and corrosion resistance properties (Athawale and Pandit, 2019).

1.3.2 PREPARATION OF UNSATURATED POLYESTER RESINS

In general, polyester resins can be synthesized from a dibasic organic acid and dihydric alcohol. They are classified as saturated polyesters, such as PET, and unsaturated polyesters (Deopura et al., 2008). Unsaturated polyesters contribute to about 75% consumption of all polyester resins used worldwide. It is advantageous due to its fast curing, good chemical resistance, electrical and mechanical properties, and comparatively low price to that of epoxy resin for various application areas. Due to its valuable properties, polyester has achieved continuous growth in its consumption worldwide at a rate of 4%–5% gain annually. It is one of the highest consumed polymer matrices for composite applications. Unsaturated polyester is the mostly used thermosetting resin for PMC. It has a low-melting-point, viscous and low-molecular-weight resin. It contains ester linkages in the main backbone chain. The presence of C=C makes it unsaturated and acts as a functional site for cross-linking. The C=C bond does not play a part in the polymerization reaction but takes part during cross-linking.

UPRs are condensation polymers formed by the reaction of polyols (also known as polyhydric alcohols), organic compounds with multiple alcohols or hydroxy functional groups, with saturated or unsaturated dibasic acids at 180°C–220°C. The commonly used polyol is ethylene glycol; the acids used are phthalic and maleic acid. Water is eliminated as a by-product in this esterification reaction, which drives towards completion. Styrene is used as a diluent in UPRs to reduce its viscosity. The solidification of liquid resin takes place by a cross-linking mechanism with the help of curing agents. The curing mechanism includes forming a new free radical at unsaturated sites (C=C), which leads to propagation in a chain reaction to another unsaturated site in the adjacent molecule and linking it in the process. The addition of catalyst decomposes into free radicals which help to complete the cross-linking of unsaturated polyester.

Typically, organic peroxides such as benzoyl peroxide (BPO) or methyl ethyl ketone peroxide (MEKP) are used as a hardener to cure UPRs (Worzakowska, 2009). Benzoyl peroxide or MEKP was used at 1%–2% of the total amount of UPR to cure it exothermally. The curing agents should be used in a calculated amount only, as excessive quantity can lead to high exothermicity and impart brittleness in the material. The brittleness may cause fractures in the cured resin even at low load conditions (Hashemi et al., 2018) (see Figure 1.1).

To obtain the desired cross-linking in UPRs, styrene is used as a comonomer and contributes to the final properties of the resin. Curing causes a shrinking in the volume of UPR by 2%–5%. Generally, UPR to styrene ratio is maintained at 0.5 to obtain the best properties. Styrene acts as a diluent for UPR and avoids its gelling while storage. The solvent increases the curing rate after the addition of the initiator and accelerator. Initiator and accelerator are added with a concentration of 1%–2%. Cobalt octoate is the most widely used accelerator for curing UPR. The initial curing rate is slow and then increases rapidly to obtain the final cured structure (Figure 1.2).

The post-curing at higher temperatures may be required to increase the glass transition temperature by ensuring complete cross-linking and ultimately enhancing the mechanical properties of polyesters. Several factors such as molecular weight,

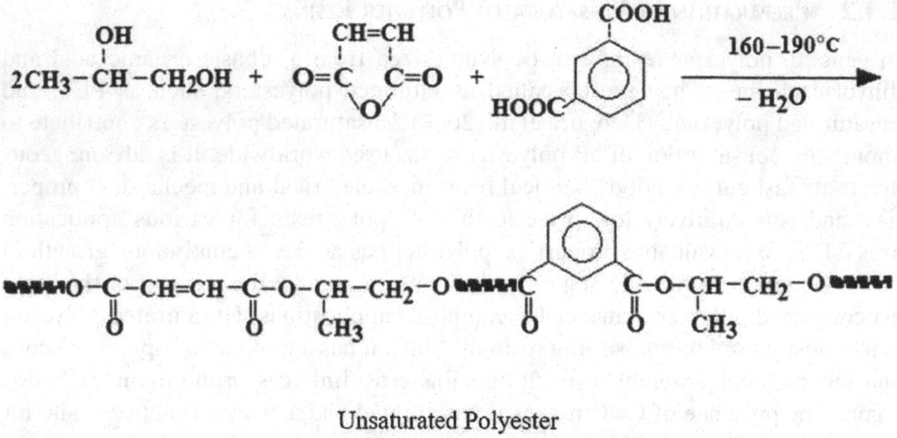

Unsaturated Polyester

FIGURE 1.1   Synthesis of unsaturated polyester (Panda and Behera, 2019).

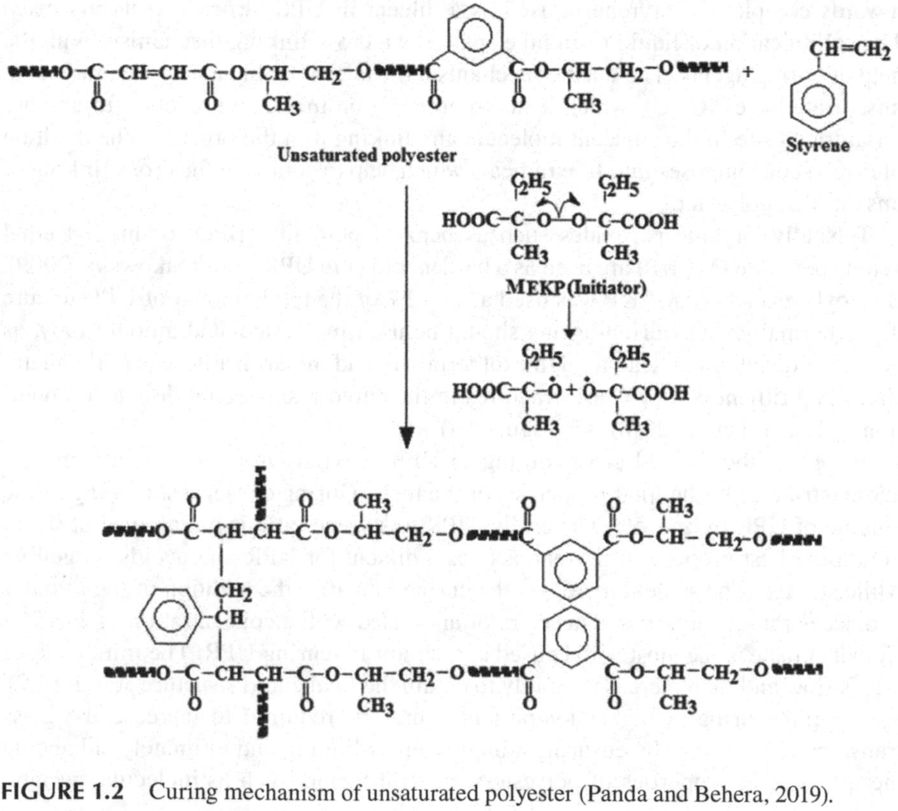

FIGURE 1.2   Curing mechanism of unsaturated polyester (Panda and Behera, 2019).

initiator and accelerator concentration affect the final properties of cured resins. The higher the molecular weight, the higher will be the point of unsaturation and the higher will be the strength of cured resins.

Various additives such as UV stabilizers, antioxidants and colourants can be added to UPRs to enhance properties. UPR is highly flammable and generates toxic gases after combustion. Halogenated compounds can be added to improve their fire resistance properties. Fillers such as nanoclay, carbon nanotubes, magnesium hydroxide, nickel hydroxide, molybdenum disulphide, antimony oxide, gypsum particles and graphene can be used to modify the properties of unsaturated polyesters (Hai et al., 2018).

The composition of monomers affects the final properties of polyester resins such as viscosity, reactivity and curing time, and mechanical and thermal properties. The advantageous properties of UPRs are good dimensional stability, chemical and thermal resistance, ease of processing, corrosion resistance and low cost (Mustafa et al., 2014; Kanimozhi et al., 2014). In a nutshell, UPRs have good structural and performance abilities at an affordable cost. Table 1.1 depicts the important mechanical properties of unsaturated polyester.

1.4 MANUFACTURING OF POLYESTER COMPOSITES

The hand lay-up and compression moulding methods are the most common ones for fabricating thermosetting composite matrices. Spray-up, pultrusion, resin transfer moulding and vacuum bag moulding are some of the other methods available. These moulding processes are divided into two categories: open mould and closed mould.

1.4.1 OPEN MOULDING

1.4.1.1 Spray-Up Method

The method of manufacture inspired the procedure's name. A spray cannon is utilized to place both the fibre and the resin into the mould in this production procedure. Rolling of the pre-layed resin and fibre preform is carried out to eliminate the bubbles entrapped within the resin since the likelihood of air bubble trapping is quite high. The type of resin used determines the curing process, which is usually done at room temperature. This fabrication technique is limited to composites.

TABLE 1.1

Mechanical Properties of Unsaturated Polyester

S. No.	Properties	Value
1	Tensile	33.3 MPa
2	Elongation at break	2.25%
3	Toughness	0.36 MPa
4	Impact strength	1.21×10^{-2} J/mm^2
5	Hardness Shore, D	88
6	Water absorption	0.21%

Source: Athawale and Pandit (2019).

The composites' characteristics, particularly the tensile strength and elastic modulus, are anisotropic in this process. The skill of the operators determines the product's quality (Kikuchi et al., 2014).

1.4.1.2 Hand Lay-Up

It is the most basic and well-known composites fabrication technique (Kozlowski et al., 2010). In comparison to other moulding or manufacturing methods, it requires a low-cost setup.

As a result, the initial investment in this process is minimal. The materials are fed into an open mould and left there until they reach the desired stable shape. The amount of time required is determined by the polymer's characteristics. At room temperature, 24 hours is required for UPR manufacturing (Yahaya et al., 2015). This is a common fabrication method for thermosetting matrices. It is utilized to make a variety of car components (Kim et al., 2014).

1.4.2 CLOSED MOULDING

1.4.2.1 Pultrusion

Pultrusion is a method of composite synthesis that produces a constant cross-section and production rate (Wiedmer and Manolesos, 2016). In this technique, fibres are drawn through a heated die. As the resin travels through the die, constant pressure is applied, allowing it to melt and impregnate the fibre reinforcement. The method of preheating, the temperature of the die, and the speed of the fibre passage all affect the quality of the produced goods.

1.4.2.2 Resin Transfer Moulding

In resin transfer moulding, the resin is transferred into a closed die under pressure. This method can create intricate parts with a smooth finish all over (Wulfsberg et al., 2014). A technique with a short cycle time in a temperature-controlled system can achieve any required thickness. A vacuum inside the mould cavity controls the resin flow. Resin transfer mouldings can be used to successfully produce hybrid composites (Pearce et al., 2000).

1.4.2.3 Compression Moulding

Compression moulding is used to make UPR-based composites. The reinforcement and thermosetting resin are sandwiched within two matching die/moulds (one male, one female), and then the pressure (up to 2000 psi) and temperature (up to 200°C) are applied, followed by the system being left for a particular amount of time to cure the material. This approach also includes room-temperature curing. The type of curing is determined by the resin's nature, as well as the sample's shape, size and thickness. The moulded product is taken from the mould after curing. For easy composite release, a releasing agent is applied before the preform is placed in the die (Wulfsberg et al., 2014).

1.4.2.4 Vacuum Bag Moulding

A variant of the hand lay-up method, vacuum bag moulding is a process that provides mechanical pressure to a laminate throughout its cure cycle. This procedure is important for removing trapped air within the laminates, which helps to keep the fibres in the appropriate orientation during curing and reduces the moisture content to improve the fibre to resin ratio within the composite (Levy and Hubert, 2019).

1.5 REINFORCEMENT USED IN UNSATURATED POLYESTER RESIN

It is one of the fundamental constituents of the PMC. Reinforcement is a discontinuous phase and adds extra strength to the composite. Reinforcement distributes the transfer load provided by the matrix. Different types of organic and inorganic reinforcing agents are used to prepare PMC. The PMC can be classified based on the reinforcement as shown in Figure 1.3.

1.5.1 NATURAL FIBRE REINFORCEMENT

Natural fibre-reinforced unsaturated polyester exhibits comparable mechanical properties with synthetic reinforcing agents. The main drawback of the natural fibre is that it is hydrophilic due to the presence of cellulosic content. Also, they need some chemical treatment to improve the interfacial properties with the polymer matrix and obtain enhanced mechanical properties (Da Silva et al., 2008). Different natural fibres such as banana, jute, hemp, sisal and coir are used effectively with UPR to obtain its composite.

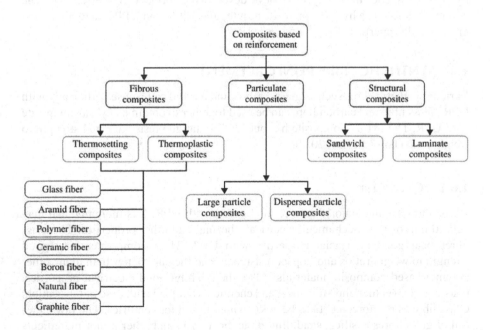

FIGURE 1.3 Classification of composites based on reinforcement (Devaraju and Alagar, 2019).

1.5.2 BANANA FIBRE

It is a natural biodegradable reinforcing agent. It contains cellulose and hemicellulose. It requires chemical treatment to remove lignin and unwanted components which can affect the adhesion with the unsaturated polyester. Banana fibres can be used up to 40% volume fraction of UPRs. Composites based on banana fibres can be fabricated using the hand lay-up technique and resin transfer moulding. They exhibits good mechanical properties (Pothan et al., 2016).

1.5.3 COIR FIBRE

It is obtained from coconut shells. It is also biodegradable and possesses cellulosic content of about 36%. It also contains 41%–45% of lignin and therefore requires alkali treatment to remove it. Due to the high microfibril angle, it has limited use and application with UPR. However, it requires surface treatment to achieve mechanical properties (Biswas et al., 2019).

1.5.4 JUTE FIBRE

It is biodegradable and contains cellulose. It can be reinforced up to 60% with UPR. Jute fibre-based composites are fabricated by hand lay-up or compression moulding. Jute along with glass fibres can be used to prepare a hybrid composite of thermosetting polyester. Jute fibres also require alkali treatment to improve the wettability with UPR and ultimately to achieve developed interfacial properties with the polymer matrix. The hybrid composite of jute, glass fibre and UPR can exhibit good mechanical properties.

1.6 SYNTHETIC FIBRE REINFORCEMENT

Various synthetic fibres such as glass, carbon and aramid can be used efficiently with UPR. Glass fibre and carbon fibre can be used together to obtain its hybrid composite with UPR. The hybrid composite has outstanding fatigue resistance and strength to weight ratio (Jain and Lee, 2012).

1.6.1 GLASS FIBRE

Glass fibres are the majorly used reinforcement with UPR. It is inorganic and may help to improve its mechanical, electrical, thermal and other properties. The glass fibres bear good interfacial properties with UPR. Their composites have good strength to weight ratios and impact resistance, and they are better than many other polymer-based composite materials. Also, they exhibit good electrical properties, resistance to weather, moisture, heat and chemicals coupled with ease of fabrication. Glass fibre as reinforcement has been known and used for centuries. The raw material of glass fibres is silica, sand, limestone, boric acid and other minor ingredients (e.g. clay, coal and fluorspar); the temperature of this melt is different for each glass type but is generally about 2300°F (1260°C). It has different types such as A, E, S,

C and D. A-glass fibre is a soda-lime silicate containing fibre possessing excellent strength, durability and good electrical resistance properties. E-glass fibres have a calcium aluminoborosilicate composition with maximum alkali content of 2%. They are used when strength and high electrical resistivity are required and are the most common fibre glasses used in composites. E-glass is inexpensive in comparison to the other glasses used for composites. S-glass has better retention of properties at elevated temperatures. S-glass is mostly used in advanced composites where strength is a premium. C-glass has a soda-lime borosilicate composition and because of its chemical stability, it is used in corrosive environments. D-glass fibres contain borosilicate having low dielectric constant and are used for electrical applications (Tables 1.2 and 1.3).

TABLE 1.2
Chemical Composition of Different Glass Fibres

Composition	A-glass	C-glass	D-glass	E-glass	S2-glass
SiO_2%	63–72	64–68	7275	52–56	64–66
Al_2O_3%	0–6	3–5	0–1	1216	24–25
B_2O_3%	0–6	4–6	21–24	5–10	-
CaO%	6–10	1115	01	16–25	0–0.2
MgO%	0–4	2–4	-	0–5	9.5–10
BaO%	-	0–1	-	-	-
Na_2O+K_2O%	14–16	7–10	0–4	0–2	0–0.2
TiO_2%	0–0.6	-	-	0–4	-
Fe_2O_3%	0–0.5	0–0.8	00.3	0–0.8	0–0.1

Source: Bagherpour (2012).

TABLE 1.3
Mechanical Properties of Glass Fibres

Properties	A-glass	C-glass	D-glass	E-glass	S2-glass
Density (g/cm³)	2.44	2.52	2.11–2.14	2.58	2.46
Tensile strength (MPa) at −196°C	-	5380	-	5310	8275
Tensile strength at 23°C	3310	3310	2415	3445	3445
Tensile strength at 371°C	-	-	-	2620	4445
Tensile strength at 538°C				1725	2415
Modulus of elasticity (GPa) at 23°C	68.9	68.9	51.7	72.3	86.9
Modulus of elasticity at 538°C				81.3	88.9
Elongation (%)	4.8	4.8	4.6	4.8	5.7

Source: Bagherpour (2012).

1.6.2 ARAMID FIBRES

They are obtained from aromatic polyamides. They are also known as Kevlar fibre. Kevlar fibres have different grades such as Kevlar 29, Kevlar 49 and Kevlar 149. Kevlar fibres have excellent creep and fatigue resistance. UPRs with Kevlar fibre can be used for structural applications. To improve the interfacial properties between aramid fibres and UPRs, coupling agents are used (Tuttle, 2003).

1.6.3 CARBON FIBRES

Carbon fibres are available in different shapes and sizes, i.e. short or continuous. Its structure can be crystalline or partially crystalline. The carbon atoms are bonded by a covalent bond. They possess high modulus. Commercially, carbon fibres are synthesized using polyacrylonitrile and contain 92% carbon. Carbon fibres have a tensile strength in the range of 1310–2480 MPa, whereas the modulus of elasticity varies from 230 to 620 GPa. Their density is in the range of 1.8–2.1 gm/cm^3 (Chung, 1994).

1.7 PROPERTIES OF POLYESTER COMPOSITES

1.7.1 THERMAL PROPERTIES OF POLYESTER COMPOSITES

The availability of a wide variety of raw materials along with the manufacturing techniques makes it convenient to tune the properties of the polyester composites to a greater extent. For various applications such as electrical, naval, automotive and electronics, their thermal properties are of utmost importance. Therefore, studying the thermal parameters such as their degradation behaviour, thermal stability, techniques to improve their thermal stability and factors affecting it helps in tailoring the properties of the polyester composite materials.

Several studies showed that the thermal degradation of UPRs takes place in two to three stages (Evans et al., 1996; Skinner et al., 1984; Satdive et al., 2020; Kicko-Walczak, 2004; Anderson and Freeman, 1959; Ravey, 1983). During the first stage, at around 200°C, phthalic anhydride is eliminated and is present in the resins which have ortho-phthalic moieties in their structure (Skinner et al., 1984; Kicko-Walczak, 2004; Satdive et al., 2020). Also, at elevated temperatures, phthalic anhydride continues to form (Bautista et al., 2017). Generally, Anderson and Freeman's mechanism for the decomposition of polyester is considered over the mechanism proposed by Ravey and Sivasamy (Ravey, 1983; Anderson and Freeman, 1959; Sivasamy et al., 1992). The Anderson and Freeman's mechanism states that the degradation initiates closer to the ester group via homolytic cleavage and the formation of free radicals. Later, bond rearrangement and hydrogen reaction yield phthalic anhydride and hydroxy ester molecules as shown in Figure 1.4a and b.

In the course of the second and third stages of UPR degradation, the cleavage of cross-links as well as weak links present in the linear chain takes place and subsequently linear chain scission (Evans et al., 1996; Skinner et al., 1984; Satdive et al., 2020; Kicko-Walczak, 2004; Anderson and Freeman, 1959; Ravey, 1983; Bautista et al., 2017). The derivative thermogravimetry curves depict these stages to be around 350°C and 390°C.

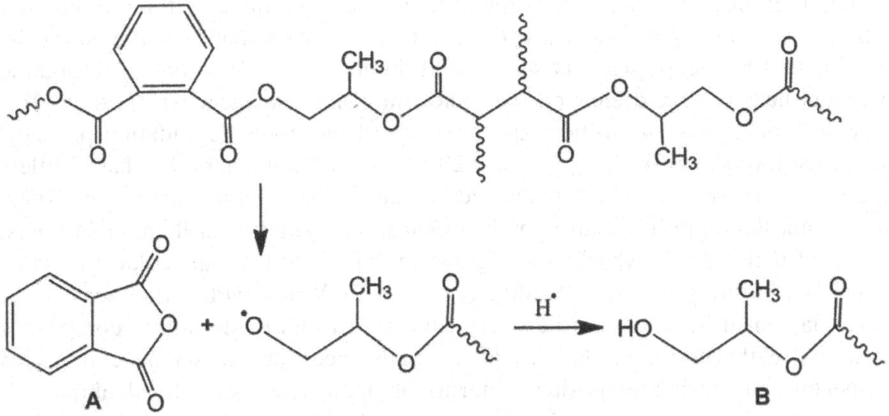

FIGURE 1.4 Anderson and Freeman's mechanism explaining the phthalic anhydride formation (Spasojevic, 2019).

Several studies show that as a result of UPR thermal degradation styrene emerges as the most abundant product while phthalic anhydride emerges as the prevalent degradation component of the polyester backbone (Evans et al., 2000). For the relevant commercial polyester thermosets, the degradation products were found to be 55%–70% for these two items. The degradation of polystyrene cross-links produces ethylbenzene, methyl styrene, benzaldehyde, benzene, 1,3-diphenyl propane, 1,2-diphenylethane/1,2-diphenylethene, toluene, styrene dimer and trimer, and phenylacetaldehyde, according to most research. These products were found in concentrations of up to 3%. Propylene glycol, maleic anhydride, acetaldehyde, acetone, allyl alcohol, benzaldehyde, etc. were discovered as polyester degradation products (Spasojevic, 2019).

The structure of prepolymer, curing procedure, amount of styrene, type and filler content along with other additives, climatic conditions, etc. all affect the thermal properties of UPR composite materials. By increasing the length of the glycol's carbon chain, UPR's thermal stability improves marginally. Straight-chain glycol-based UPRs were shown to be more stable than branched glycol-based UPRs. In glycols with β-hydrogen, the thermal stability of the resins was also reduced (propylene glycol and 1,3-butanediol). The addition of the ether group reduced thermal stability, but ether glycol resins were still more stable than branched glycols with β-hydrogen. When compared to ortho-phthalate, both the iso- and terephthalate esters demonstrated improved thermal stability. The UPR made from maleic anhydride, neopentyl glycol and isophthalic acid had the best thermal stability, implying that the lack of β-hydrogen and ortho-phthalate structures improves thermal stability (Evans et al., 1996; Hiltz, 1988; Bansal et al., 1989). Bansal et al. investigated how the curing regime affected the thermal characteristics of three UPRs (general-purpose resin, isophthalic acid-based and bisphenol-A-based). The resins were cured using three different methods: (1) 1% BPO at 120°C for half an hour, (2) 1% BPO at 85°C for 4 hours and (3) 1% MEKP at room temperature for 24 hours followed by heating at 85°C for 3 hours. The resins treated with MEKP were found to be more stable than the resins cured with BPO. Resins cured with BPO at 120°C were also more stable than those cured at 85°C.

The presence of clay fillers in the UPR matrix also impacts the thermal stability of the composite materials. George et al. showed that heat is resisted by intercalated nanoclay platelets scattered throughout the UPR matrix. Inorganic platelets help to create char on the nanocomposite's surface, which slows further decomposition due to heat and raises the barrier for degradation products to escape into the bulk (George et al., 2016). Several studies on bio-based fillers and reinforcements have been published in recent years. Natural fibre composites' significant flammability, caused by hydrocarbon polymers and flammable fibres, is one of their key drawbacks for engineering applications, particularly in aerospace and building interiors (Satdive et al., 2022). When UPR and other natural fibre (flax, sisal, jute) composites were compared to UPR glass fibre composites in terms of thermal degradation and fire resistance, the glass fibre composites outperformed the biocomposites in terms of flame resistance. Sisal fibre composites had the highest fire risk among natural fibre composites, while jute fibre composites had a short-duration fast-growing fire and flax fibre composites had a slow-growing long-duration fire (Manfredi et al., 2006).

1.7.2 Rheological Properties of Polyester Composites

Many intricate and intriguing flow parameters governed by the UPRs limit their possible application. Thus, it is critical to comprehend the link between the UP composite structure and its rheological properties. Some manufacturing techniques, such as sheet moulding compound (SMC) and bulk moulding compound (BMC), require increasing the viscosity of the input material or thickening the feed material. High viscosity improves sheet-like structure stability, eases handling before moulding, improves fibre carrying during moulding and lowers fibre segregation and polymerization shrinkage (Kia, 1993). To ensure good mixing and reinforcement wetting, the feed material's viscosity should be minimal during the early mixing process. During thickening, the compound viscosity should rapidly increase. To provide a long shelf life and dimensional consistency, it should remain constant. A higher viscosity is preferred throughout the moulding process than at the start to allow for better fibre-carrying capability. Alternatively, lower viscosity is required near the end of the mould-filling step to ensure proper filling of the mould cavity's details and corners. Group II metal oxides and hydroxides are the oldest and most commonly used thickening systems. A tiny dose of these agents could raise the viscosity of the system above 10^5 mPa s (Rodriguez, 1987). The fundamental disadvantage of alkali metal thickeners is that the thickening process is quite slow. Alkali metal thickeners are also susceptible to humidity. As a result, alternative thickening techniques have been developed previously.

Diisocyanates formed within the UPR can also be thickened the SMC or BMC paste, generating a network capable of greatly increasing paste viscosity (Gawdzik et al., 2000; Valette and Hsu, 1999; Xu et al., 1994). When compared to metal oxides or hydroxides, diisocyanates thicken much faster. The viscosity of the thickened compound is stable and reaches maximum viscosity in hours rather than days (Orgéas and Dumont, 2012). Polyurethane molecules improve the toughness of the cured SMC components as well (Xu et al., 1994).

1.7.3 MECHANICAL PROPERTIES OF POLYESTER COMPOSITES

The indentation properties of UPR-based composites are excellent. The primary quality achieved through indentation is hardness. Indentations are divided into two categories: macro/micro indentation and nanoindentation. Vickers hardness and Shore hardness are the most common indentations for polymer composites. Many parameters can be measured in the case of nanoindentation, including storage modulus, loss factor, coefficient of friction, loss factor, recovery index, creep, coefficient of friction and so on. The nanoindentation method can be used to determine the mechanical properties of thermosetting matrices and cell wall layers of natural fibre (Wang et al., 2015). Nanoindentation on the nanoscale can be used to assess the interfacial mechanical properties of multilayered natural fibre-reinforced UPR composites. The plant or natural fibre, unlike homogenous synthetic fibre, is a multi-layered structure that can be successfully subjected to nanoindentation experiments (Diederichs and Bykov, 2010).

The dispersed phase geometry has a big impact on the tensile characteristics of natural fibre-reinforced composites (Jiang et al., 2018). According to a study, jute/polyester composites offer greater strength and stiffness than wood composites and some polymers, although their mechanical properties are not comparable to those of typical composites (Weil and Levchik, 2016). The alteration of the dispersed phase to make it compatible with thermosetting matrices is another crucial factor. Chemical treatment of the matrix and the dispersed phase is possible, but in most situations, the latter is transformed by appropriate surface treatment. The tensile characteristics of thermosetting polyester resin composites are greatly improved by this modification (Andre et al., 1993). The impact qualities are primarily determined by the filler's toughness, interfacial adhesion between the filler and matrix, and filler loading (Yu et al., 2016). The impact strength of fibre/clay/UPR composites is influenced by the interaction between the constituent phases. Fibre/clay/UPR composites have a higher impact strength than fibre/UPR composites. Filler inclusion greater than 25% causes agglomeration, which reduces the ability of load transmission from matrix to reinforcement, lowering the overall impact strength (Song et al., 2008).

1.7.4 CURING PROPERTIES OF POLYESTER COMPOSITES

Depending on the initiator type used, UPRs can be cross-linked via thermal or photoinitiation. During curing, unsaturated monomers (typically styrene) function as diluent and viscosity modifiers while polyester monomers or oligomers react with one another. Curing occurs via several exothermal reaction pathways that result in the creation of a three-dimensional network and solidification, including (1) the unsaturated monomer (usually styrene) and the polyester are cross-linked intermolecularly, (2) reactions between the unsaturated monomers, (3) unsaturated polyester branching using an unsaturated monomer, and (4) unsaturated monomer homopolymerization. All of these activities have an impact on curing kinetics, but their effects on the final resin structure differ (Gonçalves et al., 2017; Davis, 2019). By connecting nearby polyester molecules in reaction 1, a macroscopic network is formed. Reaction 2 enhances cross-link density but has little effect on the creation of macroscopic networks. Reactions 3 and 4 enhance the size of the polymer coil but have minimal effect on the total macroscopic network (Gonçalves et al., 2017). Understanding curing mechanisms as a function of composition and curing circumstances is critical to accomplishing optimum curing, process control during fabrication, stresses on reinforcing components and die part surfaces, and optimum bulk characteristics in the completed part (Torre et al., 2013; Baran et al., 2014). One of the most prevalent ways of monitoring curing is in Differential Scanning Calorimeter (DSC) Rheology, however, is very beneficial because the development of viscoelastic properties while curing is related to both dynamic changes in chemical structure and the mechanical properties of the cured resin. During isothermal curing, the cross-linked polymer's viscosity increases until it forms an infinite network and gelation, or the transition from a viscous liquid to a rubbery solid occurs. Knowing the gelation time for a certain material and set of curing conditions is critical since any fibre impregnation or shaping activities must take place before gelation. Reactions continue after gelation but are severely constrained due to the high viscosity of the system until it reaches the vitrification point, at which point it becomes a stiff glass with few to no more reactions possible (Halley and Mackay, 1996).

1.8   CONCLUSIONS AND FUTURE TRENDS

UPRs are known to have less flame-retardant efficiency and compatibility with traditional flame retardants. This makes it difficult to make the synthesized matrix withstand extreme conditions such as high temperatures and radiation exposure. These materials can offer better properties than conventional structural materials in certain applications. The evolved matrix based on UPR materials has low maintenance costs and can be recycled their whole life. UPRs are the choice materials in the applications of fibres, plastics, composites, coatings, transportation industry, shipbuilding and construction engineering, among many others. The growing demands in the automotive sector have led to increasing investment. This growth along with the increasing adoption of electric vehicles warrants the need for lightweight materials. These materials need to overcome the hurdles of petroleum-derived dependency via exploring sustainable synthesizing routes. The replacement of conventional composite materials is gaining importance with the use of bio-based UPRs and other blended formulations. Newer recipes of UPRs using bio-modifiers and synthesizing biocomposites are providing promising results. Specific applications such as UPR-based products satisfying the desired mechanical properties demonstrating flame-retardant properties, UPR-based 4D printing, and enhancing mechanical properties per unit density for aerospace and automotive applications are some of the contemporary research directions. The bio-based and nanocomposite UPRs competing with the products realized from petrochemical routes in the parameters such as glass transition temperatures, tensile modulus, strength and viscosity are being reported. The dependence on the properties of cured resin as a function of thermal ageing is critical from the application point of view. The environmental concerns to reducing or eliminating health hazards associated with involved emissions are significant. Such scientific efforts will help in contributing to fuel efficiency and reducing carbon footprint. Recycling polymer waste to derive UPRs and retain their employable properties is a research avenue that needs to be explored.

REFERENCES

Abral, Hairul, Rahmat Fajrul, Melbi Mahardika, Dian Handayani, Eni Sugiarti, Ahmad Novi Muslimin, and Santi Dewi Rosanti. 2020. "Improving Impact, Tensile and Thermal Properties of Thermoset Unsaturated Polyester via Mixing with Thermoset Vinyl Ester and Methyl Methacrylate." *Polymer Testing* 81 (January): 106193. https://doi.org/10.1016/J.POLYMERTESTING.2019.106193.

Alothman, Othman Y., Mohammad Jawaid, Senthilkumar Krishnasamy, Chandrasekar Muthukumar, Basheer A. Alshammari, Hassan Fouad, Mohamed Hashem, and Suchart Siengchin. 2020. "Thermal Characterization of Date Palm/Epoxy Composites with Fillers from Different Parts of the Tree." *Journal of Materials Research and Technology* 9: 15537–15546.

An, Wenli, Xiong Lei Wang, Xuehui Liu, Gang Wu, Shimei Xu, and Yu Zhong Wang. 2022. "Chemical Recovery of Thermosetting Unsaturated Polyester Resins." *Green Chemistry* 24 (2): 701–12. https://doi.org/10.1039/D1GC03724B.

Anderson, David A., and Eli S. Freeman. 1959. "The Kinetics of the Thermal Degradation of the Synthetic Styrenated Polyester, Laminac 4116." *Journal of Applied Polymer Science* 1 (2): 192–99. https://doi.org/10.1002/APP.1959.070010210.

Andre, F., P. A. Cusack, A. W. Monk, and R. Seangprasertkij. 1993. "The Effect of Zinc Hydroxystannate and Zinc Stannate on the Fire Properties of Polyester Resins Containing Additive-Type Halogenated Flame Retardants." *Polymer Degradation and Stability* 40 (2): 267–73. https://doi.org/10.1016/0141-3910(93)90215-5.

Askeland, Donald R., and Pradeep P. Fulay. 2003. *The Science and Engineering of Materials*, 4th ed. Pacific Grove, CA: Brooks/Cole-Thomson Learning.

Athawale, Anjali A., and Jyoti A. Pandit. 2019. "Unsaturated Polyester Resins, Blends, Interpenetrating Polymer Networks, Composites, and Nanocomposites: State of the Art and New Challenges." In *Unsaturated Polyester Resins: Fundamentals, Design, Fabrication, and Applications*, pp. 1–42. Elsevier. https://doi.org/10.1016/B978-0-12-816129-6.00001-6.

Bagherpour, Salar. 2012. "Fibre Reinforced Polyester Composites." In Polyester. London: IntechOpen. https://doi.org/10.5772/48697.

Bansal, Raj K., Jagjiwan Mittal, and Prakash Singh. 1989. "Thermal Stability and Degradation Studies of Polyester Resins." *Journal of Applied Polymer Science* 37 (7): 1901–8. https://doi.org/10.1002/APP.1989.070370713.

Baran, Ismet, Remko Akkerman, and Jesper H. Hattel. 2014. "Material Characterization of a Polyester Resin System for the Pultrusion Process." *Composites Part B: Engineering* 64 (August): 194–201. https://doi.org/10.1016/J.COMPOSITESB.2014.04.030.

Bautista, Yolanda, Ana Gozalbo, Sergio Mestre, and Vicente Sanz. 2017. "Thermal Degradation Mechanism of a Thermostable Polyester Stabilized with an Open-Cage Oligomeric Silsesquioxane." *Materials* 11: 22. https://doi.org/10.3390/MA11010022.

Biswas, Bhabatosh, Nil Ratan Bandyopadhyay, and Arijit Sinha. 2019. "Mechanical and Dynamic Mechanical Properties of Unsaturated Polyester Resin-Based Composites." In *Unsaturated Polyester Resins: Fundamentals, Design, Fabrication, and Applications*, pp. 407–34. Elsevier. https://doi.org/10.1016/B978-0-12-816129-6.00016-8.

Chandrasekar, Muthukumar, Irulappasamy Siva, Senthil Muthu Kumar Thiagamani, Krishnasamy Senthilkumar, Suchart Siengchin and Nagarajan Rajini. 2020. "Influence of Fibre Inter-ply Orientation on the Mechanical and Free Vibration Properties of Banana Fibre Reinforced Polyester Composite Laminates." *Journal of Polymers and the Environment* 28: 2789–2800.

Chandrasekar, Muthukumar, Krishnasamy Senthilkumar, Mohammad Jawaid, Salman Alamery, Hassan Fouad, and Mohamad Midani. 2022. "Tensile, Thermal and Physical Properties of Washightonia Trunk Fibres/Pineapple Fibre Biophenolic Hybrid Composites." *Journal of Polymers and the Environment* 30: 4427–4434.

Chung, Deborah D. L. 1994. "Carbon-Matrix Composites." In *Carbon Fiber Composites*, pp. 145–76. https://doi.org/10.1016/B978-0-08-050073-7.50012-9.

Da Silva, Rosana Vilarim, Eve Maria Freire de Aquino, Lena Patrícia Souza Rodrigues, and Alessandro Rossano de Freitas Barros. 2008. "Development of a Hybrid Composite with Synthetic and Natural Fibers." *Matéria* 13 (1): 154–61. https://doi.org/10.1590/S1517-70762008000100019.

Davis, Virginia A. 2019. "Rheological and Curing Properties of Unsaturated Polyester Resin Nanocomposites." In *Unsaturated Polyester Resins: Fundamentals, Design, Fabrication, and Applications*, pp. 471–88. Elsevier. https://doi.org/10.1016/B978-0-12-816129-6.00018-1.

Deopura, B. L., Ramasamy Alagirusamy, Mangala Joshi, and Bhuvanesh Gupta Bhuvanesh. 2008. *Polyesters and Polyamides*. Cambridge, England: Woodhead Publishing.

Devaraju, S., and M. Alagar. 2019. "Unsaturated Polyester—Macrocomposites." In *Unsaturated Polyester Resins: Fundamentals, Design, Fabrication, and Applications*, pp. 43–66. Elsevier. https://doi.org/10.1016/B978-0-12-816129-6.00002-8.

Diederichs, Jan and Yana Bykov. 2010. "Innovative Flame-Retardants in E&E Applications Non-Halogenated Phosphorus, Inorganic and Nitrogen Flame Retardants." In Proceedings Diederichs Innovative FR. https://www.semanticscholar.org/paper/Innovative-Flame-Retardants-in-E%26E-Applications-and-Diederichs-Bykov/6c9f4a725a69a40ee70725e9ac08ad53f6e26ce2.

Edwards, K. L. 1998. "An Overview of the Technology of Fibre-Reinforced Plastics for Design Purposes." *Materials & Design* 19 (1–2): 1–10. https://doi.org/10.1016/S0261-3069(98)00007-7.

Evans, S. J., P. J. Haines, and G. A. Skinner. 1996. "The Effects of Structure on the Thermal Degradation of Polyester Resins." *Thermochimica Acta* 278 (1–2): 77–89. https://doi.org/10.1016/0040-6031(95)02851-X.

Evans, S. J., P. J. Haines, and G. A. Skinner. 2000. "Pyrolysis–Gas-Chromatographic Study of a Series of Polyester Thermosets." *Journal of Analytical and Applied Pyrolysis* 55 (1): 13–28. https://doi.org/10.1016/S0165-2370(99)00064-9.

Gawdzik, Barbara, Tadeusz Matynia, and Joanna Osypiuk. 2000. "Influence of TDI Concentration on the Properties of Unsaturated Polyester Resins." *Journal of Applied Polymer Science* 79(7): 1201–6. https://onlinelibrary.wiley.com/doi/abs/10.1002/1097-4628%2820010214%2979%3A7%3C1201%3A%3AAID-APP70%3E3.0.CO%3B2-A.

George, Manoj, George Elias Kochimoolayil, and Haridas Jayadas Narakathra. 2016. "Mechanical and Thermal Properties of Modified Kaolin Clay/Unsaturated Polyester Nanocomposites." *Journal of Applied Polymer Science* 133 (13). https://doi.org/10.1002/APP.43245.

Gonçalves, Filipa A. M. M., Ana Clotilde Fonseca, Marco Domingos, Antonio Gloria, Armacnio Coimbra Serra, and Jorge F. J. Coelho. 2017. "The Potential of Unsaturated Polyesters in Biomedicine and Tissue Engineering: Synthesis, Structure-Properties Relationships and Additive Manufacturing." *Progress in Polymer Science* 68 (May): 1–34. https://doi.org/10.1016/J.PROGPOLYMSCI.2016.12.008.

Gupta, Prashant, Bhagwan Toksha, and Mostafizur Rahaman. 2022. "A Review on Biodegradable Packaging Films from Vegetative and Food Waste." *The Chemical Record*, e202100326. https://doi.org/10.1002/TCR.202100326.

Hai, Yun, Saihua Jiang, Xiaodong Qian, Shuidong Zhang, Ping Sun, Bo Xie, and Ningning Hong. 2018. "Ultrathin Beta-Nickel Hydroxide Nanosheets Grown along Multi-Walled Carbon Nanotubes: A Novel Nanohybrid for Enhancing Flame Retardancy and Smoke Toxicity Suppression of Unsaturated Polyester Resin." *Journal of Colloid and Interface Science* 509 (January): 285–97. https://doi.org/10.1016/J.JCIS.2017.09.008.

Halley, Peter J., and Michael E. Mackay. 1996. "Chemorheology of Thermosets—an Overview." *Polymer Engineering & Science* 36 (5): 593–609. https://doi.org/10.1002/PEN.10447.

Hashemi, Mohammad J., Masoud Jamshidi, and J. Hassanpour Aghdam. 2018. "Investigating Fracture Mechanics and Flexural Properties of Unsaturated Polyester Polymer Concrete (UP-PC)." *Construction and Building Materials* 163 (February): 767–75. https://doi. org/10.1016/J.CONBUILDMAT.2017.12.115.

He, Siyao, Nicholas D. Petkovich, Kunwei Liu, Yuqiang Qian, Christopher W. Macosko, and Andreas Stein. 2017. "Unsaturated Polyester Resin Toughening with Very Low Loadings of GO Derivatives." *Polymer* 110 (February): 149–57. https://doi.org/10.1016/J. POLYMER.2016.12.057.

Hiltz, John A. 1988. "Low Temperature Thermal Degradation Studies of Styrene Cross-Linked Vinyl Ester And Polyester Resins." Defence Research Establishment Atlantic Dartmouth (Nova Scotia).

Jain, Ravi, and Luke Lee. 2012. *Fiber Reinforced Polymer (FRP) Composites for Infrastructure Applications: Focusing on Innovation, Technology Implementation and Sustainability.* Netherlands: Springer. https://doi.org/10.1007/978-94-007-2357-3.

Jiang, Mengwei, Yuan Yu, and Zhiquan Chen. 2018. "Environmentally Friendly Flame Retardant Systems for Unsaturated Polyester Resin." *IOP Conference Series: Earth and Environmental Science* 170: 032116. https://doi.org/10.1088/1755-1315/170/3/032116.

Kanimozhi, K., Pichaimani Prabunathan, Vaithilingam Selvaraj, and Muthukaruppan Alagar. 2014. "Thermal and Mechanical Properties of Functionalized Mullite Reinforced Unsaturated Polyester Composites." *Polymer Composites* 35 (9): 1663–70. https://doi. org/10.1002/PC.22819.

Khan, Akhtar S., Ozgen U. Colak, and Prabhakaran Centala. 2002. "Compressive Failure Strengths and Modes of Woven S2-Glass Reinforced Polyester Due to Quasi-Static and Dynamic Loading." *International Journal of Plasticity* 18 (10): 1337–57. https://doi. org/10.1016/S0749-6419(02)00002-5.

Kia, Hamid G. 1993. *Sheet Molding Compounds: Science and Technology.* Hanser, Ohio.

Kicko-Walczak, Ewa. 2004. "Studies on the Mechanism of Thermal Decomposition of Unsaturated Polyester Resins with Reduced Flammability." *Polymers and Polymer Composites* 12 (2): 127–34. https://doi.org/10.1177/096739110401200204.

Kikuchi, Tetsuo, Yuichiro Tani, Yuka Takai, Akihiko Goto, and Hiroyuki Hamada. 2014. "Mechanical Properties of Jute Composite by Spray up Fabrication Method." *Energy Procedia* 56 (C): 289–97. https://doi.org/10.1016/J.EGYPRO.2014.07.160.

Kim, Sang Young, Chun Sik Shim, Caleb Sturtevant, Dave Dae Wook Kim, and Ha Cheol Song. 2014. "Mechanical Properties and Production Quality of Hand-Layup and Vacuum Infusion Processed Hybrid Composite Materials for GFRP Marine Structures." *International Journal of Naval Architecture and Ocean Engineering* 6 (3): 723–36. https://doi.org/10.2478/IJNAOE-2013-0208.

Kozlowski, Ryszard, Maria Wladyka-Przybylak, Malgorzata Helwig, and Krzysztof J. Kurzydloski. 2010. "Composites Based on Lignocellulosic Raw Materials." 418. https:// doi.org/10.1080/15421400490479217.

Kuppusamy, Raghu Raja Pandiyan, Satyajit Rout, and Kaushik Kumar. 2020. "Advanced Manufacturing Techniques for Composite Structures Used in Aerospace Industries." In *Modern Manufacturing Processes*, pp. 3–12. Woodhead Publishing. https://doi. org/10.1016/B978-0-12-819496-6.00001-4.

Lee, Yoo Jin, Ji-Hyun Kim, and Jong Ryang Kim. 2016. "Preparation Method of Polyester Resin." U.S. Patent 9,284,405. https://patents.google.com/patent/ US9284405B2/en.

Levy, Arthur, and Pascal Hubert. 2019. "Vacuum-Bagged Composite Laminate Forming Processes: Predicting Thickness Deviation in Complex Shapes." *Composites Part A: Applied Science and Manufacturing* 126 (November): 105568. https://doi.org/10.1016/J. COMPOSITESA.2019.105568.

Manfredi, Liliana B., Exequiel S. Rodríguez, Maria Wladyka-Przybylak, and Analía
    Vázquez. 2006. "Thermal Degradation and Fire Resistance of Unsaturated
    Polyester, Modified Acrylic Resins and Their Composites with Natural Fibres."
    Polymer Degradation and Stability 91 (2): 255–61. https://doi.org/10.1016/J.
    POLYMDEGRADSTAB.2005.05.003.
Matthews, Frank L., and Rees D. Rawlings. 1999. *Composite Materials : Engineering and
    Science*, 1st ed. Sawston, Cambridge: Woodhead Publishing.
Miskolczi, N. 2013. "Polyester Resins as a Matrix Material in Advanced Fibre-Reinforced
    Polymer (FRP) Composites." In *Advanced Fibre-Reinforced Polymer (FRP) Composites
    for Structural Applications*, pp. 44–68. https://doi.org/10.1533/9780857098641.1.44.
Mustafa, Hauwa Mohammed, Benjamin Dauda, and Mustafa Hauwa Mohammed. 2014.
    "Unsaturated Polyester Resin Reinforced With Chemically Modified Natural Fibre."
    IOSR Journal of Polymer and Textile Engineering 1 (4): 31–38. www.iosrjournals.org-
    www.iosrjournals.org.
Nasimudeen, Nadeem Ahmed, Sharwine Karounamourthy, Joshua Selvarathinam, Senthil
    Muthu Kumar Thiagamani, Harikrishnan Pulikkalparambil, Senthilkumar Krishnasamy,
    and Chandrasekar Muthukumar. 2021. "Mechanical, Absorption and Swelling
    Properties of Vinyl Ester Based Natural Fibre Hybrid Composites." Applied Science and
    Engineering Progress. https://doi.org/10.14416/j.asep.2021.08.006.
Orgéas, Laurent, and Pierre J. J. Dumont. 2012. "Sheet Molding Compounds." In *Wiley
    Encyclopedia of Composites*, pp. 1–36. https://doi.org/10.1002/9781118097298.
    WEOC222.
Panda, Shivkumari, and Dibakar Behera. 2019. "Unsaturated Polyester Nanocomposites." In
    Unsaturated Polyester Resins: Fundamentals, Design, Fabrication, and Applications,
    pp. 101–24. Elsevier. https://doi.org/10.1016/B978-0-12-816129-6.00004-1.
Pearce, Neil R. L., John Summerscales and Felicity J. Guild. 2000. "Improving the Resin
    Transfer Moulding Process for Fabric-Reinforced Composites by Modification of the
    Fabric Architecture." *Composites Part A: Applied Science and Manufacturing* 31 (12):
    1433–41. https://doi.org/10.1016/S1359-835X(00)00140-8.
Plastics-the Facts 2021 An Analysis of European Plastics Production, Demand and Waste Data,
    Plastics Europe, Rue Belliard 40, box 16 1040 Brussels – Belgium. https://plasticseu-
    rope.org/knowledge-hub/plastics-the-facts-2021/
Polymer Resin Market Size, Share, Growth | Report, 2030. n.d. Accessed May 23, 2022.
    https://www.marketresearchfuture.com/reports/polymer-resin-market-1702.
Pothan, Laly A., Sabu Thomas, and N. R. Neelakantan. 2016. "Short Banana Fiber
    Reinforced Polyester Composites: Mechanical, Failure and Aging Characteristics."
    Journal of Reinforced Plastics and Composites 16 (8): 744–65. https://doi.
    org/10.1177/073168449701600806.
Rabnawaz, Muhammad, Ian Wyman, Rafael Auras, and Shouyun Cheng. 2017. "A Roadmap
    towards Green Packaging: The Current Status and Future Outlook for Polyesters in
    the Packaging Industry." *Green Chemistry* 19 (20): 4737–53. https://doi.org/10.1039/
    C7GC02521A.
Ravey, Manny. 1983. "Pyrolysis of Unsaturated Polyester Resin. Quantitative Aspects." *Journal
    of Polymer Science: Polymer Chemistry Edition* 21 (1): 1–15. https://doi.org/10.1002/
    POL.1983.170210101.
Rodriguez, Ernesto L. 1987. "Thickening Reaction of Unsaturated Polyester Resins with
    Inorganic Oxides and the Rubber Elasticity Theory." *Journal of Applied Polymer Science*
    34 (2): 881–86. https://doi.org/10.1002/APP.1987.070340237.
Satdive, Ajinkya, Saurabh Tayde, and Aniruddha Chatterjee. 2022. "Flammability
    Properties of the Bionanocomposites Reinforced with Fire Retardant Filler." In
    Polymer Based Bio-Nanocomposites, pp. 69–86. Singapore: Springer. https://doi.
    org/10.1007/978-981-16-8578-1_4.

Satdive, Ajinkya, Siddhesh Mestry, Pavan Borse, and Shashank Mhaske. 2020. "Phosphorus- and Silicon-Containing Amino Curing Agent for Epoxy Resin." *Iranian Polymer Journal* 29 (5): 433–43. https://doi.org/10.1007/s13726-020-00808-6.

Senthilkumar K, Siva I, Rajini N, Jeyaraj P. 2015. "Effect of Fibre Length and Weight Percentage on Mechanical Properties of Short Sisal/Polyester Composite." *International Journal of Computer Aided Engineering and Technology* 7: 60.

Senthilkumar, K., N. Saba, M. Chandrasekar, M. Jawaid, N. Rajini, Suchart Siengchin, Nadir Ayrilmis, Faruq Mohammad, and Hamad A. Al-Lohedan. 2021. "Compressive, Dynamic and Thermo-Mechanical Properties of Cellulosic Pineapple Leaf Fibre/Polyester Composites: Influence of Alkali Treatment on Adhesion." *International Journal of Adhesion and Adhesives* 106: 102823.

Senthilkumar, Krishnasamy, Saravanasankar Subramaniam, Thitinun Ungtrakul, Thiagamani Senthil Muthu Kumar, Muthukumar Chandrasekar, Nagarajan Rajini, Suchart Siengchin, and Jyotishkumar Parameswaranpillai. 2022. "Dual Cantilever Creep and Recovery Behavior of Sisal/Hemp Fibre Reinforced Hybrid Biocomposites: Effects of Layering Sequence, Accelerated Weathering and Temperature." *Journal of Industrial Textiles* 51: 2372S–2390S.

Shahroze, Rao Muhammad, Muthukumar Chandrasekar, Krishnasamy Senthilkumar, Thiagamani Senthil Muthu Kumar, Mohamad Ridzwan Ishak, Nagarajan Rajini, Sikiru Oluwarotimi Ismail. 2021. "Mechanical, Interfacial and Thermal Properties of Silica Aerogel-Infused Flax/Epoxy Composites." *International Polymer Processing* 36: 53–59.

Sheng, Liping, Kerui Xiang, Rong Qiu, Yuxuan Wang, Shengpei Su, Dulin Yin, and Yongming Chen. 2020. "Polymerization Mechanism of 4-APN and a New Catalyst for Phthalonitrile Resin Polymerization." *RSC Advances* 10 (64): 39187–94. https://doi.org/10.1039/D0RA07581G.

Silva, Marco P., Paulo Santos, João M. Parente, Sara Valvez, Paulo N. B. Reis, and Ana P. Piedade. 2020. "Effect of Post-Cure on the Static and Viscoelastic Properties of a Polyester Resin." *Polymers* 12: 1927. https://doi.org/10.3390/POLYM12091927.

Sivasamy, Palanichamy, Mallayan Palaniandavar, Chinnaswamy Thangavel Vijayakumar, and Klaus Lederer. 1992. "The Role of β-Hydrogen in the Degradation of Polyesters." *Polymer Degradation and Stability* 38 (1): 15–21. https://doi.org/10.1016/0141-3910(92)90017-Y.

Skinner, George A., Peter J. Haines, and Trevor J. Lever. 1984. "Thermal Degradation of Polyester Thermosets Prepared Using Dibromoneopentyl Glycol." *Journal of Applied Polymer Science* 29 (3): 763–76. https://doi.org/10.1002/APP.1984.070290305.

Skrifvars, Mikael, Thomas Mackin, and Bert Skagerberg. 1998. "An Application of Experimental Design to the Development of Glass Fibre Reinforced Polyester Laminates with Enhanced Mechanical Properties." *Polymer Testing* 17 (5): 345–56. https://doi.org/10.1016/S0142-9418(97)00062-7.

Song, Lei, Qingliang He, Yuan Hu, Hao Chen, and Lei Liu. 2008. "Study on Thermal Degradation and Combustion Behaviors of PC/POSS Hybrids." *Polymer Degradation and Stability* 93 (3): 627–39. https://doi.org/10.1016/J.POLYMDEGRADSTAB.2008.01.014.

Soutis, Costas. 2005. "Carbon Fiber Reinforced Plastics in Aircraft Construction." *Materials Science and Engineering: A* 412 (1–2): 171–76. https://doi.org/10.1016/J.MSEA.2005.08.064.

Spasojevic, Pavle M. 2019. "Thermal and Rheological Properties of Unsaturated Polyester Resins-Based Composites." In *Unsaturated Polyester Resins: Fundamentals, Design, Fabrication, and Applications*, pp. 367–406. Elsevier. https://doi.org/10.1016/B978-0-12-816129-6.00015-6.

Tabatabai, Habib, Morteza Janbaz, and Azam Nabizadeh. 2018. "Mechanical and Thermo-Gravimetric Properties of Unsaturated Polyester Resin Blended with FGD Gypsum." *Construction and Building Materials* 163 (February): 438–45. https://doi.org/10.1016/J.CONBUILDMAT.2017.12.041.

Thiagamani, Senthil Muthu Kumar, Harikrishnan Pulikkalparambil, Suchart Siengchin, Rushdan Ahmad Ilyas, Senthilkumar Krishnasamy, Chandrasekar Muthukumar, A. M. Radzi, and Sanjay Mavinkere Rangappa. 2022. "Mechanical, Absorption, and Swelling Properties of Jute/Kenaf/Banana Reinforced Epoxy Hybrid Composites: Influence of Various Stacking Sequences." *Polymer Composites*. https://doi.org/10.1002/pc.26999.

Thomas, Seena K., Jyotishkumar Parameswaranpillai, Senthilkumar Krishnasamy, PM Sabura Begum, Debabrata Nandi, Suchart Siengchin, Jinu Jacob George, Nishar Hameed, Nisa V. Salim, and Natalia Sienkiewic. 2021. "A Comprehensive Review on Cellulose, Chitin, and Starch as Fillers in Natural Rubber Biocomposites." *Carbohydrate Polymer Technologies and Applications* 2: 100095.

Torre, Luigi, Debora Puglia, Antonio Iannoni, and Andrea Terenzi. 2013. "Modeling of the Chemorheological Behavior of Thermosetting Polymer Nanocomposites." In *Modeling and Prediction of Polymer Nanocomposite Properties*, pp. 255–87. John Wiley & Sons, Ltd. https://doi.org/10.1002/9783527644346.CH12.

Towo, Arnold N., and Martin P. Ansell. 2008. "Fatigue Evaluation and Dynamic Mechanical Thermal Analysis of Sisal Fibre–Thermosetting Resin Composites." *Composites Science and Technology* 68 (3–4): 925–32. https://doi.org/10.1016/J.COMPSCITECH.2007.08.022.

Tuttle, Mark E. 2003. *Structural Analysis of Polymeric Composite Materials*. CRC Press. https://doi.org/10.1201/9780203913314.

Valette, Ludovic, and Chih Pin Hsu. 1999. "Polyurethane and Unsaturated Polyester Hybrid Networks: 2.: Influence of Hard Domains on Mechanical Properties." *Polymer* 40 (8): 2059–70. https://doi.org/10.1016/S0032-3861(98)00428-5.

Wang, Youchuan, Le Zhang, Yunyun Yang, and Xufu Cai. 2015. "The Investigation of Flammability, Thermal Stability, Heat Resistance and Mechanical Properties of Unsaturated Polyester Resin Using AlPi as Flame Retardant." *Journal of Thermal Analysis and Calorimetry* 122 (3): 1331–39. https://doi.org/10.1007/S10973-015-4875-7/FIGURES/8.

Weil, Edward D., and Sergei V. Levchik. 2016. "Flame Retardants in Commercial Use or Development for Unsaturated Polyester, Vinyl Resins, Phenolics." In *Flame Retardants for Plastics and Textiles*, pp. 187–203. Hanser. https://doi.org/10.3139/9781569905791.008.

Wiedmer, S., and Marinos Manolesos. 2016. "An Experimental Study of the Pultrusion of Carbon Fiber-Polyamide 12 Yarn." *Journal of Thermoplastic Composite Materials* 19 (1): 97–112. https://doi.org/10.1177/0892705706055448.

Worzakowska, Marta. 2009. "Thermal and Dynamic Mechanical Properties of IPNS Formed from Unsaturated Polyester Resin and Epoxy Polyester." *Journal of Materials Science* 44 (15): 4069–77. https://doi.org/10.1007/S10853-009-3587-4/TABLES/6.

Wulfsberg, Jens, Axel Herrmann, Gerhard Ziegmann, Georg Lonsdorfer, Nicole Stöß, and Marc Fette. 2014. "Combination of Carbon Fibre Sheet Moulding Compound and Prepreg Compression Moulding in Aerospace Industry." *Procedia Engineering* 81 (January): 1601–7. https://doi.org/10.1016/J.PROENG.2014.10.197.

Xu, Mei Xuan, Jing Song Xiao, Wen Hua Zhang, and Kang De Yao. 1994. "Synthesis and Properties of Unsaturated Polyester Diol–Polyurethane Hybrid Polymer Network." *Journal of Applied Polymer Science* 54 (11): 1659–63. https://doi.org/10.1002/APP.1994.070541109.

Yahaya, Ridwan B., Mohd Salit Sapuan, Mohammad Jawaid, Zulkiflle Leman, and Edi Syams bin Zainudin. 2015. "Effect of Layering Sequence and Chemical Treatment on the Mechanical Properties of Woven Kenaf–Aramid Hybrid Laminated Composites." *Materials & Design* 67 (February): 173–79. https://doi.org/10.1016/J.MATDES.2014.11.024.

Yeung, Celine W. S., Jerald Y. Q. Teo, Xian Jun Loh, and Jason Y. C. Lim. 2021. "Polyolefins and Polystyrene as Chemical Resources for a Sustainable Future: Challenges, Advances, and Prospects." *ACS Materials Letters* 3 (12): 1660–76. https://doi.org/10.1021/ACSMATERIALSLETT.1C00490/ASSET/IMAGES/LARGE/TZ1C00490_0004.JPEG.

Yu, Xiaojuan, Dong Wang, Bihe Yuan, Lei Song, and Yuan Hu. 2016. "The Effect of Carbon Nanotubes/NiFe$_2$O$_4$ on the Thermal Stability, Combustion Behavior and Mechanical Properties of Unsaturated Polyester Resin." *RSC Advances* 6 (99): 96974–83. https://doi.org/10.1039/C6RA15246E.

Zhang, Ning, Xianglin Hou, Xiaojing Cui, Lin Chai, Hongyan Li, Hui Zhang, Yingxiong Wang, and Tiansheng Deng. 2021. "Amphiphilic Catalyst for Decomposition of Unsaturated Polyester Resins to Valuable Chemicals with 100% Atom Utilization Efficiency." *Journal of Cleaner Production* 296 (May): 126492. https://doi.org/10.1016/J.JCLEPRO.2021.126492.

2 Pineapple Fibre-Reinforced Polyester Composites

Shiji Mathew Abraham
Temple University, Philadelphia, PA- 19122

CONTENTS

2.1 INTRODUCTION

The innumerable problems created by the production, disposal and recycling of synthetic fibres have urged researchers to find natural fibres as alternatives. Natural fibres are gaining vast acceptance due to their eco-friendly nature, admirable physical and mechanical properties, ease of availability, safe handling and biodegradable nature. Natural fibres especially lignocellulosic fibres are mostly used as excellent reinforcements in thermoplastic resin matrices such as polyester (PE), polypropylene, polyethylene and polyvinyl chloride (Venkata Deepthi et al., 2019). This is because of their attractive properties such as relatively low density and abrasiveness, high filling levels offering high stiffness, recyclability and high bending resistance. Based on the site of extraction, fibres exhibit differing inherent properties. Natural fibres are usually extracted from leaves, stems, seeds, fruits and bast. The fibres extracted from the leaves offer improved toughness in composite matrices, whereas the fibres obtained from seeds or fruit provide elastomeric properties to the matrix (Sisti et al. 2018). The natural fibres used for such purposes includes kenaf, banana, jute, oil palm, cotton, flax, hemp, sisal and pineapple leaf fibres (PALFs).

Among these natural fibres, PALF obtained from the leaf of the pineapple plant possesses the highest cellulose content and low microfibrillar angle which attributes to its increased tensile strength. Pineapple is a herbaceous and perennial plant which bears the most unique and exotic fruit (d'Eeckenbrugge et al., 2011). One of the interesting facts about this plant is that its cultivation yields more leaves than fruits. Hence, during post fruit harvesting procedures of pineapple, huge bundles of pineapple leaves are discarded as agro-wastes, which is later exploited for manufacturing textiles, papers and composites utilizing leaf fibres (Senthilkumar et al., 2015, 2021, 2022; Alothman et al., 2020).

PALF is one of the most common and abundantly available waste materials in South-east Asia, India and South America and is the least exploited natural fibre. PALF is an excellent material which can be used as a reinforcing agent in thermoset, thermoplastic, rubber and biodegradable polymer composites. Also, research has revealed that its abundance and ease of availability, acoustic insulation, good thermal, mechanical and physical properties and chemical constituents are attractive features ensuring their intensive application in textile industry, paper making, food packaging and development of composites (Shahroze et al., 2021; Nasimudeen et al., 2021; Thomas et al., 2021). This chapter gives an overview of pineapple plant morphology, species and cultivation, along with a brief description of the extraction process and present and future applications of PALFs. This chapter also discusses on the various factors of pineapple leaves that favours development of PALF-reinforced PE composites with improved properties. Also, a section briefing the various applications of PALF-reinforced PE composites is also included.

2.2 PINEAPPLE SPECIES AND PLANT MORPHOLOGY

Pineapple is one of the most common commercially grown fibre crops. Thailand and Philippines are the leading producers of PALF followed by Brazil, Hawaii, Indonesia, West Indies and India. Pineapple plant, also known as *Ananas comosus* L. Merr., belongs to the subclass monocotyledons and family Bromeliaceae (Collins, 1949). It is a non-climacteric, perennial plant, which is cultivated in tropical and subtropical countries under humid climate in abundant amount. The plant has a height of 0.75–1.5 m and 0.9–1.2 m leaf spread and the central stem contains flower buds developing into fruit at the tip. The pineapple fruit has phyllotaxy leaves on top, thick skin with scales, and juicy pulp inside (Bartholomew et al., 2003; Pandit et al., 2020). Figure 2.1 shows the features of a pineapple plant.

Pineapple fruit is generally used for fresh consumption and can also be consumed in canned, frozen or dried form (Siow and Lee, 2016). Pineapple can also be used for making juice and concentrates and can also be processed into a variety of value-added products such as jam, preserve and candy, sauce, jelly, vinegar, squash, Ready-to-Serve (RTS) beverage, chutney and pickles (Dorta and Sogi, 2016). Based on the plant growth, phenotypic characteristics, sweetness and flavours of the fruit, four important categories of pineapple cultivars include Smooth Cayenne, Queen, Spanish and Abacaxi. Different cultivars of the plant display variation in fruit sweetness, flavour and phenotypic characteristics. Smooth Cayenne variety is characterised by smooth spineless leaves and is resistant to bugs, fruit collapse and heart rot disease.

FIGURE 2.1 Pineapple plant with leaves and fruit.

Queen is the most sweet and flavourful variety and hence cultivated for fresh consumption. Spanish cultivar is characterised by spiny purplish leaves, bearing small, aromatic fruits weighing 1–2 kg and is the best for canning purpose, as the fruits are highly sweet and aromatic (Thiagamani et al., 2022; Chandrasekar et al., 2022; Chandrasekar et al., 2020). This cultivar is mainly cultivated in the coastal areas of Central and South America. The pineapple variety seen in Brazil is the Abacaxi, which has tall/long fruits with narrow spiny leaves (Bartholomew et al., 2003).

Each adult plant will have around 80 leaves, of varying size and shape, measuring approximately 90–150 cm length and 2–3-inch width. The leaves will be dark green in colour, waxy, long, thin, sword-shaped and point-tipped with sharp spines at the border (Mishra et al., 2004; Pandit et al., 2020; Senthilkumar et al., 2019b). So, when cultivated in large scale, there would be large volumes of post-harvest waste that are usually discarded as agro-waste.

2.3 LIFE CYCLE ASSESSMENT (LCA) OF PALF

PALF, being a natural fibre, is biodegradable and sustainable. Figure 2.2 shows the life cycle of PALF. Pineapple is cultivated in large scale, and after harvest, the leaves are decorticated in the decorticating machine to separate the PALF from the biomass. The leaf fibre obtained is then purified to undergo industrial processing using needlepunch

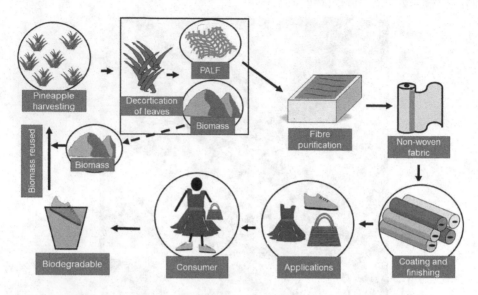

FIGURE 2.2   Life assessment cycle of PALF. Processing of pineapple leaf fibre (PALF) and development of PALF-based composites for various applications. (Source: Modified from https://www.ananas-anam.com/ and Todkar and Patil (2019).)

technology and is converted into nonwoven textiles. This nonwoven pineapple fibre textiles can be easily dyed with direct, reactive, vat and azo dyes with much fastness compared to cotton. This textile is used to make fashionable apparels, accessories, home interiors, automotive and aeronautics interior parts. These commodities being biodegradable can be easily disposed of without causing any degradation problems. At the same time, the biomass left can be used as a compost for future pineapple cultivation.

2.4 PHYSICAL AND CHEMICAL FEATURES OF PINEAPPLE LEAF FIBRE

Pineapple cultivation yields more pineapple leaves besides pineapple fruit. So, it is very necessary to find a solution to make this agro-waste productive. PALF is the most delicate fibre among all vegetable fibres. It has a ribbon-like structure, and has a vascular bundle system. Pineapple fibres are 60 cm long and can easily retain different classes of dyes. It is ten times coarser than cotton fibre. About 2.5–3.3 wt % of fibre can be extracted from green leaves. PALF is made up of 70%–82% α-cellulose, hemicelluloses, 5%–12% lignin and 1.1% ash. It is lightweight, easy to care and looks like linen (Pandit et al., 2020). The high-cellulose type I content and low microfibrillar angle impart to its better mechanical properties (Satyanarayana et al., 1986). Fine PALF possess excellent mechanical properties such as tensile strength and tensile modulus. They are highly hygroscopic in nature and are also abrasion-resistant. Other attractive features of PALF include excellent thermal and acoustic insulation, and high toughness (www.ananas). Table 2.1a summarises the physical properties of PALF, Table 2.1b lists the chemical composition of PALF and Table 2.1c lists the chemical composition of processed PALF (Franck, 2005).

TABLE 2.1A
Physical Characteristics of PALF

Single Cell

Length (mm)	3–8
Diameter (μm)	718
Fineness (tex)	2.5–4

Fibre Bundle

Length (mm)	10–90
Fineness (tex)	2.5–5.5
Tenacity (cN/tex)	30–40
Elongation (%)	2.4–3.4
Initial modulus (cN/tex)	570–700
Density (g/cm^3)	1.543

Source: Adapted and reprinted from Franck (2005).

TABLE 2.1B
Chemical Characteristics of PALF

Constituent	Content (%)
Cellulose	55–68
Hemicellulose	15–20
Pectin	2–4
Lignin	8–12
Water-soluble material	1–3
Fat and wax	4–7
Ash	2–3

Source: Adapted and reprinted from Franck (2005).

TABLE 2.1C
Chemical Composition of Processed PALF (in percentage)

Type of PALF	Alpha Cellulose	Hemicellulose	Lignin	Ash	Alcohol, Benzene
Decorticated PALF	79.36	13.07	4.25	2.29	5.73
Retted PALF	87.36	4.58	3.62	0.57	2.27
Degummed PALF	94.21	2.26	2.75	0.37	0.77

Source: Adapted and reprinted from Franck (2005).

2.5 METHODS OF EXTRACTION OF PALF

Removal of epidermal tissue of pineapple leaves can yield strong, white lustrous silky fibres that can be spun into fine textile yarn on jute and on cotton spinning system (Vincent et al., 2016). Usually, freshly harvested green leaf bundles are used for fibre extraction. The Perolera cultivar of pineapple in the Caribbean area possesses long leaves, which is considered highly suitable for fibre extraction. PALF can be extracted either by manual methods such as scraping/hand stripping process or mechanically using a decorticator. The choice of the extraction process is extremely important as it determines the quality and quantity of fibres (Kengkhetkit and Amornsakchai, 2012). Each extraction method influences the structure, chemical composition and physical properties of the fibres.

The manual method of extraction is considered more preferable than extraction by decorticator, but the yield of fibre is very less. The decorticated fibres may usually contain waxy substance and fleshy leafy parts. Hence, after extraction with decorticator, the fibrous strands have to be split out by retting and degumming methods. The difference in the yield and quality of PALF are listed in Table 2.2.

2.5.1 MANUAL EXTRACTION METHODS

2.5.1.1 Hand Stripping/Scraping

Hand stripping or scraping process is a conventional method of PALF extraction, which is done manually by scraping the fibres of small pineapple leaf using a scrapping tool called 'ketam' or using broken porcelain plate or coconut shell in case of longer leaves (Kannojiya et al., 2013). An expert worker can scrap over 500 leaves per day and extract fibres, which can be water-washed and dried in air. The fibres are later waxed to remove any entanglements and knot them. The knotting process ends up in separation of fibres from each other and is knotted at the tips to form a long continuous yarn. Even though good-quality fibres are yielded through this process, it is a highly laborious process and

---

TABLE 2.2

The Yield and Quality of PALF Obtained from Various Extraction Procedures

Method	Fibre Yield (%)	Range of Fibre Diameter (μm)	Average Fibre Diameter (μm)	Merits	Demerits
Retting	1.8	5–166	58.98	High yield	Time-consuming, water-consuming
Scraping	1.4	5–129	57.36	Good-quality fibres	Less yield
Ball milling	2.9	3–95	8.66	Simple method, easily scaled up for large-scale production of short PALF	Low yield of coarse fibres
Milling	2.8	3–68	18.70	–	–
Milling of dried leaf	3.0	5–194	63.43	–	–

Source: Reprinted and reused with permission from Kengkhetkit and Amornsakchai (2012).

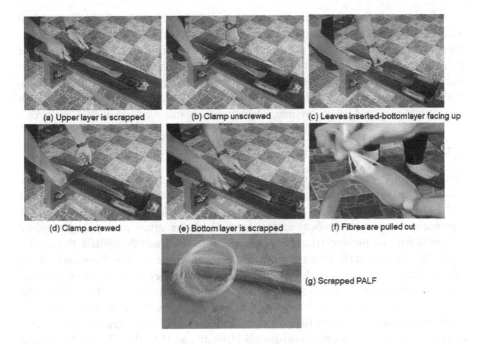

(a) Upper layer is scrapped   (b) Clamp unscrewed   (c) Leaves inserted-bottom layer facing up

(d) Clamp screwed   (e) Bottom layer is scrapped   (f) Fibres are pulled out

(g) Scrapped PALF

FIGURE 2.3   Hand stripping/scrapping method of PALF extraction. (Source: Adapted and modified from Yusof et al. (2015). Copyright © 2015 The Authors. Published by Elsevier B.V.)

needs manpower of 30 persons per ton of leaves (Jose et al., 2016). Figure 2.3 shows the different steps of hand scraping method for PALF extraction.

2.5.1.2   Retting Process

The next step in manual process of PALF extraction is retting process, which is of three types: water, bacterial and dew retting.

Water retting includes soaking the PALFs in water for about 20 days until it gets saturated. This retting process is usually preformed in water bodies such as ponds, ditches, tanks or rivers. During this process, the pectic substances are dissolved by the action of microorganisms, which leads to separation of the fibre from the cortex, resulting in clean fibres. The bacteria such as *Bacillus* and *Clostridium* are found to enhance water retting process. The extracted fibre is then washed under running tap water and then sun-dried (Yusof et al., 2015). This is a time-consuming process, and moreover, during this process, large amount of water is consumed; hence, it is not considered as an eco-friendly process.

In bacterial retting, the crushed leaves are placed in bacterial culture solution. As a result of the bacterial action, the fibres can be easily removed from the fleshy mass of leaves and extraction can be completed within 4–5 days. Also, during bacterial retting, the non-fibrous matter like hemicellulose is converted to α-cellulose, leading to the yield of high-quality fibres with good tensile strength. So, this process is less time-consuming compared to water retting process, but bacterial retting method is not commercialised yet (Cueto et al., 1978).

Another method is dew retting, fungi such as *Rhizomucor pusillus* and *Fusarium lateritium* are found to be effective. Using this process, good-quality fibres can be produced, but the method possess demerits such as time consumption. Moreover, in order to maintain homogeneity in retting process, the leaves have to be turned over at least one time. This process also needs constant monitoring to prevent over-retting of fibres (Mukherjee and Radhakrishnan, 1972). This method is applicable only in regions with heavy night dews and warm daytime, where it usually takes a duration of 3–6 weeks.

2.5.2 Mechanical Extraction

Since mechanical extraction of fibres includes many processes such as breaking, scotching and hackling, specially designed devices are needed for the extraction of PALF (Joffe et al., 2003). This is done with the help of a decorticator by leaves crushing mechanism. In mechanical extraction method, the green pineapple leaves are manually fed into the decorticating machine called the Raspador, where the revolving blades scrape out the fibres (Figure 2.4). This machine is a combination of three rollers: feed roller, leaf scratching roller and serrated roller. The leaves are fed into the feed roller, and then passed through the scratching roller, which scratches the upper layer and removes the waxy layer of leaves. Finally, the leaves reach the serrated roller where the dense blades crush the leaves and make several breaks on it to enable the entry of microorganisms (N Debasis and D Sanjoy, 2009). The decorticated fibres are then washed out and sun-dried. The fibres generated here are a mixture of liniwan and bastos. The last step is knotting the fibres by hand and then spinning it with a charkha. The final product will be a single thread, which can be used for making fabric or other materials such as ropes, carpets or sponge seats. A study conducted by Yusof et al. (2015) stated that the pineapple fibres obtained from decorticating machines were more soft, bright and creamy white colour when compared to conventional methods.

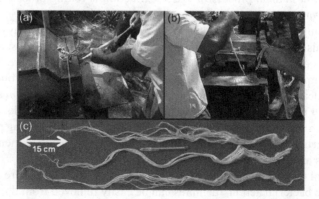

FIGURE 2.4 Extraction of PALF by mechanical process using a decorticator machine. (a) Pineapple leaves are loaded into the decorticating machine, (b) the fibres obtained are being pulled out from the decorticator and (c) sun-dried PALFs. (Source: Reused with permission from Sena Neto et al. (2013). Copyright 2014, Elsevier, B.V.)

2.5.3 DEGUMMING OF PALF

Even after extraction and retting, the fibres may remain bunched together due to the presence of gummy substances such as pectin, lignin and pentosan. Hence, it is very essential to remove this gum, as it will inhibit the further wet processing of the fibre. The process of removing gum from fibres is called degumming. Degumming further improves physical properties such as moisture regain, fitness, lustre, brightness, whiteness and other characteristics of fibres. This gum can be removed by chemical means, enzymes or with microorganisms. Chemical degumming methods primarily include alkaline treatment.

2.6 APPLICATIONS OF PALF FIBRES

PALF can be used for various applications such as for fabrication of clothing, bags, footwears, furniture and papers. It also has significant potential applications in construction and building, automotive, aerospace, marine, wind energy and consumer products. Figure 2.5 summarises the various applications of PALF. In the automotive industry, PALFs are used for manufacturing door panels, seat backs, dashboards, package trays, headliners, and many other interior parts (Reddy et al., 2020). Chemically treated PALFs can be used to prepare V-belt cord, conveyer belt cord, transmission belt cord, air-bag tying cords and industrial textiles. The following section cites certain examples and commercially available products developed from PALF.

PALF yarns and handlooms
PALF is considered as a potential commercial-grade textile natural fibre. The most common use of PALF is development of yarns and can also be combined with silk or PE for manufacturing textile fabrics. PALF being lightweight are highly suitable for making curtains, fashionable clothing and bags. Trendy apparels made of PALF has a huge demand in the global market, as socioeconomically sound people give more preference to eco-friendly, sustainable natural fibre-based fashionable dressings (Padzil

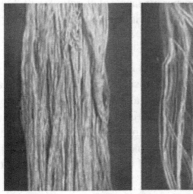

Decorticated PALF   Degummed PALF

FIGURE 2.5   Various applications of PALF. (Source: Reprinted with permission from Asim et al. (2015). Copyright © 2015 M. Asim et al. Open access journal.)

et al., 2020). In a recent study, Jose et al., 2019 reported the potentiality of PALF extracted from Indian pineapples to be used for the manufacturing of fashionable apparels. In the Philippines, PALFs are commercially marketed as Piña fabric (Emily Lush, 2019) and in Spain, it's called the Piñatex (Carmen Hijosa, 2017). Spreading more awareness about the advantages of fabrics made of PALF over synthetic fibres can motivate people to cultivate more pineapple and extract more fibres from it.

PALF footwears and baggage

PALF has been recently considered as a leather alternative for footwears and bags (ITFN, 2015). The Spanish company Pinatex® markets PALF footwears in Philippines (Dan and Mez, 2019). PALF can also be used for fabricating footwears, which are flexible, durable, smooth, odourless and chemical-resistant. Recently, PALF-made sneakers and shoes have been available in market in different international brand names such as Hugo Boss (Boss, 2018), Nike (Sidney, 2021) and H&M boots (H&M, 2019). PALF is also used for fabricating baggage and furniture.

Paper from PALF

Another important application of PALF is in pulp and paper production (Mat Nayan et al., 2014). Many studies have reported the potential of PALF in the development of papers (Laftah and Abdul Rahaman, 2015; Sibaly and Jeetah, 2017). In Thailand, Bangkok-based company Yothaka International is leading in manufacturing eco-friendly chairs, stools and benches using pineapple paper fibre (Inhabitat, 2008; Wan Nadirah et al., 2012).

In a study, Sibaly and Jeetah (2017) investigated the feasibility of using PALFs for paper production. PALFs were mixed with cane-bagasse in different ratios such as 20:80, 40:60. 60:40, 80:20 and 100:0. Similarly, PALF was also mixed with wastepaper in the same ratios and the physical and mechanical properties of the papers were analysed. The papers made of 100% PALF were found to be the most absorbent with the highest Tensile Index and Burst Index of 6.5 Nm/g and 0.84 kPam^2/g, respectively. The most abrasion-resistant paper was found to be bagasse-PALF with ratio 40:60 and the most crease-resistant paper was wastepaper-pineapple with ratio 80:20. Figure 2.6 illustrates the papers made from PALF after mixing with bagasse and wastepaper in different ratios.

PALF composites

Another important application of PALF is in the development of polymer composites. Till date, many kinds of thermoset resins such as bisphenol (Vinod and Sudev, 2018), epoxy (Doddi et al., 2020; Nagarajan et al., 2016), vinyl ester (Mazlan et al., 2019; Mohamed et al., 2014, 2010) and PE (Glória et al., 2017; SaravanaKumar et al., 2021; Senthilkumar et al., 2019b) have been reinforced with PALF and their improved properties investigated. In general, reinforcement of PALF into these polymer matrices will improve the mechanical, thermal and vibrational properties of the composites.

2.6.1   PALF-Reinforced PE Composites

PALF is found to be an important reinforcing agent in PE composites, as it is found to improve the physical, mechanical and thermal properties of the PE composites. Several studies have investigated the role of PALF in enhancing the mechanical properties of

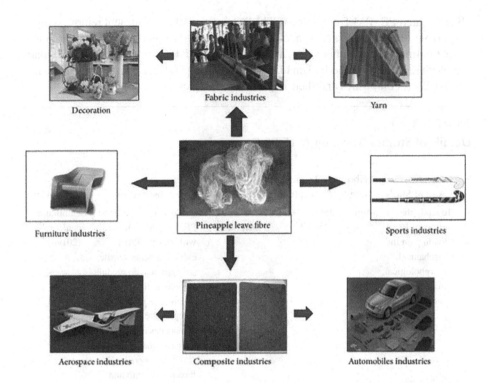

Decoration

Fabric industries

Yarn

Furniture industries

Pineapple leave fibre

Sports industries

Aerospace industries

Composite industries

Automobiles industries

FIGURE 2.6 PALF-bagasse and PALF wastepaper paper samples. (Source: Adapted and reprinted with permission from Sibaly and Jeetah (2017).)

PE composites (B a et al., 2020; B A et al., 2020; Glória et al., 2017; Keerthi Gowda et al., 2021; Rajesh et al., 2018; Venkata Deepthi et al., 2019). In a study, the loading of 30% weight of PALF has shown to improve the flexural strength, tensile strength and impact strength (Asim et al., 2015). In addition, in order to improve the fibre quality of leaf, several surface alterations can be performed, which include alkali treatment, cyanoethylation, dewaxing and grafting acetonitrile to dewaxed fibres (Mishra et al., 2001).

In another study the influence of loading different concentrations of PALF on the mechanical properties of PALF/ PE composites was studied. An increase of 23% and 25% was noticed in the tensile strength and tensile modulus, respectively, of PALF in PE -PALF composites loaded with 35% PALF compared to composites with 25% PALF. This was explained to be due to the increased region of improved interfacial adhesion between the fibre and matrix, which resulted in the enhancement of cumulative stress transfer (Senthilkumar et al., 2019b).

The effect of different surface modifications on the mechanical properties of PALF-reinforced PE composites was studied by Mishra et al. (2004). In this work, detergent washed PALF were subjected to four types of surface modification, viz., first was defatted, second was alkali-treated (5% NaOH), and third was subjected to graft copolymerization of acrylonitrile and the final sample of PALF was cyanoethylated. All four sets of surface-modified PALF were taken into non-woven mat form and used for preparing composites and the properties were compared. The results indicated that a maximum tensile strength value of 44.77 MPa was shown by alkali-treated PE -PALF composites. This is because alkali treatment removes impurities from the material surface and enhances better fibre–matrix adhesion than any other treatments. In another study, an improvement in PE composites with the loading of PALF was reported by Uma Devi et al. (2004). Table 2.3 lists various studies on PALF/PE composites and their improved properties.

TABLE 2.3
Details of Studies Based on PALF/PE Composites

S. No.	Aim of Study	Fibre Length	Loading Concentration	Method of Composite Preparation	Observation	Reference
1	To study the effect of PALF loading on the mechanical, morphological, free vibrational and damping properties	3 mm	25%, 35% and 45 wt%	Hand lay-up method	*Tensile properties:* PALF/ PE composites loaded with 35 wt% PALF exhibited better tensile strength, tensile modulus and elongation at break *Compressive properties:* compressive strength of composites improved with increased loading of PALF *Flexural properties:* flexural strength and modulus was highest at 35 wt% loaded PALF/PE composite *Free vibrational properties:* the fibre loading of 45 wt% showed the highest free vibrational properties *Damping ratio:* maximum damping value was observed for composites composed of 25 wt% PALF	Senthilkumar et al. (2019b)
2	To study the effect of alkali treatment (1 N NaOH and I N KOH) on the mechanical and morphological properties of PALF/PE composites	3 mm	25, 35, 45 wt%	Compression moulding technique	*Mechanical properties:* both 1 N NaOH and KOH-treated 45 wt% PALF-loaded composites showed highest tensile strength, tensile modulus and flexural strength *Impact strength:* NaOH-treated 25 wt% PALF-reinforced PE composites showed highest impact strength of 70 J/m	Senthilkumar et al. (2019a)

(Continued)

TABLE 2.3 (*Continued*)
Details of Studies Based on PALF/PE Composites

S. No.	Aim of Study	Fibre Length	Loading Concentration	Method of Composite Preparation	Observation	Reference
3	To study the effect of alkali and silane treatment on the mechanical properties of PALF-reinforced polypropylene composites		1%, 3%, 5% and 7% of NaOH and silane	Heat compression technique using hydraulic hot press machine	*Mechanical properties*: 5% NaOH and 3% silane-treated PALF-reinforced composites exhibited highest mechanical strength *Flexural properties*: 5% NaOH and 3% silane-treated PALF-reinforced composites exhibited improved flexural properties *Impact strength:* 5% NaOH and 3% silane-treated PALF-reinforced composites exhibited highest impact strength	Feng et al. (2020)
4	To investigate the tensile properties of PALF-reinforced PE composites	0.09–0.30 mm	10, 20, 30 volume %	-	Tensile properties improved with increase in PALF concentration	Glória et al. (2017)
5	To study the physical, mechanical and morphological behaviour of PALF-reinforced PE resin composite		10, 20, 30 and 40%	Hand lay-up technique	40 wt% of PALF loading resulted in improvement in physical and mechanical properties	B a et al. (2020)
6	To study the effect on silane-treated PALF loading on mechanical properties of phenolic composites	0.8–1 mm	40, 50, 60 wt%	Hand lay-up technique	Composites with 50% PALF loading exhibited better tensile, flexure and impact properties	Asim et al. (2018a)

(Continued)

TABLE 2.3 (*Continued*)
Details of Studies Based on PALF/PE Composites

S. No.	Aim of Study	Fibre Length	Loading Concentration	Method of Composite Preparation	Observation	Reference
7	To study the thermal, physical and flammability of silane-treated PALF/phenolic composites	0.8–1 mm	15, 25, 35%	Heat compression technique using hydraulic hot press machine	PALF loading improved the thermal, physical and flammability of the composites	Asim et al. (2018b)
8	To study the effect of PALF reinforcement in tensile, flexural and water absorption behaviour on PE composites	1 m	10, 20, 30, 40 wt%	Hand lay-up technique	Improved tensile, flexural and water absorption behaviour were observed with PE composites loaded with 40 wt% PALF	B A et al. (2020)
9	To study the mechanical, water absorption and fire-resistant properties of PALF/PE composites	10±1 mm	10, 15, 20, 25, 30, 40, 45%	Compression moulding technique	Mechanical properties of PALF/PE composites increased with PALF loading of 40%	Keerthi Gowda et al. (2021)
10	To study the effect of alkali-treated PALF/PE composites	180 mm	5–30%	Hand lay-up technique	Alkali-treated PALF in 30 wt% loading showed maximum tensile strength	Rajesh et al. (2018)
11	To study the thermal properties of PALF-loaded PE composites		0.112, 0.166, 0.213, 0.274 and 0.346	Hand lay-up technique	PALF loading was found to improve the thermal properties of PALF/PE composites compared to plain PE	Santosha et al. (2018)

2.7   CONCLUSION

Pineapple cultivation is now widespread throughout the world due to its ease in propagation and economic importance. PALFs are one of the most useful and highly exploited natural fibres. Pineapple fibres can be extracted from fresh leaves by different manual and mechanical processes. Both these methods possess merits and demerits. The quality and quantity of the PALF fibres mainly depend on the type of extraction process. So, various factors have to be considered before the fibre is extracted from the pineapple leaves. PALF is an excellent material which can be used as a reinforcing agent in thermoset, thermoplastic, rubber and biodegradable polymer composites. Also, research has revealed that its abundance and ease of availability, acoustic insulation, good thermal, mechanical and physical properties, and chemical constituents are attractive features ensuring their intensive application in textile industry, paper making, food packaging and development of composites. This chapter details on the cultivation and species varieties of pineapple. This chapter also discusses on the physical and chemical composition of PALF and explains the various methods of extracting PALF from pineapple leaves. The later sections elaborate on the applications of PALF with examples and the development of PALF/ PE -based composites. The role of PALF in improving the physical, mechanical and thermal properties of PE composites is also explained.

BIBLIOGRAPHY

Alothman, O. Y., Jawaid, M., Senthilkumar, K., Chandrasekar, M., Alshammari, B. A., Fouad, H., Hashem, M., Siengchin, S. 2020. Thermal Characterization of Date Palm/Epoxy Composites with Fillers from Different Parts of the Tree. *J. Mater. Res. Technol.* 9, 15537–15546 https://doi.org/10.1016/j.jmrt.2020.11.020.

Asim, M., Abdan, K., Jawaid, M., Nasir, M., Dashtizadeh, Z., Ishak, M.R., Hoque, M.E., 2015. A Review on Pineapple Leaves Fibre and Its Composites. *Int. J. Polym. Sci.* 2015, 950567. https://doi.org/10.1155/2015/950567

Asim, M., Jawaid, M., Abdan, K., Ishak, M.R., 2018a. The Effect of Silane Treated Fibre Loading on Mechanical Properties of Pineapple Leaf/Kenaf Fibre Filler Phenolic Composites. *J. Polym. Environ.* 26, 1520–1527. https://doi.org/10.1007/s10924-017-1060-z

Asim, M., Paridah, M.T., Saba, N., Jawaid, M., Alothman, O.Y., Nasir, M., Almutairi, Z., 2018b. Thermal, Physical Properties and Flammability of Silane Treated Kenaf/ Pineapple Leaf Fibres Phenolic Hybrid Composites. *Compos. Struct.* 202, 1330–1338. https://doi.org/10.1016/j.compstruct.2018.06.068

Bartholomew, D.P., Paull, R.E., Rohrbach, K.G. (Eds.), 2003. *The Pineapple: Botany, Production and Uses.* CABI, Wallingford. https://doi.org/10.1079/9780851995038.0000

Boss, H., 2018. Innovative Shoes Made of Pineapple Leaves for BOSS Menswear [WWW Document]. URL https://www.prnewswire.com/news-releases/innovative-shoes-made-of-pineapple-leaves-for-boss-menswear-300660293.html (accessed 4.28.22).

Chandrasekar, M., Senthilkumar, K., Jawaid, M., Alamery, S., Fouad, H., Midani, M. 2022. Tensile, Thermal and Physical Properties of Washightonia Trunk Fibres/Pineapple Fibre Biophenolic Hybrid Composites. *J. Polym. Environ.* 30, 4427–4434. https://doi.org/10.1007/s10924-022-02524-z.

Chandrasekar, M., Siva, I., Kumar, T. S. M., Senthilkumar, K., Siengchin, S., Rajini, N. 2020. Influence of Fibre Inter-ply Orientation on the Mechanical and Free Vibration Properties of Banana Fibre Reinforced Polyester Composite Laminates. *J. Polym. Environ.* 28, 2789–2800 https://doi.org/10.1007/s10924-020-01814-8.

Collins, J.L., 1949. History, Taxonomy and Culture of the Pineapple. *Econ. Bot.* 3, 335–359. https://doi.org/10.1007/BF02859162

Cueto, C.U., Quintos, A.G., Peralta, C.N., Palmario, M.S., 1978. Pineapple Fibers. The Retting Process. II. *NSDB Technol. J.* 1, 73–79.

d'Eeckenbrugge, G.C., Sanewski, G.M., Smith, M.K., Duval, M.-F., Leal, F., 2011. Ananas, in: Kole, C. (Ed.), *Wild Crop Relatives: Genomic and Breeding Resources*. Springer, Berlin, Heidelberg, pp. 21–41. https://doi.org/10.1007/978-3-642-20447-0_2

Dan, M., 2019. Pinatex - Pineapple Leather the Fabric of the Future [WWW Document]. DAN MÉZ - Sustain. Vegan Watch. URL https://danandmez.com/blog/pinatex/ (accessed 4.28.22).

Debasis, N., Sanjoy, D., 2009. A Pineapple Leaf Fibre Decorticator Assembly - Patent [WWW Document]. URL https://www.quickcompany.in/patents/a-pineapple-leaf-fibre-decorti-cator-assembly (accessed 4.10.22).

Doddi, P.R.V., Chanamala, R., Dora, S.P., 2020. Effect of Fiber Orientation on Dynamic Mechanical Properties of PALF Hybridized with Basalt Reinforced Epoxy Composites. *Mater. Res. Express* 7, 015329. https://doi.org/10.1088/2053-1591/ab6771

Dorta, E., Sogi, D.S., 2016. Value added Processing and Utilization of Pineapple by-Products, in: Lobo, M.G., Paull, R.E. (Eds.), *Handbook of Pineapple Technology*. John Wiley & Sons, Ltd, Chichester, UK, pp. 196–220. https://doi.org/10.1002/9781118967355.ch11

Emily Lush, 2019. Piña (Pineapple) Cloth, Philippines [WWW Document]. Text. Atlas. URL https://www.thetextileatlas.com/craft-stories/pina-cloth-philippines (accessed 4.23.22).

Feng, N.L., Malingam, S.D., Razali, N., Subramonian, S., 2020. Alkali and Silane Treatments towards Exemplary Mechanical Properties of Kenaf and Pineapple Leaf Fibre-Reinforced Composites. *J. Bionic Eng.* 17, 380–392. https://doi.org/10.1007/s42235-020-0031-6

Franck, R.E. (Ed.), 2005. 9- Pineapple, curauá, craua (caroá), macambira, nettle, sunn hemp, Mauritius Hemp and Fique, in: *Bast and Other Plant Fibres*, Woodhead Publishing Series in Textiles. Woodhead Publishing, pp. 322–344. https://doi.org/10.1533/9781845690618.322

Glória, G.O., Teles, M.C.A., Lopes, F.P.D., Vieira, C.M.F., Margem, F.M., de Almeida Gomes, M., Monteiro, S.N., 2017. Tensile Strength of Polyester Composites Reinforced with PALF. *J. Mater. Res. Technol.* 6, 401–405. https://doi.org/10.1016/j.jmrt.2017.08.006

H&M, 2019. Jacquard-patterned boots - Light beige/Silver-coloured - Ladies | H&M IN [WWW Document]. H&M. URL https://www2.hm.com/en_in/productpage.0742934001.html (accessed 4.28.22).

Hijosa, C., 2017. Piñatex [WWW Document]. Piñatex. URL https://www.ananas-anam.com/ (accessed 4.10.22).

Inhabitat, 2008. Pineapple Paper Furniture by Yothaka [WWW Document]. URL https://inhab-itat.com/pineapple-paper-furniture-by-yothaka/ (accessed 4.29.22).

ITFN, 2015. CANADA: Pineapple Now a Leather Alternative for Shoes – TFNet – International Tropical Fruits Network. URL https://www.itfnet.org/v1/2015/12/canada-pineapple-now-a-leather-alternative-for-shoes/ (accessed 4.29.22).

Joffe, R., Andersons, J., Wallström, L., 2003. Strength and Adhesion Characteristics of Elementary Flax Fibres with Different Surface Treatments. *Compos. Part Appl. Sci. Manuf.* 34, 603–612. https://doi.org/10.1016/S1359-835X(03)00099-X

Jose, S., Das, R., Mustafa, I., Karmakar, S., Basu, G., 2019. Potentiality of Indian Pineapple Leaf Fiber for Apparels. *J. Nat. Fibers* 16, 536–544. https://doi.org/10.1080/15440478.2018.1428844

Jose, S., Salim, R., Ammayappan, L., 2016. An Overview on Production, Properties, and Value Addition of Pineapple Leaf Fibers (PALF). *J. Nat. Fibers* 13, 362–373. https://doi.org/10.1080/15440478.2015.1029194

Kannojiya, R., Kumar, G., Ranjan, R., Tiyer, N.R., Pandey, K.M., 2013. Extraction of Pineapple Fibres for Making Commercial Products. *J. Environ. Res. Dev.* 7, 1385–1390.

Keerthi Gowda, B.S., Naresh, K., Ilangovan, S., Sanjay, M.R., Siengchin, S., 2021. Effect of Fiber Volume Fraction on Mechanical and Fire Resistance Properties of Basalt/Polyester and Pineapple/Polyester Composites. *J. Nat. Fibers* 1–15. https://doi.org/10.1080/15440478.2021.1904479

Kengkhetkit, N., Amornsakchai, T., 2012. Utilisation of Pineapple Leaf Waste for Plastic Reinforcement: 1. A Novel Extraction Method for Short Pineapple Leaf Fiber. *Ind. Crops Prod.* 40, 55–61. https://doi.org/10.1016/j.indcrop.2012.02.037

Laftah, W.A., Abdul Rahaman, W.A.W., 2015. Chemical Pulping of Waste Pineapple Leaves Fiber for Kraft Paper Production. *J. Mater. Res. Technol.* 4, 254–261. https://doi.org/10.1016/j.jmrt.2014.12.006

Mat Nayan, N.H., Wan Abdul Rahman, W.A., Abdul Majid, R., 2014. The Effect of Mercerization Process on the Structural and Morphological Properties of Pineapple Leaf Fiber (PALF) Pulp. *Malays. J. Fundam. Appl. Sci.* 10. https://doi.org/10.11113/mjfas.v10n1.63

Mazlan, A.A., Sultan, M.T.H., Shah, A.U.M., Safri, S.N.A., 2019. Thermal Properties of Pineapple Leaf/Kenaf Fibre Reinforced Vinyl Ester Hybrid Composites. *IOP Conf. Ser. Mater. Sci. Eng.* 670, 012030. https://doi.org/10.1088/1757-899X/670/1/012030

Mishra, S., Misra, M., Tripathy, S.S., Nayak, S.K., Mohanty, A.K., 2001. Potentiality of Pineapple Leaf Fibre as Reinforcement in PALF-Polyester Composite: Surface Modification and Mechanical Performance. *J. Reinf. Plast. Compos.* 20, 321–334. https://doi.org/10.1177/073168401772678779

Mishra, S., Mohanty, A.K., Drzal, L.T., Misra, M., Hinrichsen, G., 2004. A Review on Pineapple Leaf Fibers, Sisal Fibers and Their Biocomposites. *Macromol. Mater. Eng.* 289, 955–974. https://doi.org/10.1002/mame.200400132

Mohamed, A.R., Sapuan, S.M., Khalina, A., 2014. Mechanical and Thermal Properties of Josapine Pineapple Leaf Fiber (PALF) and PALF-Reinforced Vinyl Ester Composites. *Fibers Polym.* 15, 1035–1041. https://doi.org/10.1007/s12221-014-1035-9

Mohamed, A.R., Sapuan, S.M., Shahjahan, M., Khalina, A., 2010. Effects of Simple Abrasive Combing and Pretreatments on the Properties of Pineapple Leaf Fibers (Palf) and Palf-Vinyl Ester Composite Adhesion. *Polym.-Plast. Technol. Eng.* 49, 972–978. https://doi.org/10.1080/03602559.2010.482072

Mukherjee, R.R., Radhakrishnan, T., 1972. Long Vegetable Fibres. *Text. Prog.* 4, 1–75. https://doi.org/10.1080/00405167208688974

Nagarajan, T.T., Babu, A.S., Palanivelu, K., Nayak, S.K., 2016. Mechanical and Thermal Properties of PALF Reinforced Epoxy Composites. *Macromol. Symp.* 361, 57–63. https://doi.org/10.1002/masy.201400256

Nasimudeen, N. A., Karounamourthy, S., Selvarathinam, J., Kumar Thiagamani, S. M., Pulikkalparambil, H., Krishnasamy, S., Muthukumar, C. 2021. Mechanical, Absorption and Swelling Properties of Vinyl Ester Based Natural Fibre Hybrid Composites. *Appl. Sci. Eng. Progress.* https://doi.org/10.14416/j.asep.2021.08.006.

Padzil, F.N.M., Ainun, Z.M.A., Abu Kassim, N., Lee, S.H., Lee, C.H., Ariffin, H., Zainudin, E.S., 2020. Chemical, Physical and Biological Treatments of Pineapple Leaf Fibres, in: Jawaid, M., Asim, M., Tahir, P.Md., Nasir, M. (Eds.), *Pineapple Leaf Fibers, Green Energy and Technology.* Springer, Singapore, pp. 73–90. https://doi.org/10.1007/978-981-15-1416-6_5

Pandit, P., Pandey, R., Singha, K., Shrivastava, S., Gupta, V., Jose, S., 2020. Pineapple Leaf Fibre: Cultivation and Production, in: Jawaid, M., Asim, M., Tahir, P., Nasir, M. (Eds.), *Pineapple Leaf Fibers, Green Energy and Technology.* Springer, Singapore, pp. 1–20. https://doi.org/10.1007/978-981-15-1416-6_1

Praveena, B.A., P Shetty, B., Vinayaka, N., Srikanth, H.V., Singh Yadav, S.P., AvinashL., 2020. Mechanical Properties and Water Absorption Behaviour of Pineapple Leaf Fibre Reinforced Polymer Composites. *Adv. Mater. Process. Technol.* 1–16. https://doi.org/10.1080/2374068X.2020.1860354

Praven, B. A., P Shetty, B., Sachin, B., Singh Yadav, S.P., Avinash, L., 2020. Physical and Mechanical Properties, Morphological Behaviour of Pineapple Leaf Fibre Reinforced Polyester Resin Composites. *Adv. Mater. Process. Technol.* 1–13. https://doi.org/10.1080/2374068X.2020.1853498

Rajesh, G., Siripurapu, G., Lella, A., 2018. Evaluating Tensile Properties of Successive Alkali Treated Continuous Pineapple Leaf Fiber Reinforced Polyester Composites. *Mater. Today Proc.* 5, 13146–13151. https://doi.org/10.1016/j.matpr.2018.02.304

Reddy, B.S., Rajesh, M., Sudhakar, E., Rahaman, A., Kandasamy, J., Sultan, M.T.H., 2020. Pineapple Leaf Fibres for Automotive Applications, in: Jawaid, M., Asim, M., Tahir, P., Nasir, M. (Eds.), *Pineapple Leaf Fibers, Green Energy and Technology*. Springer, Singapore, pp. 279–296. https://doi.org/10.1007/978-981-15-1416-6_14

Sani, I.K., Marand, S.A., Alizadeh, M., Amiri, S., Asdagh, A., 2021. Thermal, Mechanical, Microstructural and Inhibitory Characteristics of Sodium Caseinate Based Bioactive Films Reinforced by ZnONPs/Encapsulated *Melissa officinalis* Essential Oil. *J. Inorg. Organomet. Polym. Mater.* 31, 261–271. https://doi.org/10.1007/s10904-020-01777-2

Santosha, P.C.R., Gowda, A.S.S.S., Manikanth, V., 2018. Effect of Fiber Loading on Thermal Properties of Banana and Pineapple Leaf Fiber Reinforced Polyester Composites. *Mater. Today Proc.* 5, 5631–5635. https://doi.org/10.1016/j.matpr.2017.12.155

SaravanaKumar, M., Kumar, S.S., Babu, B.S., Chakravarthy, CH.N., 2021. Influence of Fiber Loading on Mechanical Characterization of Pineapple Leaf and Kenaf Fibers Reinforced Polyester Composites. *Mater. Today Proc.* 46, 439–444. https://doi.org/10.1016/j.matpr.2020.09.804

Satyanarayana, K.G., Ravikumar, K.K., Sukumaran, K., Mukherjee, P.S., Pillai, S.G.K., Kulkarni, A.G., 1986. Structure and Properties of Some Vegetable Fibres: Part 3Talipot and Palmyrah Fibres. *J. Mater. Sci.* 21, 57–63. https://doi.org/10.1007/BF01144699

Sena Neto, A.R., Araujo, M.A.M., Souza, F.V.D., Mattoso, L.H.C., Marconcini, J.M., 2013. Characterization and Comparative Evaluation of Thermal, Structural, Chemical, Mechanical and Morphological Properties of Six Pineapple Leaf Fiber Varieties for Use in Composites. *Ind. Crops Prod.* 43, 529–537. https://doi.org/10.1016/j.indcrop.2012.08.001

Senthilkumar, K., Rajini, N., Saba, N., Chandrasekar, M., Jawaid, M., Siengchin, S., 2019a. Effect of Alkali Treatment on Mechanical and Morphological Properties of Pineapple Leaf Fibre/Polyester Composites. *J. Polym. Environ.* 27, 1191–1201. https://doi.org/10.1007/s10924-019-01418-x

Senthilkumar, K., Saba, N., Chandrasekar, M., Jawaid, M., Rajini, N., Alothman, O.Y., Siengchin, S., 2019b. Evaluation of Mechanical and Free Vibration Properties of the Pineapple Leaf Fibre Reinforced Polyester Composites. *Constr. Build. Mater.* 195, 423–431. https://doi.org/10.1016/j.conbuildmat.2018.11.081

Senthilkumar, K., Saba, N., Chandrasekar, M., Jawaid, M., Rajini, N., Siengchin, S., Ayrilmis, N., Mohammad, F., Al-Lohedan, H. A. 2021. Compressive, Dynamic and Thermo-Mechanical Properties of Cellulosic Pineapple Leaf Fibre/Polyester Composites: Influence of Alkali Treatment on Adhesion. *Int. J. Adhes. Adhes.*106, 102823. https://doi.org/10.1016/j.ijadhadh.2021.102823.

Senthilkumar, K., Siva, I., Rajini, N., Jeyaraj, P. 2015. Effect of Fibre Length and Weight Percentage on Mechanical Properties of Short Sisal/Polyester Composite. *Int. J. Comput. Aided Eng. Technol.* 7, 60. https://doi.org/10.1504/IJCAET.2015.066168.

Senthilkumar, K., Subramaniam, S., Ungtrakul, T., Kumar, T. S. M., Chandrasekar, M., Rajini, N., Siengchin, S., Parameswaranpillai, J. 2022. Dual Cantilever Creep and Recovery Behavior of Sisal/Hemp Fibre Reinforced Hybrid Biocomposites: Effects of Layering Sequence, Accelerated Weathering and Temperature. *J. Industr. Text.* 51, 2372S–2390S. https://doi.org/10.1177/1528083720961416.

Shahroze, R. M., Chandrasekar, M., Senthilkumar, K., Senthil Muthu Kumar, T., Ishak, M. R., Rajini, N., Siengchin, S., Ismail, S. O. 2021. Mechanical, Interfacial and Thermal Properties of Silica Aerogel-Infused Flax/Epoxy Composites. *Int. Polym. Process.* 36, 53–59 https://doi.org/10.1515/ipp-2020-3964.

Sibaly, S., Jeetah, P., 2017. Production of Paper from Pineapple Leaves. *J. Environ. Chem. Eng.* 5, 5978–5986. https://doi.org/10.1016/j.jece.2017.11.026

Sidney, P., 2021. Nike's "Happy Pineapple" Pack Is Made With Pineapple Leaf Waste [WWW Document]. Futurevvorld. URL https://futurevvorld.com/footwear/nike-happy-pine-apple-pinatex-cork-sneakers-air-force-1-air-max-90-95-zoom-type-free-trail-run/ (accessed 4.28.22).

Siow, L.-F., Lee, K.-H., 2016. Canned, Frozen and Dried Pineapple, in: Lobo, M.G., Paull, R.E. (Eds.), *Handbook of Pineapple Technology*. John Wiley & Sons, Ltd, Chichester, UK, pp. 126–139. https://doi.org/10.1002/9781118967355.ch7

Sisti, L., Totaro, G., Vannini, M., Celli, A., 2018. Retting Process as a Pretreatment of Natural Fibers for the Development of Polymer Composites, in: Kalia, S. (Ed.), *Lignocellulosic Composite Materials*, Springer Series on Polymer and Composite Materials. Springer International Publishing, Cham, pp. 97–135. https://doi.org/10.1007/978-3-319-68696-7_2

Sneakers Made From Pineapple Leaves, 2020. Hum. Are Vain. URL https://humansarevain.com/sneakers-made-from-pineapple-leaves/ (accessed 4.28.22).

Thiagamani, S. M. K., Pulikkalparambil, H., Siengchin, S., Ilyas, R. A., Krishnasamy, S., Muthukumar, C., Radzi, A. M., Rangappa, S. M. 2022. Mechanical, Absorption, and Swelling Properties of Jute/Kenaf/Banana Reinforced Epoxy Hybrid Composites: Influence of Various Stacking Sequences. *Polym. Compos.* https://doi.org/10.1002/pc.26999.

Todkar, S.S., Patil, S.A., 2019. Review on Mechanical Properties Evaluation of Pineapple Leaf Fibre (PALF) Reinforced Polymer Composites. *Compos. Part B Eng.* 174, 106927. https://doi.org/10.1016/j.compositesb.2019.106927

Uma Devi, L., Joseph, K., Manikandan Nair, K.C., Thomas, S., 2004. Ageing Studies of Pineapple Leaf Fiber-Reinforced Polyester Composites. *J. Appl. Polym. Sci.* 94, 503–510. https://doi.org/10.1002/app.20924

Venkata Deepthi, P., Sita Rama Raju, K., Indra Reddy, M., 2019. Dynamic Mechanical Analysis of Banana, Pineapple Leaf and Glass Fibre Reinforced Hybrid Polyester Composites. *Mater. Today Proc.* 18, 2114–2117. https://doi.org/10.1016/j.matpr.2019.06.484

Vincent, O.A., Rachael, T.B., Oyeniyi, O.S., 2016. Assessment of Feeding Value of Vegetable-Carried Pineapple Fruit Wastes to Red Sokoto Goats in Ogbomoso, Oyo State of Nigeria. *Afr. J. Biotechnol.* 15, 1648–1660. https://doi.org/10.5897/AJB2016.15257

Vinod, B., Sudev, L.J., 2018. Study on Electrical Properties of PALF Reinforced Bisphenol-A Composite. *MATEC Web Conf.* 144, 02006. https://doi.org/10.1051/matecconf/201814402006

Wan Nadirah, W.O., Jawaid, M., Al Masri, A.A., Abdul Khalil, H.P.S., Suhaily, S.S., Mohamed, A.R., 2012. Cell Wall Morphology, Chemical and Thermal Analysis of Cultivated Pineapple Leaf Fibres for Industrial Applications. *J. Polym. Environ.* 20, 404–411. https://doi.org/10.1007/s10924-011-0380-7

Yusof, Y., Yahya, S.A., Adam, A., 2015. Novel Technology for Sustainable Pineapple Leaf Fibers Productions. *Procedia CIRP* 26, 756–760. https://doi.org/10.1016/j.procir.2014.07.160

3 Jute Fibre-Reinforced Polyester Composites

M. Ramesh
KIT-Kalaignarkarunanidhi Institute of Technology

D. Balaji and L. Rajeshkumar
KPR Institute of Engineering and Technology

CONTENTS

3.1 INTRODUCTION

Natural fibre does have many benefits over artificial fibres as a reinforcing agent, including renewability (as shown in Figure 3.1), low cost, lightweight, as well as other satisfactory standards [1]. Ecologically responsible alternative solutions to traditional reinforcement materials (synthetic fibres such as polyester as well as rayon) include natural fibres. Due to the hollow as well as cellular nature of natural fibres, they have

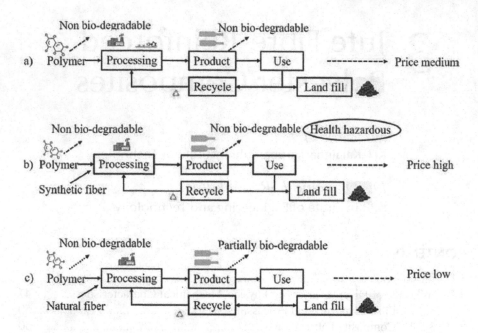

FIGURE 3.1 Environmental impact of various fibres [62].

a higher surface area as well as lower bulk density, making them less fuel-intensive to process [2]. Synthetic fibrils such as glass, but there are more, have a higher fibre density (40% greater than natural materials) and therefore are problematic in terms of machinability and recycling. Because of this, natural fibre-reinforced composites are becoming increasingly popular in various industries, including automotive and construction. New design flexibility, enhanced corrosion, resistance to abrasion as well as fatigue life are some of the advantages of composites [2]. The cellulose fibre element of fibre-reinforced polymeric materials could provide considerable low-temperature preparation, cost benefits, renewable as well as environmentally friendly properties. Organic fibre-reinforced research focuses on a variety of areas, such as the substitution of conventional compositional or aesthetic components with cost-effective alternative solutions, lighter weight as well as high strength. Polymeric substance was used to make 380 million metric tonnes of plastic products, which were consumed in 2013, as per Sayeed and co-researchers [3]. These plastic materials are also not compostable and produce enormous amounts of waste when disposed of in landfills, emit carbon dioxide even during predominant material preparation instead recycling processes [4]. Therefore, the next era would then inherit a healthier environment thanks to the addition of organic fibres to polymers.

Many countries have access to abundant supplies of jute, making it one of the world's leading fibre-producing crops [5]. Jute fabrics are stronger and more ductile than other fibres and yarns. As for the matrix materials, epoxy, phenolic, and polyester have indeed been extensively utilized. Despite this, polyester matrix is inexpensive, easy to process, as well as easily obtainable in the industry [6,7]. Even though thermosetting polymer production can be done in a variety of ways, the hand lay-up method is the most convenient, energy-efficient, and cost-effective method [3]. The effectiveness of various layers of jute fibre, mat, as well as the combined effect with these other natural fibres [8,9], synthetic fibres [9], and metals [6] on polypropylene composites has already been studied extensively. Nevertheless, because it is unclear which jute variety was used, the influence of jute diverse array on the thermosetting composite material cannot be determined with certainty. Besides that, the Bangladesh Jute Research Institute (BJRI) has recognized approximately 2000 jute genotypes from either the country as well as abroad [10] along with the requirements for jute fibre transformation as the retting method progresses. BJRI developed Tossa (O-9897) jute variety and distributed through the Department of Agriculture for Extension to all of the country's agricultural producers. The physico-mechanical effects of this jute variety on polyester composite must be carefully studied. In addition, jute sheet layers in the composite increase the final product's cost, but it is not clear how much of an impact this has on their mechanical properties [11–13]. Jute fabrics manufactured from Tossa jute diverse array were used in the polyester binder and their physical and mechanical properties were studied.

3.2 INFLUENCE OF DIVERSE LAYERS ON PHYSICO-MECHANICAL CHARACTERISTICS

The jute fibre in the composite materials is optimized to meet mechanical requirements. Comparatively, the physical and mechanical properties of layered jute composites were superior to those of other polyester–jute composites. Nevertheless, a two-layer jute fabric material was found to be more cost-effective in terms of sustainability. This jute material has a broad range of applications, including furniture and kitchen interior design materials [14]. Composite materials with high-performance properties were created by incorporating fibre-reinforced polymers into the matrix. Discontinuous and continuous fibres were used as reinforcement agents by scientists. Despite the fact that continuous fibres are stronger, discontinuous fibres are less so. Therefore, researchers used continuous fibre reinforcement to increase the composite's axial sturdiness. For continuous fibre composite materials, laminate configuration is a common one. When two or more layers of reinforcement materials have been layered together in a reinforcement polymer, the result is a laminated composite [15,16].

For sporting, aeronautic, automotive, as well as marine applications, laminated composites seem to be fascinating since these composites produce substantial mechanical performance and seem to be lightweight [17,18]. Numerous laminated fabrics, including unidirectional, woven, nonwoven, knitted, as well as braided laminate fabrics [19–21], have been used as reinforcement fabrics to fix engineering challenges. Amongst some of the various laminated fabric types, woven structures have been used as reinforcement materials by several investigators. Weaved fabrics have a number of advantages over unidirectional fabrics that are well known [22], and they are easy to handle and extremely resistant to impact. A woven fabric seems to be reliable because of the yarn fibres there in warp as well as weft directions. Load can be distributed in whatsoever longitudinal or transverse direction, and it retains the same shape or dimensions. It is also fascinating that the knitted fabric is easy to work with during the manufacturing process. Orthogonally angles 0° and 90° are the most commonly used in weaving. Weaved fabric comes in a variety of forms, including plain, twill, satin, as well as basket weaves. The much more frequently utilized laminated composite material is a woven fabric with something like a simple architectural design. The tensile, flexural, as well as impact characteristics of woven laminated composites are superior to those of unlaminated composites when viewed out of plane [23,24].

Composite materials made from petroleum-based as well as renewable matrix polymers can be made with natural fibre reinforcement. Bio-based plastics and natural fibres are combined to create green composites [25]. Using natural fibre-based composites instead of synthetic fibres is a more environmentally friendly option. A life cycle assessment has also been conducted to evaluate long viability of natural fibres and their composite materials as reinforcing materials in previous studies [26,27]. A significant increase of lightweighting can indeed be attained by simply replacing artificial materials with organic fibres, especially in automotive as well as structural applications, which have a number of environmental benefits, such as recyclability as well as renewable resources. Inconsistent fibre dimensions and chemical composition are two of the many problems with natural fibres. When developing innovative laminated structures, the mechanical characteristics of organic woven fibres are critical [28–31].

These natural fibre-reinforced composites are difficult to manufacture due to their physical characteristics, resulting in lengthy and labour-intensive mechanisms as well as increase in cost. The woven and laminated variables of organic knitted laminated composites are just two of the many variables at play. Laminated variables include volume/weight fraction, layering sequence, layering size, as well as fabric orientation, whilst knitted variables include yarn crimp, yarn twist, linear density, as well as fabric structures. Knitted laminated composites are more difficult to model than continuous fibre composites because of these factors. The in-plane characteristics of woven composite material can be determined mathematically using a variety of micromechanical modelling techniques. Unidirectional composite materials are used in many micromechanical models, including Halpin–Tsai, inverse rule of mixture (IROM), Hirsch, as well as rule of mixture (ROM). According to one such research, the tensile modulus of unsaturated polyester composites made of single and dual-reinforced jute and ramie was determined using micromechanical models such as Halpin–Tsai and Hirsch. A graph was used to compare experimental as well as micromechanical model values to determine which theory values have been extremely consistent with the experimental value systems [32,33].

Using natural fibres as reinforcements in polymer matrices has resulted in lightweight, renewable, and non-hazardous products. Polymer matrix composites have been prepared to reinforce natural fibres including *Acacia leucophloea*, elephant grass, isora, jute, fish tail palm, and banana, but also date fibres as well as *Ipomoea staphylina*, wild cane grass, and bamboo fibres [34–37]. The physical, chemical, and thermal characteristics of abaca/bamboo composite materials were studied. Resin transfer moulding was used to create the composite plates. Using transverse thermal conductivity tests, researchers discovered two distinct patterns of rising conductivity for bamboo and abaca fibre composites, respectively. E-glass and hemp fibre composites were compared. Researchers discovered that hemp fibre has superior reinforcement characteristics for composite materials than flax because it is more environmentally friendly [38,39].

Previously, a study on the fish tail palm-fibre polymer composites was conducted. At a volume fraction of 0.50, the mechanical characteristics of composites were already at their best [40]. An investigation of the golden cane grass fibre polymer composites under various volume fractions was carried out. Compared to pure polyester-based composites, composites' mean tensile strength and modulus increased [41]. The coconut sheath/polyester composites were discussed and found that the composites had a higher flexural as well as impact strength than other composites, according to the tests. The influence of silane treatment on coconut sheath-reinforced polyester composites was studied and found that the strength of fibre composites that have undergone surface treatment is higher than that of similar composites made with untreated fibres. The mechanical characteristics of hemp as well as palmyra fibre-reinforced polyester composites were predicted. Palmyra and hemp fibres had higher tensile properties when the fibres were 30 mm long [42,43]. Sun hemp fibre/polymer composite materials were investigated in terms of fibre weight ratio as well as length at 55%, and fibres of 30 mm length were found to have the greatest strength [44,45]. When the volume fraction (V_f) of snake grass natural fibre-reinforced polymer composites was varied between 10% and 30%, the best tensile strength was accomplished at 25% V_f, while the best flexural strength was accomplished at 25% V_f for 120 and 150 mm length. Few researchers investigated the influence of fibre loading on polyester composites made of banana fibres. Fibre loading was found to have a positive impact on composites strength, according to the findings. Compression moulding was used to create the sisal/hemp/jute/coir-reinforced composite materials by some researchers. Composites made using the compression moulding technique were documented and the various characteristics of the composite materials were analysed. The laminates were found to have improved mechanical as well as thermal properties as a consequence of the chemical treatment [46–48].

Using a hand lay-up method, researchers fabricated natural fibre hybrid composites to study their properties. The mechanical properties of hybrid composites were superior to those of non-hybrid composites. Banana and jute fibre-reinforced epoxy composites were investigated by few researchers, and the mechanical properties of the banana and jute fibres reinforced with epoxy matrix were increased by up to 30% weight percent [49–51]. Some researchers studied bamboo fibre-reinforced polyester composites containing between 10% and 60% bamboo fibres. For example, voids, fibre bending, and matrix failure were discovered by scanning electron microscopy (SEM) analysis. Natural fibre composites made from jowar as well as polyester resin were developed. Composites based on the jowar fibre, as opposed to sisal as well as bamboo fibres, demonstrated superior tensile and flexural strength. Some researchers investigated composites made from short, randomly oriented sansevieria cylindrical fibres and polyester. According to some other researchers, a critical fibre length of 30 mm and an optimum fibre weight percent of 40% were found. Some researchers used waste grass broom fibres to create green composites. Variations in fibre content and temperature were used to study thermal properties [52–54]. They focused on random as well as unidirectional oriented polymer composites using natural fibres such as banana, jute, bamboo, elephant grass, snake grass, as well as wild cane grass in a review performed by them. By ranging the weight percent of Indian mallow fibre-polyester composites from 10% to 50%, the various thermal and mechanical characteristics have been evaluated [55,56].

Jute, a natural fibre, is in high demand in a wide range of industries. Jute fabrics are becoming increasingly popular, both among furniture designers and consumers, for their aesthetic appeal. Jute fabrics-reinforced composite materials have been therefore developed using jute fabrics with different types of matrix materials to improve their efficiency according to their final use. Layered jute–polyester composites were studied to determine their physical as well as mechanical properties. The jute fabrics have been incorporated into unsaturated polyester resin as well as hardener by the easy hand lay-up method at room temperature as well as humidity situations to create different layers of composites. Analysis of the fracture surface of the two-layer, four-layer, and six-layer jute–polyester composites was carried out for determining physico-mechanical characteristics. These composites provided significant cost savings, adequate strength, and environmental-friendly attributes [57–60].

3.2.1 FIBRE HANDLING AND PROCESSING

When treated with emulsion (25% of weight), jute fibres were allowed to rest for 48 hours before being spun in BJRI's mechanical processing department, where they were then passed through a series of steps culminating in the creation of a 10-count yarn. A piconol double rapier (Belgium) loom was used to weave 15 ends per inch and 15 peaks per inch hessian fabrics in the weaving division of BJRI after beam preparation [61]. Inside the textile physics section of BJRI, composite materials made of jute and polyester have been created (Figure 3.2).

3.2.2 COMPOSITE FABRICATION

To reduce the moisture content prior to actual preparation of the composite, 12×12 jute fabrics were cut as well as baked for 6 hours at 105°C to 110°C. At room temperature and humidity, polyester matrix was used to create jute–polyester composites. Unsaturated polyester resin and hardener are being used in the hand lay-up technique to make jute–polyester composites with two, four, and six layers. Methyl ethyl ketone peroxide (MEKP) was used as a hardener. Jute composite materials have been cured at 25°C for 24 hours at room temperature [62]. Characterization could begin well after the curing process (Figure 3.3).

Hand lay-up and compression moulding were used to create an unsaturated polyester composite material of woven jute with ramie reinforcement. $300 \times 300\,mm^2$ of

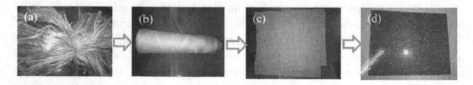

FIGURE 3.2   Jute fibre at various stages [62].

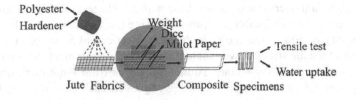

FIGURE 3.3 Schematic for composite specimen preparation [62].

plain-woven jute and ramie fabric are cut. Through the use of an analytical balance, we calculated a 1:50 mixture of unsaturated polyester and MEKP. When the composite plate had dried, a discharge agent was sprayed upon that surface of the mould to aid removal. Above the mould, weaved fabric was placed. The woven fabric was therefore sprayed within the resin mixture. Excess air was removed from fabric's surface using a roller, and the resin was distributed evenly throughout the fabric. A load somewhere between 100 and 200 N was applied to the mould. Curing appears to take about 8 hours at ambient temperature of 22°C–24°C, and ASTM 3039-compliant specimen was then cut from the finished composite material [63]. The specimen is of 25 mm (length)×25 mm (width)×3 mm (thickness). An universal testing machine (UTM) (INSTRON 3369) fitted with a 10 kN load cell was used to carry out the tensile testing. The crosshead speed was set to 1 mm/min on the UTM machine. There in the direction of such warp, the fabric was evaluated. Seven samples were given to each group. The results of the screening were subjected to statistical analysis (Analysis of Variance). The four micromechanical frameworks used in this study include the ROM, the IROM, Halpin–Tsai, as well as Hirsch models [64–66].

A scissor was used to chop the jute fibres into 15 mm lengths, as suggested by Venkateshwaran et al. [67]. The polyester matrix was created by combining polyester resin with 2% catalyst as well as 2% accelerators. After that, the mixture was thoroughly whisked to ensure even mixing. Hand lay-up and static compression were used to create the composites, which contained a constant 16% of total fibres in each one, following preparation of reinforcement as well as matrix materials. A stainless steel mould with size of 300 mm×200 mm×3 mm was used to create the 3-mm thick composite laminate during the fabrication method. Before being eliminated from the mould, thus every composite's cast was given 24 hours to cure under a weight of 50 kg. After curing, the laminates were easily removed from the mould with the use of silicon spray. The mechanical, thermal, and water absorption characteristics of the samples have been examined after they have been cut to suitable ASTM sizes with a diamond cutter.

3.3 PROPERTIES

3.3.1 MECHANICAL PROPERTIES

Tensile strength can be used to describe the tensile behaviour because all parameters, including the sample size as well as thickness, have been kept constant throughout tensile strength testing. Tensile test specimens made of jute composites have been cut to ASTM D3039 specifications from each of the three materials

[55]. The Instron universal testing machine was used to test the tensile strength of various layers of jute–polyester composite. During the tensile testing, the load cell had a capacity of 50 kN and the cross head had a speed of 5 mm per min. Strain to failure and tensile strength were calculated based on the machine data. ASTM D3039 was then used to calculate Young's modulus from the strain–stress plot. Accuracy was determined by testing five samples in each case [68,69]. Fracture surfaces of layered composite materials (Figure 3.4) were examined following tensile testing.

Zinc oxide (ZnO) nanoparticles have been shown to enhance the interfacial adhesion among resin, reinforcement, as well as filler in a tensile test of jute fibre treatment. So the composite's tensile strength is improved due to strong adhesion. There is little difference in the tensile strength of fabricated composites. According to these results, the scoured process had lower hydrogen bonding interaction and therefore lower tensile strength than other process, which increased the size of the fibre's OH groups. The chain movement is constrained by the increased hydrogen bonding. As a result, chains that are subjected to tensile force have a difficult time sliding. There are three other sets of specimens that exhibit the same tensile strength behaviour. When tested against other materials, the best tensile strength was found in 40 g/L treated jute-reinforced composite. Similarly, the finest tensile strength characteristics of composite materials were obtained by adding ZnO nanoparticles to the matrix. As a result, the highest tensile characteristics were found in composites made with 40 g/L treated reinforcement as well as ZnO particles in the matrix. The tensile stress–strain contours of all specimens were not obtained by plotting, because they all exhibited the same behaviour [70,71].

The tensile properties of untreated as well as treated jute composites in the aspects of tensile strength, tensile modulus, and percentage of strain at break are determined. Due to an efficient surface modification, all treated jute composite materials have superior tensile properties than the untreated jute composites. The poor fibre–matrix adhesion in untreated jute composites will have the least tensile characteristics: (1) alkali-treated jute composite seem to have a tensile strength of 12% and tensile modulus of 17% higher than jute composite. Several researchers have found that alkali treatment increased the tensile characteristics of jute composites [72–75]. This enhanced interfacial adhesion seen between matrix and fibres is due to alkali treatment increasing the surface roughness of the fibre by eliminating hemicelluloses and lignin. (2) PLA-coated jute composite does have higher strength than jute composite but lower strength than other treated composites owing to mean bonding among fibres as well as matrix. (3) To put it simply, the treated one is the strongest and most

FIGURE 3.4 Tensile fracture of (a) two-layer, (b) four-layer, and (c) six-layer jute–polyester composite [62].

elastic of all jute composites, with tensile strength, modulus and % of strain at breaks all 20% greater than jute composite. (4) The combination of alkali treatment as well as PLA coating may be responsible for the composite's high adhesion.

3.3.2 WATER ABSORPTION TEST

Using distilled water and the ASTM D570-98 water uptake criterion of 24, 48, and 72 hours at room temperature, different jute fabric–polyester composites were investigated [64]. Five samples from each category were subjected to the testing process. Following Fan et al. [76], the impact of water absorption on jute fabrics-reinforced composite materials was examined. When the specimens have been weighed, they were cut into the 76.2×25.4 mm dimensions and dried in an oven at 50°C before cooling to ambient temperature in desiccators. This procedure was carried out over and over again until the specimen's mass was stabilized. The specimens were subjected to a series of timed water absorption tests in a beaker filled with distilled water. Tissue papers were used to wipe samples from the water, then weighed and re-inserted into the water. Within one minute of being removed from the water, the samples were reweighed to the nearest 0.1 mg. As a result of the process, equilibrium was achieved. The percentage of water absorption was calculated using equation (3.1):

$$\Delta M(t) = \frac{m_t - m_o}{m_o} \times 100, \tag{3.1}$$

where weight of the specimens after soaking inside the solution and their initial weights are m_t and m_o in this example.

Young's modulus is the ratio of stress to strain during tensile testing, and the maximum load is referred to as tensile strength. Young's modulus and tensile strength were found to enhance with increasing layers of the load-bearing component. In addition, the load is transferred from one fibre to the next by the matrix material. There in jute–polyester composites, the strain to failure enhanced in direct proportion to the increase in jute fabric layers. It was found that the natural fibre inhibited/protected ductility of polyester, as the jute fibre layers or bundles debonded or delaminated under tensile stress. In addition, the thickness of the composite increased slightly with the addition of jute fabrics. Both the warp and the weft yarn counts were equal in plain-woven fabrics. Therefore, the tensile criteria would then be the same irrespective of whether the fabric is warped or wefted. Prediction of physical and mechanical properties of various jute–polyester composite layers is the focus of the current study. Using jute materials, particularly Tossa jute fabrics, can improve the tensile strength of polyester. Increased jute content in the polyester reduces the water absorption of jute–polyester composites. Jute–polyester composites should indeed be coated in which moisture contact is essential. Because of the degradation of mechanical strength due to swelling of jute fibres at the fibre–matrix interface, further research is needed. Moisture uptake is reduced when fibre content is low, but mechanical criteria are increased when fibre content is high. Using a two-layered composite instead of a four-layered composite for a moderately weighted product is preferable [77–79].

The existence of hydroxyl groups in natural fibres causes water absorption, which is one of the greatest disadvantages of natural fibres. Micro-gaps in the polymer matrix and interfacial bonding among fibres along with matrix are the primary mechanisms for water absorption in natural fibre-reinforced polymer composites. In addition, the bulging of the fibres caused micro-cracks to propagate, which affected water absorption. Jute composites show a linear rate of water assimilation at the outset. Over a specified length of time of soaking, the fraction of water assimilation slows down and reaches saturation. The Fickian diffusion method can be used to model the conduct of jute composites [65]. The slope of something like the water absorption versus the time graph could be used to determine the rate of water assimilation, which is noticed to be smallest for such composites because of strong adhesion that severely prohibits the advancement of micro-voids. Due to the hydrophilic existence of jute fibres as well as micro-voids in the composite, the jute composite shows the highest percentage of water absorption. Treated jute composites seem to have 20% lower water absorption compared with untreated, alkali-treated jute composites. Alkali treatment may be the cause for this. Separation of hemicellulose and lignin in alkali treatment result in a decrease in fibre diameter, which results in an increase in the aspect ratio, which means that more fibres seem to be readily accessible for adhesion [80,81].

Sampathkumar et al. [66] found that alkali treatments convert the hydrophilic existence of jute fibre into the hydrophobic existence, which lowers absorption of water. Both untreated and treated jute composites have lower water penetration than PLA-coated jute composite, which is better at resisting water absorption because the PLA coating isolates the jute fibres out from water getting absorbed. As a result, treated composite does have the least fraction of a percent of water absorption (4.02), which is 42% lesser than that of the untreated composite. This may be attributable to the combination of alkali treatment as well as PLA coating that offers the greatest fibre–matrix adhesion, negligible micro-voids, converts the hydrophilic nature into the hydrophobic nature, and keeps the fibres separated from water. Diffusion, sorption, and permeability coefficient measurements were made on untreated as well as treated jute composites. The Fick's model's main output is the diffusion coefficient that demonstrates the capacity of water molecules to diffuse into composite materials. Due to the hydrophilic existence of jute fibre as well as micro-voids in the composite, the natural jute composite had the greatest diffusion coefficient. However, characterized composites had lower values because hemicellulose and lignin, which are the primary causes of water assimilation, have been removed. Diffusion coefficients derived from this research seem to be very similar to those already identified in the literature. A further important result of the Fick's model is the sorption coefficient that also provides the capability of composites to resist water molecule diffusion. Pure jute composite was found to have the lowest value, while treated jute composite materials had higher values. It must have been expected that jute composite would have the highest permeability coefficient because of its water absorption capacity [68,69].

3.3.3   RULE OF MIXTURE (ROM) ALONG WITH INVERSE ROM

The mechanical properties of composites can be calculated and predicted using the ROM. Depending on the direction of the fibres, the mixture rule can be altered. In spite of this, there are a number of issues with the ROM. To predict exactly the tensile performance, numerous considerations, including stress and strain raisers from embedded reinforcements, interface failure along with statistical dispersion effect, presence of void as well as unevenness [], must be taken into account. Errors in data collection lead to an overestimation of the utmost tensile strength calculated from ROM data. Fibre orientation is presumed to be sequentially connected as well as homogeneously dispersed according to the assumptions of ROM. Although its spread and alignment appear to be consistent, the fibre exhibits a degree of non-homogeneity. ROM doesn't really help anticipate the transverse orientation of a consistent fibre [44,52,72].

3.3.4 MICROMECHANICAL MODEL FOR UNIT FIBRE-BASED COMPOSITE MATERIAL

In both single jute and ramie composite materials, modulus vs. volume fraction is plotted by investigation. They have been using ROM, Halpin–Tai, as well as Hirsch to determine the conceptual toughness of their composite materials. The elasticity modulus tends to increase the percentage of such fibre length. Studies and theoretical values follow the same patterns. There were some discrepancies between the evaluated micromechanical model's elasticity modulus and that of the investigational one. At the relatively short end of the fibre, the estimated values have been compared with the investigational value. Nevertheless, because the volume was so much smaller, the variance was much greater. Fibre–matrix interactions may be weaker at higher volume fractions. The modulus value for both jute and ramie fibre composite materials enhances with increase in volume fraction. All volume fraction distributions have higher ROM values when measured axially. Using the findings, it is clear that the axial ROM generates the highest possible results. The relatively low bound is obtained by running the ROM in the perpendicular direction. The ROM outcome is significantly different from the investigational one. The combination of mixture law anticipates a higher bound for the tensile characteristics of the blended system. The blended system's lack again for impact of fibre orientation as well as fibre interaction with matrix is to blame for this issue. Halpin–Tsai's micromechanical model characterizing elasticity modulus $x = 1$ shows nearly ideal contract to data acquired from single jute composite materials in volume fractions ranging from 20% to 25%. The Halpin–Tsai micromechanical design with an empirical variable $x = 1$ doesn't really fit the experimentally obtained value for single ramie fibre composites. The Halpin–Tsai model with empirical parameter $x = 2$ hardly appears to fit ramie fibre composite material values with volume fractions varying from 0% to 20% [38,49].

3.3.5 INFLUENCE OF ALKALI TREATMENT AS WELL AS POLY (LACTIC ACID) COATING

Immerse jute fibres in 5% NaOH solution for 30 minutes at 30°C. As a final step, the fibres were immersed in a solution of very dilute hydrochloric acid to eliminate the NaOH from their surfaces. A hot air oven was used to dry the fibres for 24 hours at 70°C after they were scrubbed multiple times with distilled water. NaOH reacts with jute fibres in the following manner (equation 3.2).

$$Fibre - OH + NaOH \rightarrow Fibre - O^-Na^+ + H_2O \qquad (3.2)$$

3.3.6 Covering of PLA-Based Fibres

Originally, PLA pellets were immersed in a chloroform solution containing 2% PLA weight per volume for eight hours. In order to ensure a uniform distribution of PLA into the chloroform solution, the solution was stirred manually as well as heated to 60°C, and then the PLA was added. Jute fibres were immersed in PLA solvent and then extracted instantaneously. As a final step, the coated fibres were dried in an oven at 40°C for 6 hours at room temperature.

3.3.7 Scanning Electron Microscopy Analysis

Prior to the creation of composites, the SEM evaluation of fibres is a critical step in predicting their adhesion to polymers. Consequently, the SEM images' failure surfaces (tensile, flexural, and impact) of jute fibres utilized as reinforcement in polyester matrix are shown in Figure 3.5. It shows the smooth texture of untreated jute fibres. Hemicelluloses and lignin may be the cause for keeping the surface smooth. Composites show a rough surface as a result of alkali treatment removing hemicelluloses and lignin. Rougher fibres allow the polymer matrix to wet a larger area of the fibre, enhancing the adhesion between the fibre and matrix [64,82–84]. A PLA coating on the exterior of fibres, acting as an outer non-uniform rough thin sheet on the fibres' surface, could also enhance the surface harshness of fibres without removing hemicelluloses and lignin. Moreover, PLA coating can change natural fibres' hydrophilic nature into a hydrophobic one, increasing the fibres' adhesion to polymers. Due to the consolidated impact of alkali treatment as well as PLA coating, composite has extremely high surface roughness.

3.3.8 Flexural Characteristics

Flexural characteristics of untreated as well as treated jute composites on aspects of flexural strength, flexural modulus, and strain at break proportion were analysed. Flexural characteristics show the same trend as tensile characteristics. The flexural strength and modulus of treated composite seem to be 35% and 49% higher than the untreated ones, respectively, while the flexural strength and modulus of composite with coated seem to be 37% and 59% higher than those of untreated. Due to good interfacial bonding among fibres and matrix, high flexural characteristics in coated composite could be due to effective stress transmission from matrix to fibres. This could be explained by the interaction between the polar jute fibres and the non-polar polymer matrix in jute fibre-reinforced polymer composites, which usually results in poor interfacial adhesion [85]. The flexural characteristics of PLA-coated jute fibres are improved because they behave more like non-polar materials and adhere to the polymeric matrix better. Flexural characteristics, however, are shown by treated composite to be in the middle range. Observe that the proportion of strain at break increases significantly after treatment, with the greatest enhancement found in the coated composite.

The composite specimens have higher flexural strength ranging from 45 to 50 MPa for specimens, which had a maximum value of 55 MPa. The ZnO nanoparticles enhance the interfacial adhesion of the matrix, reinforcement, as well as filler which

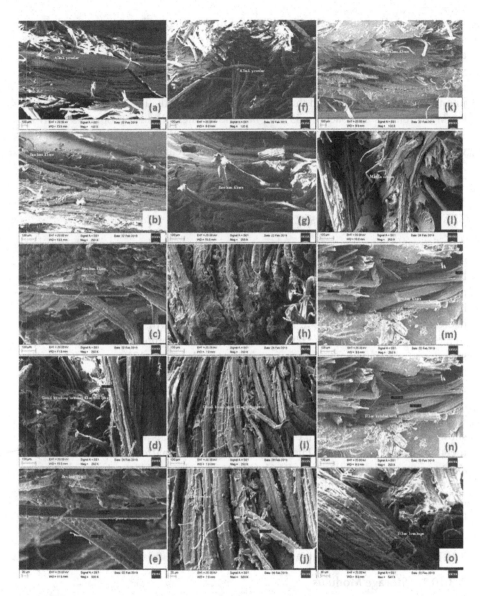

FIGURE 3.5 Fracture morphology of jute fibre-polyester composites due to (a–e) tensile fracture, (f–j) flexural failure, and (k–j) impact loading [64].

can be seen from this pattern of behaviour. Increased flexural strength can be achieved by adhering strongly to the composite. Additionally, the flexural strength behaviour of all untreated composite samples is nearly identical, as can be seen. Flexural strength may be affected by treatments that focus on reinforcement, as evidenced by this behaviour. The flexural strength of treated composite sample is superior to that of untreated one [49,52]. Because the mercerized method produced larger-OH groups on the fibres, the hydrogen bonding interactions between the polymer chains of jute were stronger, as well as this resulted in higher flexural strength. The chain can only move so far because of the tighter hydrogen bonds. As an outcome, chain flexural force applications lead to inadequate chain sliding. The flexural characteristics of composite materials are illustrated by the interaction of the matrix and composite reinforcements. When particularly in comparison to other materials, a 40 g/L HFC-treated jute-reinforced composite had the highest flexural strength. The coated samples with flexural strengths approaching 55 MPa had the greatest flexural properties of all the composites tested when ZnO nanoparticles were added to the matrix. Because of this, we can say that composites made with 40 g/L HFC-treated reinforcement as well as ZnO particle-added matrix exhibited the best flexural characteristics in this study. Tensile characteristics were also found to exhibit this phenomenon [60,64,86].

3.3.9 IMPACT CHARACTERISTICS

Alkali-treated jute composite has lower impact strength than untreated jute composite, which is fascinating. Few researchers experienced a related result. That's because fibre pull-out is an important factor in determining how much energy a composite can absorb while in an impact assessment; alkali treatment strengthens the bond among fibres as well as matrix, reducing the amount of fibre pull-out and thus decreasing the impact strength. In terms of impact strength, untreated is lower than treated one. Coated composite had the highest impact toughness (5.30 kilo Joules per square metre), beating out other composite materials by 4%, 131%, and 34%. During the sample impact experiment, the increased capacity to absorb impact energy and resist crack propagation is attributed to deformation of a ductile PLA skin wrapped around jute fibres. The composite samples' impact strength was measured at 10.01 J. Together, all composite material samples had nearly identical impact values. This shows that the impact strength of such composite materials was unaffected by treatments on the reinforcement. The impact strength was unaffected by the addition of ZnO nanoparticles. Adding only 1% ZnO nanoparticles could have had an impact on the results. Specimens' impact strength was unaffected by this small quantity. No matter what type of sample was used, the impact test yielded the same results; therefore, these experiments were just not even further examined [28,39,44].

3.3.10 DYNAMIC MECHANICAL CHARACTERISTICS

3.3.10.1 Storage Modulus

The storage modulus is indeed a measure of the quantity of elastic energy that can be sequestered in a material all through deformation. The stiffness as well as load-bearing capabilities of a material can be determined by the modulus of Young's modulus [85]. Untreated jute composite materials have lower storage modulus than those that have been treated. Those characterized composite materials have higher E'' values than untreated jute composite in the glassy region. Coated composite which had the highest storage modulus was followed by other composites. As the temperature rises, the storage modulus among all composite materials decreases, which is most likely due to the fibres losing their stiffness and elasticity. Many researchers have previously reported similar findings [87,88]. There is a gradual decrease in the values of E'' for all composites within the transition zone; however, the integration of treated as well as coated jute fibres in coated one causes a slow decrease in the value of E′. In the rubbery province, this composite possesses the greatest value of E′, accompanied by other combinations. Due to this same strong interfacial adhesion among treated jute fibres as well as the polyester matrix, E′ value is the highest in the family. Coated composite has a better load-bearing potential than only those other composite materials at higher temperatures [89–91].

3.3.10.2 Damping

We must split the loss modulus from the storage modulus in sequence to determine damping. Strength as well as stiffness of reinforcing fibres and the bonding between fibres and matrix all play a role in determining not only the strength but also the stiffness of a material. There is more of an impact with a higher damping (Tan Θ) value. It was determined that the treated composites have good damping properties, even though they have low load-bearing capacity. Coated composite seems to have the least Tan Θ value because of its high fibre–matrix adhesion and lower damping, making it the most stable material. As a result of alkali treatment and PLA coating, both fibre–matrix adhesion and load-bearing capacity have improved. Tan Θ peaks increase indicates an increment in thermal stability, which would be evaluated by Tg. The composite's thermal stability as well as maximum Tg value must be reflected in its highest Tan Θ peak shift to the right [92–94].

3.3.10.3 Loss Modulus

When a material deforms, it releases heat through thermal dissipation, which is measured as loss modulus. Dynamic Tg refers to the point on a polymer's loss modulus curve where it reaches its maximum value. The spikes of either the E″ or Tan Θ curves can be used to calculate the Tg. The value of E″ increases up to Tg and then decreases as the temperature rises. The coated composite has a higher Tg as well as loss modulus because of the addition of alkali treated with PLA-coated jute, which also promotes the progress of excellent bonding as well as reduces the movement of the polyester bonder. This composite has a higher Tg value than only some other composite materials, which indicates its superior thermal consistency [95–98].

3.3.11 Chemical Resistance

Chemical opposition has indeed been found in ZnO nanoparticles reinforced with jute. Adding ZnO nanoparticles to unsaturated polyester results in the best chemical resistance, which is almost five times lower than that of the preliminary untreated composite materials, from 5.45 to 1.51 [96]. Figure 3.6 clearly shows that ZnO nanoparticles added to CR polyester resulted in composites with said optimal outcomes.

3.3.12 Flame Retardancy

The composite materials' flame retardancy was enhanced by adding jute-reinforced ZnO nanoparticles. Samples containing ZnO nanoparticles had the highest burning time of 149 seconds, which was significantly longer than other samples. Because ZnO nanoparticles have such a bigger surface region, they are more resistant to combustibility than other nanoparticles [97,98]. There is also a noticeable difference in flame retardancy between the first two specimens. Jute fibre treatments augmented the attraction

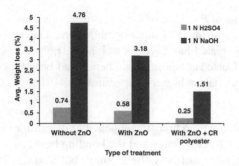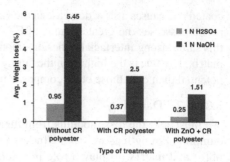

FIGURE 3.6 Influence of treatment over chemical resistance [96].

between both the jute reinforcement and the matrix, which resulted in higher strength. Third, fourth, and fifth specimen sets have the same pattern of behaviour [99–101]. Because resin accounts for 75% of the composite's weight and reinforcement seems to be hardly 25%, this consequence may be marginal. Both unsaturated polyester and ZnO nanoparticle enhanced flame resistance, as well as the highest suitable flame resistance was attained when both are used together. Whenever these two cases are evaluated independently, this same flame resistance was lowered to 182 seconds for unsaturated polyester by itself as well as 160 seconds for ZnO nanoparticles-added composites, which is significantly lower than the previous results [102–105].

3.4 CONCLUSION

Jute fibre-reinforced polyester composites have been discussed from various spectrum initiated with the influence of diverse layers on physico-mechanical characteristics. Fibre handling and processing and composite fabrication are preliminary parameters being understood. The evaluating methods including tensile test, water absorption test, and their characteristics are discussed. Concern to move forward is to be addressed for modelling of composite material for the micromechanical characteristics, which includes the ROM along with IROM, micromechanical model for the single jute fibre-based composite material, influence of alkali treatment as well as coating and covering of PLA-based jute fibres reinforcement composite material. The next realization is about the fabrication method of jute composite material and their corresponding SEM analysis. Thereby, this leads to learn about the other evaluation for their flexural characteristics, impact characteristics along with their dynamic mechanical characteristics including storage modulus, damping, loss modulus, chemical resistance and flame retardancy.

REFERENCES

1. Vimalanathan P, Venkateshwaran N, Srinivasan SP, et al. Impact of surface adaptation and *Acacia nilotica* biofiller on static and dynamic properties of sisal fiber composite. *Int J Polym Anal Charact* 2018;2:99–112.
2. Navin C, Mohammed F. Natural fibers and their composites. *Tribol Nat Fiber Polym Compos* 2008;2008:1–58.

3. Sayeed M, Sayem S, Haider J. *Encyclopedia of Renewable and Sustainable Materials: Opportunities with Renewable Jute Fiber Composites to Reduce Eco-Impact of Nonrenewable Polymers.* New York: Elsevier; 2020.

4. Shahinur S, Mahbub H. *Encyclopedia of Renewable and Sustainable Materials: Jute/Coir/ Banana Fiber Reinforced Bio-Composites: Critical Review of Design, Fabrication, Properties and Applications.* New York: Elsevier; 2020.

5. Townsend T. *Natural Fibres and the World Economy – July 2019.* Hurth: Food and Agriculture Organization of the United Nations (FAO); 2019.

6. Deb A, Das S, Mache A, et al. A study on the mechanical behaviors of jute-polyester. *Compos Procedia Eng* 2017;173:631–38.

7. Sahayaraj AF, Muthukrishnan M, Ramesh M, et al. Effect of hybridization on properties of tamarind (*Tamarindus indica* L.) seed nano-powder incorporated jute-hemp fibers reinforced epoxy composites. *Polym Compos* 2021;42(12):6611–20. https://doi.org/10.1002/pc.26326.

8. Rahman F, Eiamin MA, Hasan MR, et al. Effect of fiber loading and orientation on mechanical and thermal properties of jute-polyester laminated composing. *J Nat Fibers* 2020. https://doi.org/10.1080/15440478.2020.1788485.

9. Ramesh M, Rajeshkumar L, Balaji D, et al. Properties and characterization techniques for waterborne polyurethanes. In *Sustainable Production and Applications of Waterborne Polyurethanes* (pp. 109–23). Cham: Springer; 2021.

10. Hossain A, Miah M, Prodhan MH, et al. *Evaluation of Selected Kenaf (Hibiscus cannabinus) Germplasm.* Dhaka: Bangladesh Jute Research Institute; 2018.

11. Sweety S, Sharmin A, Zakirul I, et al. Impact of different layers on physico-mechanical criteria of jute fabrics polyester composites. *Adv Mater Process Technol* 2021. https://doi.org/10.1080/2374068X.2021.1872246.

12. Mohankumar D, Amarnath V, Bhuvaneswari V, et al. Extraction of plant based natural fibers–A mini review. *IOP Conf Ser* 2021;1145(1):012023.

13. Ramesh M, Rajeshkumar L, Bhuvaneswari V. Leaf fibres as reinforcements in green composites: A review on processing, properties and applications. *Emergent Mater* 2021. https://doi.org/10.1007/s42247-021-00310-6.

14. Mazlan N, Yusoff MZM, Ariff AHM. Investigation of alkaline surface treatment effected on flax fibre woven fabric with biodegradable polymer based on mechanical properties. *J Eng Technol Sci* 2020;52:677–90.

15. Ramesh M, Balaji D, Rajeshkumar L, et al. Tribological behavior of glass/sisal fiber reinforced polyester composites. In *Vegetable Fiber Composites and Their Technological Applications* (pp. 445–459). Singapore: Springer; 2021.

16. Rajesh M, Jayakrishna K, Sultan MTH, et al. The hydroscopic effect on dynamic and thermal properties of woven jute, banana, and intra-ply hybrid natural fiber composites. *J Mater Res Technol* 2020;9:10305–15.

17. Carmisciano S, De Rosa IM, Sarasini F, et al. Basalt woven fiber reinforced vinylester composites: Flexural and electrical properties. *Mater Des* 2011;32:337–42.

18. Rajeshkumar L. Biodegradable polymer blends and composites from renewable Resources. In *Biodegradable Polymer Blends and Composites* (pp. 527–49). Woodhead Publishing; 2021. https://doi.org/10.1016/B978-0-12-823791-5.00015-6.

19. Topalbekiroglu M, Kaynak HK. The effect of weave type on dimensional stability of woven fabrics. *Int J Cloth Sci Technol* 2008;20:281–88.

20. Ramesh M, Rajeshkumar L, Balaji D, et al. Keratin-based biofibers and their composites. In *Advances in Bio-Based Fiber: Moving Towards a Green Society,* Rangappa et al. (eds.) (pp. 315–34). Duxford, UK: Elsevier; 2021.

21. Deepa C, Rajeshkumar L, Ramesh M. Thermal properties of kenaf fiber-based hybrid composites. In *Natural Fiber-Reinforced Composites: Thermal Properties and Applications,* Senthilkumar et al., (eds.) (pp. 167–82). Germany: Wiley VCH; 2021.

22. Misra M, Pandey JK, Mohanty A. *Biocomposites: Design and Mechanical Performance*; Sawston, UK: Woodhead Publishing; 2015.

23. Dissanayake NPJ, Summerscales J. Life cycle assessment for natural fibre composites. *Green Compos Nat Resour* 2013;8:157–86.

24. Ramesh M, Rajeshkumar L. Case-studies on green corrosion inhibitors. *Sustain Corr Inhibit* 2021;107:204–21. https://doi.org/10.21741/9781644901496-9

25. Malviya RK, Singh RK, Purohit R, et al. Natural fibre reinforced composite materials: Environmentally better life cycle assessment–A case study. *Mater Today Proc* 2020;26:3157–60.

26. Hamdan MH, Siregar JP, Cionita T, et al. Water absorption behaviour on the mechanical properties of woven hybrid reinforced polyester composites. *Int J Adv Manuf Technol* 2019;104:1075–86.

27. Ramesh M, Rajeshkumar L, Saravanakumar R. Mechanically induced self-healable materials. In *Self-Healing Smart Materials and Allied Applications* (pp. 379–403). Germany: Wiley; 2021. https://doi.org/10.1002/9781119710219.ch15

28. Tezara C, Zalinawati M, Siregar JP, et al. Effect of stacking sequences, fabric orientations, and chemical treatment on the mechanical properties of hybrid woven jute–ramie composites. *Int J Precis Eng Manuf Technol* 2021;9:273–85.

29. Ramesh M, Rajeshkumar L, Balaji D, et al. Self-healable conductive materials. In *Self-Healing Smart Materials and Allied Applications* (pp. 297–319), Germany: Wiley; 2021. https://doi.org/10.1002/9781119710219.ch11.

30. Mohamad Hamdan MH, Siregar JP, Thomas S, et al. Mechanical performance of hybrid woven jute–roselle-reinforced polyester composites. *Polym Polym Compos* 2019;27:407–18.

31. Ramesh M, Rajeshkumar L, Bhoopathi R. Carbon substrates: A review on fabrication, properties and applications. *Carbon Lett.* 2021;31:557–80. https://doi.org/10.1007/s42823-021-00264-z.

32. Hamdan A, Mustapha F, Ahmad KA, et al. The effect of customized woven and stacked layer orientation on tensile and flexural properties of woven kenaf fibre reinforced epoxy composites. *Int J Polym Sci* 2016;2016:6514041.

33. Ramesh M, Rajeshkumar L, Balaji D. Mechanical and dynamic properties of ramie fiber-reinforced composites. In *Mechanical and Dynamic Properties of Biocomposites* (pp. 274–291). Germany: Wiley; 2021.

34. Sanjay MR, Madhu P, Jawaid M, et al. Characterization and properties of natural fiber polymer composites: A comprehensive review. *J Clean Prod* 2018;172:566–81.

35. Ramesh M, Rajeshkumar L, Balaji D. Aerogels for insulation applications. *Mater Res Found* 2021;98:57–76. https://doi.org/10.21741/9781644901298-4.

36. Sanjay MR, Siengchin S, Parameswaranpillai J, et al. A comprehensive review of techniques for natural fibers as reinforcement in composites: Preparation, processing and characterization. *Carbohydr Polym* 2019;207:108–21.

37. Ramesh M, Maniraj J, Rajesh Kumar L. Biocomposites for energy storage. In *Biobased Composites: Processing, Characterization, Properties, and Applications* (pp. 123–142). Germany: Wiley; 2021. https://doi.org/10.1002/9781119641803.ch9.

38. Vignesh V, Balaji AN, Karthikeyan MKV. Extraction and characterization of new cellulosic fibers from Indian mallowstem: An exploratory investigation. *Int J Polym Anal Char* 2016;21(6):504–12.

39. Ramesh M, Rajeshkumar L. *Technological Advances in Analyzing of Soil Chemistry. Applied Soil Chemistry* (pp. 61–78). Germany: Wiley; 2021. https://doi.org/10.1002/9781119711520.ch4.

40. Madhu P, Sanjay MR, Pradeep S, et al. Characterization of cellulosic fibre from *Phoenix pusilla* leaves as potential reinforcement for polymeric composites. *J Mater Sci Technol* 2019;8(3):2597–604.

41. Muralimohanrao K, Ratna Prasad AV, Rangababu MNV, et al. A study on tensile properties of Elephant grass fiber Reinforced polyester composites. *J Mater Sci* 2007;42:3266–72.
42. Madhu P, Sanjay MR, Jawaid M, et al. A new study on effect of various chemical treatments on Agave Americana fiber for composite reinforcement: Physico-chemical, thermal, mechanical and morphological properties. *Polym Test* 2020;85:106437.
43. Ramesh M, Rajeshkumar L, Balaji D. Influence of process parameters on the properties of additively manufactured fiber-reinforced polymer composite materials: A review. *J Mater Eng Perform* 2021;30:4792–807. https://doi.org/10.1007/s11665-021-05832.
44. Joshy MK, Mathew L, Joseph R. Influence of fiber surface modification on the mechanical performance of Isora-polyester composites. *Int J Polym Mater* 2009;58:2–20.
45. Ramesh M, Deepa C, Niranjana K, et al. Influence of Haritaki (*Terminalia chebula*) nano-powder on thermo-mechanical, water absorption and morphological properties of Tindora (*Coccinia grandis*) tendrils fiber reinforced epoxy composites. *J Nat Fibers* 2022;19(13):6452–68. https://doi.org/10.1080/15440478.2021.1921660.
46. Liu K, Takagi H, Yang Osugi R. Effect of physicochemical structure of natural fiber on transverse thermal conductivity of unidirectional abaca/bamboo fiber composites. *Compos A: Appl Sci Manuf* 2012;43:1234–41.
47. Ramesh M, Rajeshkumar L, Balaji D, et al. *Green Composite Using Agricultural Waste Reinforcement. Green Composites. Materials Horizons: From Nature to Nanomaterials* (pp. 21–34). Singapore: Springer; 2021. https://doi.org/10.1007/978-981-15-9643-8_2.
48. Ramesh M, Rajeshkumar L, Deepa C, et al.. Impact of silane treatment on characterization of *Ipomoea staphylina* plant fiber reinforced epoxy composites. *J Nat Fibers* 2022;19(13):5888–99. https://doi.org/10.1080/15440478.2021.1902896.
49. Scarponi C, Messano M. Comparative evaluation between E-Glass and hemp fiber composites application in rotorcraft interiors. *Compos B: Eng* 2015;69:542–9.
50. Devarajan B, Saravanakumar R, Sivalingam S, et al. Catalyst derived from wastes for biofuel production: A critical review and patent landscape analysis. *Appl Nanosci* (2021). https://doi.org/10.1007/s13204-021-01948-8.
51. Ramesh M, Deepa C, Rajeshkumar L, et al. 2021. Influence of fiber surface treatment on the tribological properties of *Calotropis gigantea* plant fiber reinforced polymer composites. *Polym Compos* 2021;42(9):4308–17.
52. Ramanaiah K, Prasad AR, Chandra Reddy KH. Mechanical and thermo-physical properties of fish tail palm tree natural fiber-reinforced polyester composites. *Int J Polym Anal Char* 2013;18:126–36.
53. Ramesh M, Rajesh Kumar L. Bioadhesives. In *Green Adhesives* (pp. 145–61). Germany: Wiley- Scrivener Publisher; 2020. https://doi.org/10.1002/9781119655053.ch7.
54. Balaji D, Ramesh M, Kannan T, et al. 2021. Experimental investigation on mechanical properties of banana/snake grass fiber reinforced hybrid composites. *Mater Today: Proc* 2021;42:350–5.
55. Ratna Prasad AV, Atluri V, Mohan Rao K, et al. Experimental Investigation of Mechanical properties of golden cane grass fiber-reinforced polyester composites. *Int J Polym Anal Char* 2013;18:30–9.
56. Ramesh M, Deepa C, Selvan MT, et al. Mechanical and water absorption properties of *Calotropis gigantea* plant fibers reinforced polymer composites. *Mater Today: Proc* 2021;46:3367–72.
57. Sreenivasan VS, Ravindran D, Manikandan V, et al. Mechanical properties of randomly oriented short *Sansevieria cylindrica* fiber/polyester composites. *Mater Des* 2011;32:2444–55.
58. Bhuvaneswari V, Priyadharshini M, Deepa C, et al. Deep learning for material synthesis and manufacturing systems: A review. *Mater Today: Proc* 2021;46:3263–9.

59. Winowlin Jappes JT, Siva I. Studies on the influence of Silane treatment on Mechanical properties of coconut sheath reinforced polyester composites. *Polym Plast Technol Eng* 2011;50:1600–5.

60. Ramesh M, RajeshKumar L, Bhuvaneshwari V. Bamboo fiber reinforced composites. In *Bamboo Fiber Composites. Composites Science and Technology* (p. 113). Singapore: Springer; 2021. https://doi.org/10.1007/978-981-15-8489-3_1.

61. Hadi AE, Hamdan MH, Siregar JP, et al. Application of micromechanical modelling for the evaluation of elastic moduli of hybrid woven jute–ramie reinforced unsaturated polyester composites. *Polymers* 2021;13(15):2572.

62. Shahinur S, Akter S, Islam Z, et al. Impact of different layers on physico-mechanical criteria of jute fabrics polyester composites. *Adv Mater Process Technol* 2022;8(2):1728–38.

63. Sathiskumar TP, Navaneetha Krishnan P, Shankar S. Tensile and flexural properties of snake grass natural fiber reinforced polyester composites. *Compos Sci Technol* 2012;72:1183–90.

64. Sajin JB, Paul RC, Binoj JS, et al. Impact of fiber length on mechanical, morphological and thermal analysis of chemical treated jute fiber polymer composites for sustainable applications. *Curr Res Green Sustain Chem* 2022;5:100241.

65. Sudha S, Thilagavathi G. Analysis of electrical, thermal and compressive properties of alkali-treated jute fabric reinforced composites. *J Ind Textil* 2018;47(6):1407–23.

66. Sampathkumar D, Punyamurthy R, Bennehalli B, et al. Effect of esterification on moisture absorption of single areca fiber. *Int J Agric Sci* 2012;4:227–29.

67. Venkateshwaran N, ElayaPerumal A, Alavudeen A, et al. Mechanical and water absorption behaviour of banana/sisal reinforced hybrid composites. *Mater Des* 2011; 32: 4017–21.

68. Dabade BM, Ramachandra Reddy G, Rajesham S, et al. Effect of fiber length weight ratio on tensile properties of sun hemp and palmyra fiber reinforced polyester composites. *J Reinforc Plast Compos* 2006;25:1733–8.

69. Udaya Kiran C, Ramachandra Reddy G, Dabade BM, et al. Tensile properties of sun hemp, banana and sisal fiber reinforced polyester composites. *J Reinforc Plast Compos* 2007;26(10):1043–50.

70. Saravana Kumar A, Maivizhi Selvi P, Rajeshkumar L. Delamination in drilling of sisal/banana reinforced composites produced by hand lay-up process. *Appl Mech Mater* 2017;867:29–33.

71. Ramesh M, Rajeshkumar L. Wood flour filled thermoset composites. In *Thermoset Composites: Preparation, Properties and Applications* (Vol. 38, pp. 33–65). Materials Research Foundations; 2018. http://dx.doi.org/10.21741/9781945291876-2.

72. Arthanarieswaran VP, Kumaravel A, Kathirselvam A, et al. Mechanical and thermal properties of acacia leucophloea fibre epoxy composites: Influence of fibre loading and alkali treatment. *Int J Polym Anal Charact* 2016; 21: 571–83.

73. El-Abbassi FE, Assarar M, Ayad R, et al. Effect of alkali treatment on Alfa fibre as reinforcement for polypropylene based eco-composites: Mechanical behaviour and water ageing. *Comp Struct* 2015; 133: 451–57.

74. Ibrahim NA, Yunus WMZW, Othman M, et al. Effect of chemical surface treatment on the mechanical properties of reinforced plasticized poly(lactic acid) biodegradable composites. *J Reinf Plast Compos* 2011; 30: 381–88.

75. Thiruchitrambalam M, Shanmugam D. Influence of pre-treatments on the mechanical properties of palmyra palm leaf stalk fiber-polyester composites. *J Reinf Plast Compos* 2012; 31: 1400–14.

76. Fan Y, Gomez A, Ferraro S, et al. The effects of temperatures and volumetric expansion on the diffusion of fluids through solid polymer. *J Appl Polym Sci*. 2017;134(31):45151.

77. Rana AK, Mandal A, Bandyopadhyay S. Short jute fiber reinforced polypropylene composite effect of compatibiliser, impact modifier and fiber loading. *Compos Sci Technol* 2006;63:801–6.

78. Ramesh M, Kumar LR, Khan A, et al. Self-healing polymer composites and its chemistry. In *Self-Healing Composite Materials* (pp. 415–27). Woodhead Publishing; 2020.
79. Ramesh M, Deepa C, Kumar LR, et al. Life-cycle and environmental impact assessments on processing of plant fibres and its bio-composites: A critical review. *J Indust Text* 2022;51(4_suppl):5518S-42S.
80. Jothibasu S, Mohanamurugan S, Vijay R, et al. Investigation on the mechanical behavior of areca sheath fibers/jute fibers/glass fabrics reinforced hybrid composite for lightweight applications. *J Ind Textil* 2018;49(8):1036–60.
81. Malingam SD, Chan KH, Subramaniam K, et al. The static and dynamic mechanical properties of kenaf/glass fibre reinforced hybrid composites. *Mater Res Express* 2018;5(9):095304.
82. Wong KJ, Zahi S, Low KO, Lim CC. Fracture characterization of short bamboo fiber reinforced polyester composites. *Mater Des* 2010;31:4147–54.
83. RatnaPrasad AV. Mechanical properties of natural fiber reinforced polyester composites: Jowar, sisal and bamboo. *Mater Des* 2011;32:658–63.
84. Ramanaiah K, Prasad AR, Reddy KHC. Thermal and mechanical properties of waste grass broom fiber-reinforced polyester composites. *Mater Des* 2012;40:103–8.
85. ASTM D3039/D3039M-17. *Standard Test Method for Tensile Properties of Polymer Matrix Composite Materials.* West Conshohocken: PA: ASTM International; 2017.
86. Gupta MK. Investigations on jute fibre-reinforced polyester composites: Effect of alkali treatment and poly (lactic acid) coating. *J Indust Text* 2020;49(7):923–42.
87. Jannah M, Mariatti M, Abu Bakar A, et al. Effect of chemical surface modifications on the properties of woven banana-reinforced unsaturated polyester composites. *J Reinf Plast Comp* 2009;28:1519–32.
88. Luo, Z.; Li, X.; Shang, J., et al. Modified rule of mixtures and Halpin–Tsai model for prediction of tensile strength of micron-sized reinforced composites and Young's modulus of multiscale reinforced composites for direct extrusion fabrication. *Adv Mech Eng* 2018;10:1–10.
89. Munoz E.; and Garcia-Manrique A. Water absorption behaviour and its effect on the mechanical properties of flax fibre reinforced bioepoxy composites. *Int J Polym Sci* 2015;6:1–10.
90. Dixit S and Verma P. The effect of surface modification on the water absorption behavior of coir fibers. *Adv Appl Sci Res* 2012;3:1463–65.
91. Goud G and Rao RN. Effect of fibre content and alkali treatment on mechanical properties of Roystonea regia-reinforced epoxy partially biodegradable composites. *Bull Mater Sci* 2011;34:1575–81.
92. Shinoj S, Visvanathan R, Panigrahi S, et al. Dynamic mechanical properties of oil palm fibre (OPF)-linear low density polyethylene (LLDPE) biocomposites and study of fibre matrix interactions. *Biosyst Eng* 2011;109:99–107.
93. Gupta MK. Thermal and dynamic mechanical analysis of hybrid jute/sisal fibre reinforced epoxy composite. *Proc Inst Mech Eng Pt L J Mater Des Appl* 2018;232:743–48.
94. Idicula M, Malhotra SK, Joseph K, et al. Dynamical mechanical analysis of randomly oriented intimately mixed short banana/sisal hybrid fibre reinforced polyester composites. *Compos Sci Technol* 2005;65:1077–87.
95. Winowlin Jappes JT, Siva I, Rajini N. The fractography Analysis of naturally woven coconut sheath reinforced polyester composites: A Novel Reinforcement. *Polym Plast Technol Eng* 2011;49:419–24.
96. Uz Zaman S, Shahid S, Shaker K, et al. Development and characterization of chemical and fire resistant jute/unsaturated polyester composites. *J Text Inst* 2022;113(3):484–93.
97. Maheswari CU, Reddy KO, Muzenda E, et al. A comparative study of modified and unmodified high-density polyethylene/borassus fiber composites. *Int J Polym Anal Charact* 2013;18:439–50.

98. ASTM D570-98 (2018). *Standard Test Method for Water Absorption of Plastics*. West Conshohocken, PA: ASTM International; 2018.
99. Akil HM, Cheng LW, Ishak ZAM, et al. Water absorption study on pultruded jute fibre reinforced unsaturated polyester composites. *Compos Sci Technol* 2009;69:1942–48.
100. Dixit S, Verma P. The effect of surface modification on the water absorption behavior of coir fibers. *Adv Appl Sci Res* 2012;3:1463–65.
101. Gupta MK, Deep V. Effect of stacking sequence on flexural and dynamic mechanical properties of hybrid sisal/glass polyester composite. *Am J Polym Sci Eng* 2017;5:53–62.
102. Rodriguez ES, Stefani PM, Vazquez A. Effects of fibers' alkali treatment on the resin transfer molding processing and mechanical properties of jute-vinylester composites. *J Compos Mater* 2007;41:1729–41.
103. Gomes A, Matsuo T, Goda K, et al. Development and effect of alkali treatment on tensile properties of curaua fiber green composites. *Compos Part A* 2007;38:1811–20.
104. Stocchi A, Lauke B, Vazquez A, et al. A novel fiber treatment applied to woven jute fabric/vinylester laminates. *Compos Part A* 2007;38:1337–43.
105. Cyras VP, Commisso MS, Mauri AN, et al. Biodegradable double-layer films based on biological resources: Polyhydroxybutyrate and cellulose. *J Appl Polym Sci* 2007;106:749–56.

4 Bamboo Fiber-Reinforced Polyester Composites

Dheeraj Kumar
National Institute of Technology Durgapur

Nadeem Faisal
Central Institute of Petrochemicals
Engineering & Technology

Ranjan Kumar Mitra and Apurba Layek
National Institute of Technology Durgapur

CONTENTS

DOI: 10.1201/9781003270980-4

4.1 INTRODUCTION

The ever-increasing costs of raw resources for structural and standard plastic products, concerns about the long-term viability of natural reservoirs, and the growing environmental risk have compelled researchers and manufacturers of polymeric materials to turn to naturally occurring, recyclable materials [1,2]. The usage of synthetic fibers had controlled the sector of reinforcements in the past years; nevertheless, the reinforcements using natural fibers had received a great push to replace this synthetic fiber in numerous fields [3]. Over the past decade, there has been an increase in interest in the production of polymer composites that are successful in the market with synthetic composites by combining natural fibers with polymer matrices derived from both non-renewable (petroleum-based) and renewable resources [4]. These composites are created by mixing natural fibers with polymeric composites.

Biodegradable polymers and bio-based polymeric goods from renewable resources may build sustainable and eco-friendly products that can compete and conquer the existing market which is occupied by petroleum-based merchandise. These types of goods can also be formed from renewable resources. Investigators have used both softwoods and hardwoods in their efforts to isolate fibers that may be used as reinforcements in a variety of composites. Natural fibers, such as cotton, jute, and sisal, are of critical significance to the economies of some underdeveloped countries. Cotton is particularly important in a number of West African countries, while jute and sisal are important in Bangladesh and Tanzania, respectively [5].

In regions of the world where there is a shortage of natural forests, farmlands have been used in research and innovations related to polymer composites. Bamboo is one of the agricultural products that has the potential to be employed in polymer composite research and innovation. Due to its benefits, bamboo is one prominent agricultural product that may be utilized [6]. The regions of Asia and South America have abundant supplies of bamboo. Although it is recognized as a natural engineering material, bamboo has not been studied to its maximum degree in many of the Asian nations. Due to the fact that it takes many months for the material to mature, it has become an essential component in determining the socioeconomic level of society. Because of its exceptional strength relative to its weight, bamboo has traditionally been used in a variety of applications, including housing and equipment. This quality is a result of the fibers being aligned in a longitudinal direction. In actual fact, it is obligatory to construct the bamboo-based composite materials in combination with the supervised extracting of bamboo fibers from bamboo trees [7,8]. This is the case even if the extraction of bamboo fibers is the primary focus. Bamboo fibers inherently contain finer mechanical qualities than other natural fibers; however, compared with other natural fibers, bamboo fibers have a more brittle nature owing to the higher lignin concentration that covers the bamboo fibers.

Currently, bamboo is thought of as an essential plant fiber, and it has a significant potential for use in the polymer composite sector. Because of the variations in its structural makeup, mechanical qualities, extraction of fibers, surface modification, and thermal characteristics, it has become a versatile material for application in automotive industries [9,10]. Some of the most interesting characteristics of bamboo are its commercial benefits, thin and light, excellent mechanical properties, and non-hazardous essence of bamboo fibers, which makes investigators work in the path of composite technology. Another attractive property of this material is that it does not contain any hazardous elements. Because of this, it is clear that composites made from bamboo fiber have capabilities to be used in the automobiles. These composites have the potential to recover the non-renewable and expensive synthetic fibers that are currently used in composite materials, especially in automotive industry and in residential sectors. Currently, there is danger to the environment, which has resulted in several nations passing regulations requiring the use of recyclable materials in automobiles to the extent of 95%. The use of natural fibers, and more specifically composites based on bamboo fiber, in day-to-day life is becoming more popular now. Extensive research is being conducted in every area, including engineering, biotechnology (also known as genetic designing), farming, and other related areas, all with the intention of achieving one objective: the improved use of bamboo fibers in composite materials.

4.2 INTERPRETATIONS OF THE SOCIAL AND ECONOMIC ENVIRONMENT RELATING TO BAMBOO AND COMPOSITE MATERIALS CONTAINING BAMBOO FIBER

There are around one thousand subspecies of bamboo widely present, which is itself a reflection of the incredible variety of this plant. Instead of saying that bamboo keeps growing incredibly quickly, it is more accurate to remark that it is an incredibly fast-growing grassland. Bamboo has been put to use in a number of Asian nations from prehistoric days, as well as in South American countries for hundreds of years. In many contexts, bamboo may be seen as an environmentally sound alternative to traditionally used wood species. When opposed to hardwood, which usually takes nearly around 20 years to achieve maturation, bamboo may be harvested after just 3 years. When bamboo has reached maturity, its tensile strength is equivalent to that of mild steel [11]. The average growth of bamboo is incredible; the renowned and fastest tends to grow upward of 2 inches per 60 minutes, and in certain species of moso bamboo, the height of 60 feet is attained in only 90 days. As a result, trying to cut down this alternative source of wood would have no impact whatsoever on environmental health. Trading in bamboo and items made from bamboo is expanding at a very quick rate. The explanation for the high selling price of bamboo is that there is lack of wood production in developing countries, and bamboo is the preferred option available to replace wood in terms of its growth factor [12]. The production and distribution of bamboo and the goods made from it, whether they be household items, panels, or decoration products, all have an impact, collectively, not only on the state of the world economy but also on the environment.

Agricultural workers are the ones who stand to profit the most from the cultivation and harvesting of bamboo. Cultivating and harvesting bamboo has provided them with a reliable source of income, which has allowed them to hone their fundamental agricultural skills, remain calm in the face of potential financial setbacks associated with bamboo sales, and increase their capacity for self-determination. Bamboo planting has a direct bearing on the entire ecosystem, not only does it aid in the prevention of mudslides and soil depletion, but it also converts the previously worthless area into farmland that can support agriculture. The ability to cultivate bamboo may help impoverished farmers in rural areas improve their standard of living [13]. Not only can the cultivation of bamboo help restore the damaged land and the ecosystem, but it can also help improve food and nutritional security by being intercropped with other crops that provide food. The cultivation of bamboo involves just a little cost, which is used to purchase bamboo reproduction, site, and labor. The social economic subsidies of cultivating raw resources or designing products (household equipment, floor coverings, bamboo-based nanocomposite, barricades, decorative items, and so on) that produce large quantities of long-lasting goods and services are, without a doubt, making a significant contribution to the growth of business attributes in many regions of the world. This is the case because these activities lead to the fabrication of basic commodities [14]. Interest in employing bamboo in composite technology has been sparked as a result of the destruction of the environment and the rapidly rising price of gasoline. Bamboo is one of the greatest resources that can be used as a reinforcement material in composite materials rather than using glass fibers, which also entirely depends on the depletion of natural energy reserves [15]. The imposition of strict regulations to construct sustainable and environment-friendly consumer goods is pushing industry sector to enhance the research methods of someone using bioremediation assets for the improvement of mechanical properties.

Organizations are looking for ways to mitigate this ecological imbalance in a more sustainable manner. The rapid expansion of the reinforced sector may be attributed to the widespread accessibility of bamboo. In addition to minimizing the necessity for the use of wood, there is a significant need for a step toward the formulation of public policies as well as the acceleration of technical efforts [16].

The development of bamboo-based composite materials for use in home furnishings, mobility, and buildings has contributed to the economy and simultaneously providing socioeconomic benefits to the general populace. New job opportunities have opened up as a result of the marketing of bamboo-based composites; regulations are currently being developed all over the community to cultivate attention among some of the general community by asserting a series of strategies; for instance, bamboo materials will no longer be subject to excise obligations [17].

4.3 BAMBOO FIBERS

4.3.1 THE AVAILABILITY OF BAMBOO AROUND THE WORLD

Bamboo is cultivated in several of the world's largest regions, as well as the industry has indeed been subdivided to reflect this fact: the Asia–Pacific bamboo area, the American hardwood territory, the African hardwood province, and the European

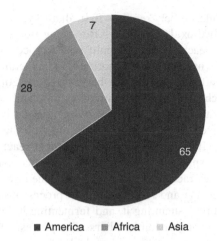

■ America ■ Africa ▨ Asia

FIGURE 4.1 Statistics on bamboo productivity at the worldwide level.

and North American bamboo geographic area. The region encompassing Asia and the Pacific is home to the greatest number of bamboo plantations in the universe. In the continents of Asia, bamboo is referred to by a variety of names. In China, it is regarded as the "companion of mankind," while in India and Vietnam, it is known as the "woods of the poor" and "the brother" [18]. The statistics on bamboo productivity at the worldwide level were given by FAO and are shown in Figure 4.1. India, China, Indonesia, the Philippines, Myanmar, and Vietnam are just few of the nations that are responsible for a significant portion of Asia's bamboo production. The monopodial kind of bamboo, as opposed to the monopodial variety, makes up the majority of the world's population [19]. The widespread promotion of bamboo cultivation across China has resulted in an approximately 30% increase in the number of pods per plant of bamboo production.

4.3.2 THE PROCESS OF EXTRACTING BAMBOO FIBERS

The organic genuine bamboo fabric and the bamboo pulp fibers are the two types of bamboo fiber that may be acquired mostly from trees. Each kind of bamboo fiber is processed in a completely different way; thus, it is possible to get both types (namely bamboo viscose fiber or regenerated cellulose bamboo fiber). Utilizing just physico-mechanical processes, the raw bamboo fiber is obtained from the bamboo plant in its purest form, without the use of any reprocessing agents. We refer to it as existing natural fibers or real natural hardwood fibers so that we can distinguish something from bamboo pulp fibers, which is also known as bamboo viscose fibers. However, bamboo pulp (viscose) fiber is a kind of chemical fiber that belongs to the category of regenerating cellulose fiber. Machinery processing and biochemical treatment are the two main methods for obtaining bamboo fibers. Mechanical treatment is the more common of the two. The separation of bamboo sticks is the first step in both procedures. Afterward, the strips are either subjected to mechanical treatment or chemical preparation, depending on the final application of bamboo fibers.

The production of cellulose fibers requires first alkali hydrolysis, which is accomplished using sodium hydroxide (NaOH). After being treated with alkali, the cellulose fibers are next bleached using a multi-phase process that involves carbon disulfide. This method is used by a vast majority of manufacturers since this requires the shortest amount of effort to produce the bamboo fibers [20].

In contrast, the mechanical technique involves the bamboo being first crushed before being subjected to enzyme treatment, which results in the production of a sponge-like mass. Isolated fibers are then recovered with the use of mechanically comb fiber technology. In comparison to the chemical reaction, this approach is environmentally safe, despite the fact that it is a lesser economically viable procedure. The procedure of extracting the fiber was explained by the investigators, and it was separated into coarse and refined bamboo processing [21]. After chopping and separating the bamboo, steaming it, and fermenting it with bamboo catalysts, we were able to acquire rough bamboo fibers. The process of heating, fermenting bamboo with just an enzymatic, washing and bleaching it, treating it with acids, immersing it in oil, and air-drying it are the processes that must be taken in order to achieve good quality bamboo.

4.3.3 THE STRUCTURE AND FUNCTION OF BAMBOO

The culm, the roots, the rhizome foundation, the shoot, the branching, and the leaf make up the anatomical components of a bamboo plant. Phyllostachys elegans in the United States and Phyllostachys edulis known as "Moso" in China are examples of bamboo species that produce blooms sporadically [22]. The term "sporadic blossoming" refers to the process through which flowers open in response to ambient factors instead of genetic information [23].

According to Zhang et al. [24] and Jiang et al. [25], the morphological structure (outside look) of bamboo may be broken down into two primary components: the rhizome and the culm networks. Additionally, the bamboo plant may be broken down into two distinct sections: the subterranean section, which is made up of rhizomes, roots, and buds, and the aboveground section (composed of stems, branches, and foliage). The purpose of extant bamboo structures is outlined below in detail. Every job that is performed out by the same structure of bamboo is connected to every other configuration of bamboo in order to guarantee that perhaps the bamboo plant is able to develop consistently [26].

4.3.4 RHIZOME

It is a horizontally and fragmented stem that projects from the source species and has the ability to penetrate underground in order to colonize new land with the aim of expanding its territory. It has a sheath that protects it from the elements. When a plant breaks the surfaces to develop a culm, the sheath serves the purpose of providing the necessary protection for the plant to do so. The rhizome is responsible for the collection, storage, and distribution of moisture and nutrients for the functions of parenchyma and conducting tissue. It also participates in the vegetative development of the plant by developing into the nodes of new growth or bamboo culms.

4.3.5 CULM

It is a cylindrical-shaped, wooden stem that is made up of nodes and lateral roots. Culms are almost always hollowed inside. In spite of this, there are varieties of bamboo which have culms that seem to be solid. A culm's sheathing, foliage, leaflets, flowering, and branching system are its recognized characteristics. The root system provides everything with stability, which enables it to endure the additional load of leaflets as they expand. After three or four years, a culm reaches its complete maturity, and as it develops, it increases exponentially and becomes more rigid.

4.3.6 ROOT

The primary purpose of the root system is to provide structural support for the culms, which enables the culms to better withstand environmental stresses such as adverse weather. Additionally, this makes it possible for the culm to support increased load, which in turn enables it to produce additional leaves across greater distances. Storage of nutrition is another function of the root. They usually don't go deeper than one foot into the ground, coming from either the rhizome node or the bottom of the culm. The morphology of roots is often rather identical, both in terms of their size and their form.

4.3.7 BRANCHES

Nearly all species of bamboo are capable of producing several branches from either a solitary bud that's also situated at a node. Nevertheless, there are genera that are capable of growing many branches at every terminal. Chusquea is one such example.

4.3.8 LEAVES

Leaves may be found on every major component of a bamboo plant, including the rhizomes, the culm, and the branches. The blade, the sheathing, and the ligule are the three components that make up the anatomical structure of a leaf. The rhizome is the first place where leaflets may be found, and at this stage they are virtually entirely made up of the sheath. At this point, the leaves act as a safety covering to encapsulate the rhizome as it moves underneath. This helps the rhizome to continue its growth. The blades will emerge as the most prominent characteristic of the plant after the rhizome has broken through the topsoil and transformed into a culm. By absorbing sunlight and turning it into usable energy, the blade performs the photosynthesis functionality of the plants. The shape of the blade may vary greatly from species to species. The way that the leaves appear is a significant factor in determining whether or not something is bamboo [27].

4.4 THE CONSTITUENT SUBSTANCES THAT INCORPORATE BAMBOO FIBER

There are three primary chemical components that make up bamboo fiber: cellulose, hemicellulose, and lignin [28,29]. Bamboo fiber is a development of biomaterial.

4.4.1 CELLULOSE

Cellulose is the structural base that constitutes the cellular structure of bamboo fiber, and it is mostly constituted of three components including carbon, hydrogen, and oxygen. Microfibrils are the most common type of cellulose that is found to be preserved inside the cell walls of plants. Following separation, bamboo fiber was examined for its chemical components by Wang et al. [31], who discovered that somehow the cellulose content of the fibers achieved 73.83%. Bamboo fiber is used in the clothing industry. Cellulose is the primary component that determines tensile properties along the grains of bamboo fiber, as well as the amount of cellulose present in bamboo is directly proportional to the maturity of the bamboo. It is important to take into account that the aging of bamboo results in a reduction in the amount of cellulose present in almost the same bamboo materials [30].

4.4.2 HEMICELLULOSE

Hemicellulose is an amorphous polymer that has moderate crystallinity and may be found in the spaces among fibers. Hemicellulose is a complicated polysaccharide that mostly consists of xylan as the primary chain and, among other things, 4-O-methyl-D-glucuronic acid, L-arabinose, and D-xylose [32]. According to the findings of Fengel et al. [33], who investigated the polysaccharides proportion of Phyllostachys makinoi removed with 5% and 17.5% NaOH, the major ingredient of this fraction was arabinoxylan, and the ratios of xylose to arabinose seemed to be 17–18:1. Following retrieval of filtered water, the majority of the bamboo polysaccharide constituents were found to be glucose. However, following removal with an alkali solution, the proportion of xylose found in the bamboo polysaccharide elements was found to be significantly higher. This research was conducted by Sun et al. [34].

4.4.3 LIGNIN

Lignin is a type of polymer that may take on many different forms and has intricate patterns. Lignin is composed of many different types of monomers, and the most important ones are guaiacyl, syringyl, and p-hydroxyphenyl monomers. The structural units that make up lignin are related to one another mostly by ether linkages and carbon–carbon solitary bonding [34]. There is not a uniform distribution of lignin across the secondary walls of the bamboo fibers. When compared, the percentage of lignin inside the wide layer is often lower, whereas the percentage of lignin in the narrower layer is typically greater [32]. The inclusion of lignin in bamboo materials contributes, in addition, to the molecule's level of resilience. There is also a correlation between the maturity of the bamboo and the amount of lignin it contains. As shown in Figure 4.2, the fundamental framework of lignin can be generally separated into three distinct types.

Contrary to its three primary components, lignin typically consists of a variety of carbohydrate, lipid, and protein compounds as well as minute amounts of ashes components [28,29,32]. Not only these chemical constituents impact the qualities of bamboo, but they also affect how bamboo fiber may be used in various applications.

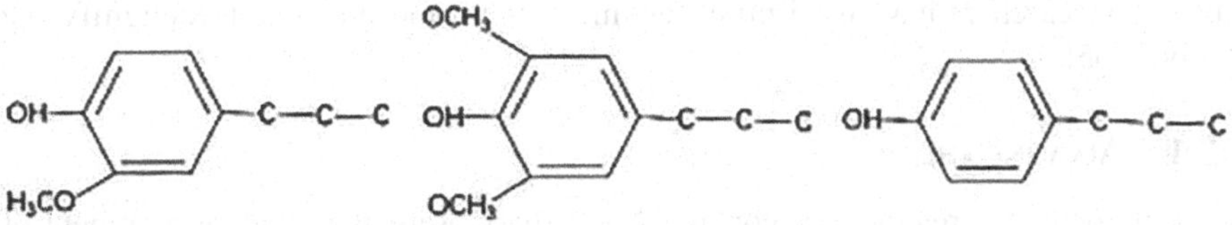

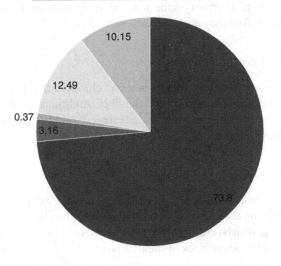

FIGURE 4.2 Fundamental framework of lignin.

Table 4.1 shows the mechanical properties of bamboo fiber. Figure 4.3 represents the chemical constituents of bamboo fiber.

4.5 BAMBOO TREATMENTS

The strands of bamboo have water-soluble quality because their synthetic constituents, such as lignin, may weaken bonds well with water-repellent framework elements. This results in the bamboo having a more water-loving character. A wide

TABLE 4.1
Mechanical Properties of Bamboo Fibers

Bamboo Fiber	Standard Values
Tensile (MPa)	290.01
Young's modulus (GPa)	17.02
Elongation at break (%)	-
Density (g/cm³)	1.251

■ Cellulose ■ Aqueous Extract ■ Pectin ▨ Hemi-Cellulose ▨ Lignin

FIGURE 4.3 Chemical constituents of Bamboo fiber.

range of researchers have used mixtures, medicines, and methods to delignify the bamboo [35].

4.5.1 ALKALINE THERAPY

When it comes to creating thermoplastics and thermosets, the alkaline treatment of natural fibers is by far the most common chemical treatment that is used. The alkaline treatment eliminates the extra lignin, wax, and oils that are present on the exterior surface of the fiber cell. The treatment with NaOH ionizes the hydroxyl group, which is the goal of the process. There are two different effects that the alkaline therapy exerts on the fibers. The first is the capability of increasing the quantity of cellulose exposed on the fiber surface, and the second is the capability of increasing the roughness over the outer surface, which in turn raises the mechanical interlocking [36].

4.5.2 THE TREATMENT WITH SILANE

The silane treatment is most noticeable when there is moisture present, and it may also lower the number of hydroxyl groups in cellulose. Moisture, when blended with a hydrolyzable alkoxy group, contributes to the creation of silanols, which respond to the hydroxyl group of the fiber to form covalent bonds with the wall. Silane treatment is most effective when there is moisture present.

4.5.3 ACRYLATION OF NATURALLY OCCURRING FIBERS

Following acetylation, also known as the incorporation of acetyl substituents into the molecule, the hygroscopic quality of the plant fibers is significantly diminished. In order to improve the fiber–matrix adhesion, the acetylation solution is frequently applied to sisal fibers before they are spun into yarn.

4.5.4 TREATMENT THROUGH BENZOYLATION

Attributed to the prevalence of benzoyl in benzoyl chloride, this procedure has the benefit of reducing the aqueous solubility of the fiber. Additionally, interactions with matrices are much improved in terms of their adherence. Additionally, it brings about a reduction in moisture absorption.

4.6 PERMANGANATE DIAGNOSIS

During the permanganate treatment procedure, solutions of permanganate are employed most of the time. The hydrophilic character of the fiber is reduced by permanganate treatments, which result in increased adhesive between both the fiber and matrix. This is one of the advantages of the diagnosis.

4.6.1 THE PROCESS OF ISOCYANATE AND PEROXIDE TREATMENTS

Agents such as benzoyl peroxide and dicumyl peroxidase, among others, are used in the process of modifying the natural fibers. After undergoing a preparation with alkaline, the fiber is given an exterior coating of (Black Phosphorus) BP or

(Dichloropropane) DCP in acetone solution during this type of therapy. The tensile properties of the fibers may be increased as a result of this treatment. There is a high degree of reactivity between the hydroxyl group of cellulose and the isocyanate functional group. The hydrophilic character of the fibers is diminished by these treatments, which also results in an increase in the qualities of adhesion.

4.7 APPLICATION

Bamboo is a natural fiber composite material that can be found growing abundantly in the majority of tropical regions. As a result of technological developments and an increase in business trend, bamboo has already been utilized extensively for the production of a variety of consumer goods as well as for industrial purposes.

4.7.1 CAPABILITIES FOR CONSTRUCTION WORK

The tensile strength that was required for the development of the bamboo link during the time period just before the First World War is the proof in this case. The pedestrian bridge made of bamboo had first been constructed so that people could cross the river and maintain their commercial ties.

4.7.2 APPLICATIONS RELATED TO INTERIOR DESIGN

Bamboo bio-composites have a significant impact on the development of interior design, and this impact contributes to the material's overall commercial value. Utilization of bio-composites in the production of various bamboo products for both exterior and interior usage has a respectable demand in markets all over the globe This is further strengthened by its perfectness as an inspirational material, which has received recognition from a variety of sources. This demonstrates that half and half's fabric can beat other building materials from a variety of perspectives, such as physiological, structural, and aesthetic [37,38].

In modern times, several kinds of mixtures based on bamboo products have been established. These mixtures have been used for the design of roofs, walls, floors, entrances, frame boundaries, staircases, and even house decorative embellishments.

4.7.3 APPLICATIONS RELATED TO FURNITURE

Many countries have recently started producing new pieces of decor; for instance, Malaysia is the world's largest transporter of furniture, as it ships its wares to over a hundred different countries. Bamboo bio-composite has been shown by a large number of researchers to have several benefits when it comes to the production of different kinds of furniture and its components. For the purpose of demonstrating that the durability of the decor may be equivalent to that of wood, new ideas have been devised to use smart materials. Investigations into bent and shaped materials such as wood, for example, may help strengthen the durability of the bamboo fibers. Work that is done throughout the world to produce more sustainable and environment-friendly (bio-based) furniture pieces has the potential to bring about a continuous shift in the level of product development accomplishment. Initiatives that aim to extend the use of bio-composites are extremely highly applauded and encouraged because of the potential of these resources to lessen their impact on the natural environment, improve development, and enable more advanced manufacturing techniques.

4.7.4 APPLICATIONS IN THE AUTOMOTIVE INDUSTRY

The development of novel bamboo-tangled polished composite cross-breeds is ongoing, and continuous research is leading to the creation of new hybrids in which all the sections of wood (core, facing, and rear tiers) are formed of bamboo mate [39]. In addition, bamboo tangles with wooden polish panels are used as components of train carriages. This is supported by a large number of researchers who have proved the usefulness and feasibility of bamboo as an alternative material in the industry that produces automobiles.

4.7.5 BAMBOO FIBER COMPOSITES: POTENTIAL BENEFITS AND POTENTIAL PITFALLS

The economic value of bamboo fiber is quite high in many different sectors, from raw materials to finished goods. Bamboo is used in everything from paper to clothing. It is anticipated that the worldwide market for bamboo fiber would reach 98.30 billion US dollars by the year 2025 [40]. Because it is such a diverse substrate for industrialized items as well as furniture, demand for it is quite strong within these sectors. As a consequence, the expansion of the bamboo industries will ultimately lead to improve in the socioeconomic standing of the surrounding communities. As a consequence, unemployment, poverty, and lack of development in local communities will be significantly reduced or eliminated entirely. A situation in which supply and demand are in equilibrium is anticipated to result in an increase in the price of bamboo fiber. The bamboo fiber-reinforced polymer composites do not seem to have any economic value at this point, which is probably also true. The polymer composites of synthetic fiber may work well with the bamboo reinforcing fibers that are being used. Despite this, there are a number of challenges that need to be addressed before bamboo-reinforced polymer composites may find widespread use [41–45].

The present moment is a great opportunity to teach the wider populace regarding bamboo composites, which would need full backing of government educational institutions. Inventions are not disseminated to the general population since there are not enough relevant venues on which they may be published. This means that they are kept inside the scientific community. It is the role of the nation to provide venues that are easily accessible for communicating research results to all levels of society. The results of research on bamboo composites may be effectively disseminated via a variety of channels, including editorials, social networks, awareness programs, and/or community events. Additional problem is that there is little financing, which diminishes the progression of research studies, restricts participation, and prevents presenters from attending academic conferences.

Unfortunately, there is a restriction placed on research grants on a worldwide scale during the period of the COVID-19 epidemic. It makes perfect sense for administrations to put the revival of the economy and society at the top of their agendas. To further the internationalization of bamboo composites in the foreseeable future, progress should be launched on a number of different fronts. As a consequence of this, this chapter compiles and disseminates the most recent information regarding bamboo-reinforced polymer composites. As a consequence of this, researchers will be able to concentrate their role to make new strategies regarding bamboo-reinforced polymer matrix composites study.

4.8 FUTURE INNOVATIONS IN BAMBOO-REINFORCED COMPOSITES

The current industrialization, which is focusing on increasing the eco-efficiency of manufacturing goods and the processes used to manufacture them, is key to ensuring a stable environment for future generations. All through using techniques for chemically produced and gasoline-related goods and services, which currently dominate the industry and are depleting natural petroleum sources, high-performance, biodegradable waste and renewable plant materials may build a stronger foundation for sustainable and eco-efficient advanced technologies. These materials can also be used to make plant materials. Natural fibers and bio-composites made from natural resources have the potential to replace the prevailing use of petroleum products soon since they combine sustainability, environmental friendliness, and highly completed products. Cellulose fibers and biomaterials are manufactured from natural resources. Bamboo fiber is derived from a source that is well known for being renewable in the sense that it has a high rate of growth and superior mechanical qualities. The exploitation of bamboo fiber in the production of biomaterials via the application of cutting-edge technology revolutionizes the future of the next generations. Products made from bamboo fiber that are thoughtfully developed and constructed have the potential to contribute to a new revolution aimed at preserving our environmental capital. As a result, based on this concise analysis, bamboo fibers have the potential to be employed for the creation of advanced and designed products for a variety of various purposes. It will be an alternate opportunity to grow biomaterials, which can be especially helpful for meeting the day-to-day requirements of citizens, such as in the manufacture of lighter weight automotive parts or sporting goods. This includes domestic tables and chairs, houses, barricades, trusses, and floor coverings, among other things. They will be the leading factor in the transition from a dependent world today to a sustainable economy because of their cheap cost, quick accessibility, and aesthetically designed products.

4.9 CONCLUSION

The utilization of bamboo fibers in a spectrum of uses has made it possible for academics and businesses to explore new routes toward the goal of developing a renewable framework for the continued use of bamboo fibers in the upcoming years. Bamboo fibers have found widespread use in the composite industry, which have contributed significantly to the socioeconomic and cultural leadership of citizens. The manufacturing of bamboo fiber-based composites using various matrices has resulted in the development of biomaterials that are both cost-efficient and environmentally benign. This has a direct influence on the stock prices of bamboo. In order to build such composites, it is required to do exhaustive research on the basic, biomechanical, and functional properties of bamboo fibers. In light of this, an effort has been made throughout this research to compile data about the fundamental characteristics of composites based on bamboo fiber and the economic applications of such composites. Researchers from all around the globe have conducted a wide range of studies and created novel ideas in order to simply provide the most basic support to people who are both employed and earning. Investigations being done now on bamboo fiber-based composites make use of both theoretical and applied engineering, whether in terms of transformation, mechano-physical, thermal, or other qualities. However, the final objective of using the bamboo fiber to its maximum extent remains a long way off from its predicted landmark, especially in Malaysia. Other nations, such as India and China, have advanced much farther in employing bamboo fiber in a socioeconomic sense.

A successful strategy for the bamboo-based composites sector would aid in using bamboo in a manner that is different from the typical and conventional use of the material. In terms of research and interpretation, there must be more progress made in order to effectively characterize bamboo fiber, as well as composite materials based on bamboo fiber. In this chapter, we have made an effort to obtain information on the many forms of experiments and investigations that have been carried out. However, scientists have indeed completed a significant amount of work on composites based on bamboo, but there is still a need for more study and invention in this field in order to address any possible difficulties that may arise in future. People living in metropolitan areas and in remote communities, who rely more on artificial composite materials, will find their lives simplified as a result of these factors.

REFERENCES

1. Senthilkumar K, Saba N, Chandrasekar M, Jawaid M, Rajini N, Siengchin S, Ayrilmis N, Mohammad F, Al-Lohedan HA. Compressive, dynamic and thermo-mechanical properties of cellulosic pineapple leaf fibre/polyester composites: Influence of alkali treatment on adhesion. *Int J Adhes Adhes* 2021; 106: 102823.
2. Puglia D, Biagiotti J, Kenny JM. A review on natural fibre-based composites – part II. *J Nat Fibres* 2005; 1:23–65.
3. Alothman OY, Jawaid M, Senthilkumar K, Chandrasekar M, Alshammari BA, Fouad H, Hashem M, Siengchin S. Thermal characterization of date palm/epoxy composites with fillers from different parts of the tree. *J Mater Res Technol* 2020; 9: 15537–15546.
4. Thomas SK, Parameswaranpillai J, Krishnasamy S, Begum PMS, Nandi D, Siengchin S, George JJ, Hameed N, Salim Nisa V, Sienkiewicz N. A comprehensive review on cellulose, chitin, and starch as fillers in natural rubber biocomposites. *Carbohydr Polym Technol Appl* 2021; 2: 100095.
5. Mohanty AK, Misra M, Drzal LT. Sustainable bio-composites from renewable resources: opportunities and challenges in the green materials world. *J Polym Environ* 2002; 10:19–26.
6. Coutts RSP, Ni Y, Tobias BC. Air-cured bamboo pulp reinforced cement. *J Mater Sci Lett* 1994; 13:283–285.
7. Senthilkumar K, Siva I, Rajini N, Jeyaraj P. Effect of fibre length and weight percentage on mechanical properties of short sisal/polyester composite. *Int J Comput Aided Eng Technol* 2015; 7:60.
8. Senthilkumar K, Subramaniam S, Ungtrakul T, Kumar TSM, Chandrasekar M, Rajini N, Siengchin S, Parameswaranpillai J. Dual cantilever creep and recovery behavior of sisal/hemp fibre reinforced hybrid biocomposites: Effects of layering sequence, accelerated weathering and temperature. *J Industr Text* 2022; 51:2372S–2390S.

9. Kitagawa K, Ishiaku US, Mizoguchi M, Hamada H. *Bamboo-Based Ecocomposites and Their Potential Applications in Natural Fibres, Biopolymers, and Biocomposites.* Amar K. Mohanty Manjusri Misra Lawrence T. Drzal; 2005 [chapter 11].

10. Nasimudeen NA, Karounamourthy S, Selvarathinam J, Kumar Thiagamani SM, Pulikkalparambil H, Krishnasamy S, Muthukumar C. Mechanical, absorption and swelling properties of vinyl ester based natural fibre hybrid composites. *Appl Sci Eng Progr* 2021; 14:680–688.

11. Mohanty AK, Misra M, Drzal LT. Sustainable bio-composites from renewable resources: opportunities and challenges in the green materials world. *J Polym Environ* 2002; 10:19–26.

12. Deshpande AP, Rao MB, Rao CL. Extraction of bamboo fibres and their use as reinforcement in polymeric composites. *J Appl Polym Sci* 2000; 76:83–92.

13. Das M, Chakraborty D. Evaluation of improvement of physical and mechanical properties of bamboo fibres due to alkali treatment. *J Appl Polym Sci* 2008; 107:522–527.

14. Shahroze RM, Chandrasekar M, Senthilkumar K, Senthil Muthu Kumar T, Ishak MR, Rajini N, Siengchin S, Ismail SO. Mechanical, interfacial and thermal properties of silica aerogel-infused flax/epoxy composites. *Int Polym Process* 2021; 36: 53–59.

15. Jain S, Kumar R, Jindal UC. Mechanical behaviour of bamboo and bamboo composite. *J Mater Sci* 1992; 27:4598–604.

16. Chandrasekar M, Senthilkumar K, Jawaid M, Alamery S, Fouad H, Midani M. Tensile, Thermal and physical properties of Washightonia trunk fibres/pineapple fibre biophenolic hybrid composites. *J Polym Environ* 2022; 30: 4427–4434.

17. Bansal AK, Zoolagud SS. Bamboo composites: material of the future. *J Bamboo Rattan* 2002; 1:119–130.

18. Farrelly D. *The Book of Bamboo: A Comprehensive Guide to This Remarkable Plant, Its Uses, and Its History.* London: Thames and Hudson; 1984.

19. Thiagamani SMK, Pulikkalparambil H, Siengchin S, Ilyas RA, Krishnasamy S, Muthukumar C, Radzi AM, Rangappa SM. Mechanical, absorption, and swelling properties of jute/kenaf/banana reinforced epoxy hybrid composites: Influence of various stacking sequences. *Polym Compos* 2022; 43(11):8297–8307.

20. Waite M. Sustainable textiles: The role of bamboo and comparison of bamboo textile properties. *J Text Appl Technol Manage* 2009; 6:1–21.

21. Yao W, Zhang W. Research on manufacturing technology and application of natural bamboo fibre. In *2011 Fourth International Conference on Intelligent Computation Technology and Automation* (Vol. 2, pp. 143–148). IEEE; 2011 March.

22. Li LJ, Wang YP, Wang G, Cheng HT, Han XJ. Evaluation of properties of natural bamboo fibre for application in summer textiles. *J. Fiber Bioeng. Inform* 2010; 3(2):94–99.

23. Understanding the anatomy of bamboo, *Green Pot Enterprise*, 2016. Available online: http://gre enpotenterprises.com/understanding-the-anatomy-of-bamboo/ [Accessed: 05-Aug-2016].

24. Zhang QS, Jiang SX and Tang YY. Industrial utilization on bamboo: Technical report 26. *The International Network for Bamboo and Rattan (INBAR), People's Republic of China.* 2002.

25. Wang, G, Innes, JL, Lei, J, Dai, S, Wu, SW. China's forestry reforms. *Science* 2007; 318(5856):1556–1557.

26. Osorio L, Trujillo E, Van Vuure AW, Verpoest I. Morphological aspects and mechanical properties of single bamboo fibres and flexural characterization of bamboo/ epoxy composites. *J Reinf Plast Compos* 2011; 30:396–408.

27. Puttasukkha J, Khongtong S, Chaowana P. Curing behavior and bonding performance of urea formaldehyde resin admixed with formaldehyde scavenger. *Wood Res* 2015; 60(4):645–654.

28. Zhang Y, Wang CH, Peng JX. *Preparation and Technology of Bamboo Fiber and Its Products.* Beijing: China Textile Publishing House; 2014.

29. Chokshi S, Parmar V, Gohil P, Chaudhary V. Chemical composition and mechanical properties of natural fibers. *J Nat Fibers*, 2020; 1–12. DOI 10.1080/15440478.2020.1848738.

30. Takahashi Y, Matsunaga H. (1991). Crystal structure of native cellulose. *Macromolecules*, 24(13), 3968–3969. DOI 10.1021/ma00013a035.

31. Wang YP, Wang G, Cheng HT, Tian GL, Liu Z. Structures of bamboo fiber for textiles. *Text Res J*, 2010; 80(4):334–343. DOI 10.1177/0040517509337633.

32. Xu W. (2006). Extraction of natural bamboo fiber and the research on its structure and chemical properties (Master Thesis). Suzhou, China: Soochow University. JRM, 2022, vol. 10, no. 3, 619.

33. Fengel D, Shao D. A chemical and ultrastructural study of the bamboo species Phyllostachys makinoi Hay. *Wood Sci Technol* 1984; 18(2), 103–112. DOI 10.1007/BF00350469.

34. Sun SN, Yuan TQ, Li MF, Cao XF, Liu QY. Structural characterization of hemicelluloses from bamboo culms (Neosinocalamus affinis). *Cell Chem Technol* 2012; 46(3–4):165–176.

35. Kushwaha P, Varadarajulu K, Kumar R. Bamboo fiber reinforced composite using non-chemical modified bamboo fibers. *Int J Adv Res Sci Technol* 2012; 1(2):95–98.

36. Abdul Khalil HPS, Bhat IUH, Jawaid M, Hermawan AZ. Bamboo fibre re-inforced biocomposites: A review. *Mater Des* 2012; 42:353–368.

37. Bovea MD, Vidal R. Increasing product value by integrating environment impact, cost and customer valuation. *Resour Conserv Recyc* 2004; 41(2):133–145.

38. Van der Lugt P, Vogtländer JG, Van Der Vegte JH, Brezet JC. Life cycle assessment and carbon sequestration; the environmental impact of industrial bamboo products. In *Proceedings of the IXth World Bamboo Congress*. 2012 April.

39. Sabeel Ahmed K, Vijayarangan S. Tensile, flexural and interlaminar shear properties of woven jute and jute-glass fabric reinforced polyester composites. *J Mater Process Technol* 2008; 207:330–335.

40. Bamboos Market Size, Share & Trends Analysis Report, By Application (Raw Materials, Industrial Products, Furniture, Shoots), By Region, And Segment Forecasts, 2019–2025. Grand View Research. 2021. Available online: https://www.grandviewresearch.com/industry-analysis/bamboos-market [Accessed: 04-Jan-2022].

41. Javadian A, Smith IFC, Hebel DE. Application of sustainable bamboo-based composite reinforcement in structural-concrete beams: design and evaluation. *Materials* 2020; 13:696.

42. Faisal N, Pandey MK, Chauhan T, Verma G, Farrukh M, Kumar D. Experimental investigation on mechanical behavior of palm based natural fibre reinforced polymer composites. *AIJR Abstracts*, 2021; 161. https://books.aijr.org/index.php/press/catalog/book/108/chapter/842

43. Kumar D, Farrukh M, Faisal N. Nanocomposites in the food packaging industry: recent trends and applications. *Res Anthol Food Waste Reduct Alter Diets Food Nutr Sec* 2021; 122–146. https:// doi/10.4018/978-1-7998-5354-1.ch006

44. Kumar D, Mitra RK, Chouhan T, Farrukh M, Faisal N. Effect of moisture absorption on interfacial shear properties of the bio-composites. In Muthukumar C, Krishnasamy S, Thiagamani SMK, Siengchin S (eds.). *Aging Effects on Natural Fiber-Reinforced Polymer Composites* (pp. 237–256). Singapore: Springer; 2022.

45. Kumar D, Faisal N, Layek A, Priyadarshi G. Enhancement of mechanical properties of carbon and flax fibre hybrid composites for engineering applications. In *AIP Conference Proceedings* (Vol. 2341, No. 1, p. 040032). AIP Publishing LLC; 2021.

5 Banana Fibre-Reinforced Polyester Composites

Abhishek Biswal R., Divyashree J. S., Dharini V., and Periyar Selvam S.
SRM Institute of Science and Technology

Sadiku Emmanuel R.
Tshwane University of Technology

CONTENTS

5.1 INTRODUCTION

Composites are generally utilised in our everyday life. Because of their light weight and can be formed for explicit use, they command significant distinctions in their applications, especially in the aviation and automotive industries, to mention a few. Nowadays, natural fibres are the source of renewable energy, and they are relatively inexpensive and abundantly available (Kiruthika and Veluraj, 2009). Natural fibre strands supplant the synthetically manufactured filaments as support in different grids. Natural fibre composites can really be used as a replacement for wood and also in other specialised fields, e.g., in the automotive industry (Nasimudeen et al., 2021; Shahroze et al., 2021; Chandrasekar et al., 2022). In any event, a modification is recommended to work on the connection between the fibre and the lattice, compound (Rashid et al., 2020). The physiochemical characteristics of banana fibre are listed in Table 5.1. The impact of compound alteration on the unique mechanical properties of banana fibre-supported polyester composites is being examined in order to explore the interfacial properties. The interfacial peculiarity was thus viewed to be connected with the inclusion of corrosive base cooperation, which happens because of the particular associations between less responsive utilitarian gatherings of each accomplice at the point of interaction (Srinivasababu et al., 2009). The fibre technology mainly focuses on the longevity, stability and inertness by increasing the usage of plant fibre-based composites. Cellulose is the most abundant component of plant fibre, followed by hemicellulose, lignin and gelatin (Thiagamani et al., 2022;

DOI: 10.1201/9781003270980-5

TABLE 5.1
Physiochemical Properties

Functional Properties	Parameters/Properties	Values
Mechanical properties	Width or diameter (μm)	80–250
	Density (kg/m³)	1350
	Cell ℓ/d ratio	150
	Microfibrillar angle	10 ± 1
	Initial modulus (GPa)	7.7–20
	Ultimate tensile strength (MPa)	54–754
	Elongation (%)	10.35
Physical properties	Moisture content (wt.%)	80–250
	Ash content (wt.%)	1350
	Carbon content (wt.%)	150
	Water absorption (wt.%)	10 ± 1
	Tensile strength (MPa)	7.7–20

TABLE 5.2
Nutrient Components in Banana Fibre

Components	Name of Compound	Values (Units)
	A_{13}^{+}	0.14
	Ca^{+}	5.72
Metal elements	Mg^{+}	1.77
	Na^{3+}	0.28
	Si^{4+}	1.41
	Lignin	9.00
Polymeric compounds (wt.%)	Cellulose	43.46
	Hemicellulose	38.54

Chandrasekar et al., 2020). Table 5.2 presents the nutrient contents in the banana fibre. The higher the perspective proportion of the fibre, the higher is the elasticity-aiding reinforcement possible. There are many studies being carried out using natural fibres such as coir, sisal, pineapple leaf fibre and banana (Venkateshwaran and Elayaperumal, 2010). The effect of fibre treatment of natural fibre improves the fibre–matrix bonding, while reinforcing themselves in the cellulose content and microfibrillar angle that manifests better mechanical behaviour to the material. One such fibre is banana fibre that has extremely notable improvement in mechanical, physical and chemical attributes (Shankar et al., 2013). About 64% of high cellulose content with low microfibrillar angle of banana has better reinforcing potential. Some other composites, e.g., cotton hybrid fabric, are useful in the preparation of low-strength materials. The continuous improvement in the characteristics of natural fibres, and the use of non-inexhaustible and non-biodegradable assets are expanding dramatically, in this modern age (Samivel and Babu, 2013). Normal strands, which have excellent mechanical qualities, are typically inexhaustible material assets all over the world, particularly in tropical regions. Regular filaments can be used as a substitute for designed strands in various ways due to their low thickness, low cost, their sustainability and biodegradability. Lately, specialists' advantage has centred on the utilisation of normal strands, as support system in polymeric network (Hanifawati et al., 2011). By considering the ecological angle, it would be exceptionally beneficial to utilise regular filaments, rather than the manufactured strands. Normal natural filaments, such as coir, hemp, sisal, pineapple leaf strands, jute and others, were tested for their ability to be components in the production of thermoset composites. The gum is commonly injected at the lowest part of the form and fills it up to reduce air entrapment (Prabu et al., 2012). At the point when the sap begins to spill into the pitch trap, the cylinder is cinched to limit tar misfortune. Since the interaction happens in a shut form, the development of styrene will be lowered. Resin transfer moulding (RTM) doesn't include huge infusion pressure. Infusion time goes from few moments for little and straightforward parts to hours for enormous, complex parts with fibre-rich content (Pothan and Thomas, 2004). This has prompted the ecological contamination and has drawn in the consideration of a lot of analysts. Over the past several years, scientists have been working to include new classifications of materials and objects that are both environmentally friendly and practical to utilise (Senthilkumar et al., 2021; Alothman et al., 2020). This chapter emphasises the physical, mechanical and tribological characteristics of banana fibre-reinforced polymer composites, as well as the factor properties of the composites, with extensive industrial applications (Kusić et al. 2020).

5.1.1 Properties of Banana Fibre

Banana fibre has better density and microfibrillar angle than sisal and pineapple leaf and has better elongation properties compared to palmyra, sisal and pineapple leaf. Figure 5.1 depicts the image of paper extracted from banana plant fibres. Fibres such as coir and bagasse have lower tensile strength compared to banana fibre. Nicollier et al. (2001) studied the actual and mechanical properties of banana fibre concrete composites. The normal strands are called building materials, and the cellulose content and the microfibril point determine these real properties. The high cellulose content and the low microfibril level of the banana filaments emphasise their obvious suitability as a material for construction. In the material plan, the hybridisation of two single strands has become an effective method (Thomas et al., 2021; Senthilkumar et al., 2015, 2022). The mechanical characteristics of the cross-link have decreased by about half when compared to the hybridised link. Some studies have shown that banana strands and sisal twine offer good execution and can be used as suitable candidates in polyester netting (Laly et al., 2003 and Kalaprasad et al., 1997). In order to improve the mechanical characteristics, banana filaments were hybridised with glass strands. As a result, it may be deduced that the test produces relative designs rather than true structures.

FIGURE 5.1   Tuxing technique to separate fibre from pith.

5.2   RESEARCH ON BANANA FIBRE-REINFORCED POLYESTER COMPOSITES

Biocomposites have emerged as a key component in the construction industry as widely used, lightweight and efficient composite designs. These biocomposites outperform their basic partners in terms of mechanical characteristics. The characteristics of the biocomposites vary depending on the additive utilised. Hybridisation of banana fibres improves ductility.

Chandu et al. (2018) examined composites with specific fibre and matrix compositions that boost the material's strength. Pin-on-disc type friction and wear monitor with data collecting equipment were employed as the testing approach. The material is a banana pineapple fibre-reinforced hybrid composite that was manually laid up with epoxy glue and hardener as additional ingredients. Different composite combinations were tested, as shown in Table 5.3.

The mechanical characteristics of these hybrid composites were investigated by using different weight fractions, viz. 2, 4 and 6 kg at 0.75 and 1 m/s of sliding velocities. The wear behaviour of the composites is affected by the impact load and chemical bonding. Since the surface of the composites is made only with fibre and the matrix smooth, the penetration of load was higher and this led to higher weight loss when compared to the banana pineapple-reinforced hybrid composites. It is noteworthy that the hybrid composite with 30% banana/70% pineapple exhibited different loads at 1 m/s sliding velocity, which resulted in lower weight loss and coefficient of friction. The fine distribution of fibres along with the chemical bonding between the matrix and reinforcements must be the reason behind the mechanical strength of this hybrid composite.

Karimi et al. (2014) employed a unidirectional glass/banana fibre epoxy laminate by using the hand laying technique and binder materials, resulting in increased tensile strength between the composites. The Unidirectional Glass Fibre (UGF) contains high tensile strength values up to 567 MPa that can be used for moderate-stability applications, especially in the construction articles, e.g., partition boards. In exploratory investigation, the creation of deserts in composites, for example, voids, breaks, framework and fibre crack, unfortunate holding, ill-advised network dispersion and

TABLE 5.3

Ratio of Banana Fibres and Pineapple Fibres in the Hybrid Composites

S. No.	Banana:Pineapple
1	10:0
2	7:3
3	5:5
5	3:7
5	0:10

nonalignment of filaments were the likely explanations for attaining lower ductile incentive for UGF and UB (Banana)/GF composites when contrasted with the theoretical/mathematical data.

Pujari et al. (2018) carried out some experiments by conducting thermal conductivity measurements according to ASTM E 122587. Because of its non-reactive nature, the thermal conductivity of the compounds reduced as the volume percentages of fenugreek increased. The fenugreek banana components were effective and had great heat resistance. The newly innovated composite material may be utilised for a variety of purposes, including steam pipes, car interior parts and electronics packages. The thermal conductivity of the composite was 0.259 W/mK at the greatest extreme volume percent of fenugreek.

Mohan et al., (2019) used banana fibres, untreated raw sodium montmorillonite clay along with some binders, e.g., epoxy materials and hardeners infused with nanoparticles. Scanning Electron Microscopy (SEM) examination and wear tests were employed in the experiments. Microscopy analysis revealed that Nanoclay Banana Fibre-supported composites indicated a moving layer between the wear test example worn surface and the counter face, resulting in improved wear qualities. Additionally, the micro-clay particles added, increased the hardness and the grating to the composites, thereby enhancing the deterioration of the composites qualities.

Wang et al. (2016) experimented with raw coconut fibre and waste banana stalk fibre along with adhesives, such as phenol formaldehyde and paraffin wax, as waterproofing agents. The material was tested for its mechanical integrity, viz. elastic modulus, thickness and elasticity. The mechanical properties of the Coir Banana (CB) molecular sheet were directly related to the fibre content of the banana. The high surface area of the banana fibres and the microstructure qualities led to a hike in the phenol formaldehyde content and the 24 h average thickness swelling ranges for CB chipboard. The appearance of the CB agglomeration was impacted by the organisation of the banana fibre component, which was ascribed to two features of the banana fibre, viz. higher cellulose and hemicellulose content and lower lignin content.

Vigneshkumar and Rajasekaran (2018) employed sisal, kenaf, banana and carbon fibre filler material in polyester resin and catalytic materials. The fibre-supported polymer composite on Pin on a plate hardware was used to accomplish the experiment. With the enlarged burden of a symmetrical display technique, the wear and frictional characters of carbon, fibre of sisal and rice husk composites were enhanced. The weight of the sisal, kenaf and banana fibre composite was reduced further as the heap and sliding speeds increased. The wear and frictional forces of the sisal and rice husk composites were more problematic when it came to the speeding up and time.

Hariprasad et al. (2013) used woven fibres from banana psuedostem with coconut to improve tensile strength. The half-and-half salt-treated banana–coconut–epoxy composite exhibited significantly higher stiffness and potency than an untreated mixed banana–coconut–epoxy composite, according to their findings. Antacid is a type of antiseptic. The flexural strength of the treated banana–coir–epoxy half-and-half composite was lower than that of the composite blended with untreated banana–coir–epoxy. This study was useful in determining the advantages of fibre mix composites in the underlying applications, as well as in identifying where stresses were fundamental and harmed the connection point under varied stacking settings. The centres of the top sections of the intermediate components employed in the Finite Element Analysis are the most concerning.

Alavudeen et al. (2010) used the banana and kenaf fibres along with the catalyst and polyester as the matrix and reported the fact that the fabric tensile test showed a maximum stress of 137.221 MPa, leading to the development of a good clutch disc compound, especially in terms of stiffness and strength. This composite material had high machinability and would extend clutch disc lifespan and decrease energy loss during power transfer. Banana and kenaf fibres have been found to be viable reinforcements in the creation of a reinforced composite alternatives and novel material for low-cost clutch disc applications.

Estrada et al. (2015) experimented with banana fibre containing 6% sodium hydroxide with epoxy resin and catalyst/network association. After fibre pre-treatment, the banana fibre/epoxy composite exhibited the greatest compressive strength of 122.11 MPa, which was 38.35% greater than that obtained without treatment. The banana fibre/eco-polyester composites had the highest compressive strength of 122.88 MPa, which was attained without fibre pre-treatment. Higher ecological clogging was observed for the epoxy/banana fibre composites with soluble pre-treatment, followed by the untreated banana/polyester fibre composites. Due to the advancement in fibre/lattice formulation, fibre pre-treatment is crucially important for epoxy composites and the weakening of the work of the successive fibre arrangements in polyester composites.

Subramanya et al. (2020) used the water tank roasting method to extract banana fibre from the pseudostem part of the plant. The weight ratio of the composite was 3:7, consisting of fibres and epoxy resin, together with 10% (w/w) grade K6 as a hardener. Two composite laminates were experimented, one with alkali-treated banana fibre and the other with untreated banana fibre. Previous studies had found that tensile strength was directly related to fibre-matrix bonding. The caustic-treated fibre composite had better tensile strength than the untreated fibre, representing a significant difference of ~15.8%. The impact strength was 27.28% higher for the treated banana compound than the untreated. SEM analysis showed that the voids in the green fibre composite were the result of poor adhesions in the fibre and epoxy resin.

The smooth surface of the untreated fibres resulted in a poor epoxy bonding to the fibres. This reduced the adhesions/interactions between them and resulted in a lower tensile strength. Therefore, this study showed that the alkali-treated banana fibre composite had better impact strength, tensile strength and fracture toughness than the untreated banana fibre composite.

Bilba et al. (2007) experimented on the banana stalk with 9% alkali and 25% of sulphidity treatments to make a cement matrix. Pulped banana fibre (Cavendish) was found to be suitable for use in a concrete network. The flexural characteristic of over 20 MPa was recorded at a stacking of between 8% and 16% by mass (as high as those found in a few engineered strands-supported concretes and up to 90% that of pulped softwood fibre concrete materials). The flexural strength was about 25 MPa and the fracture durability was 1.74 kJ m^2 at a 14% fibre stacking by mass. The mechanical tests revealed that a fibre content of ~12% by mass resulted in an excellent representation of the various waste fibre-reinforced composites, with flexural strength improvement of ~20 MPa and toughness value in the range of ~1.01.5 kJ m^2.

Banana fibres with unsaturated polyester and methyl ethyl ketone peroxide as a hardener were reported by Jannah et al. (2009). The stiffness and modulus of elasticity of the composite were extended in relation to the volumetric expansion of a portion of fibreglass due to fibreglass's closer resemblance to banana fibre with polystyrene. The impact of fibre content, fibre stacking and the blending effect on the mechanical characteristics such as elasticity, elastic modulus, elongation-at-break and flexural properties of composites were investigated. Table 5.4 comprises the density and thermal conductivity of PP (polypropylene) and banana fibre composites. In addition to the elongation-at-break, the volume fraction of the glass fibre, in terms of the absolute fibre fraction, enhances the mechanical characteristics in each situation. The elastic and flexural characteristics of the composites were shown to improve when the fibre stacking was increased (vol.%). The reticular fibre's perfect qualities were further refined when the banana fibre was modified. The additive rule of cross-breeds was used to assess the cross potency.

Paul et al. (2008) used polypropylene fibres and banana fibres to make a composite and analysed the thermophysical properties of these composites. Banana fibres have been chemically treated with alkali, potassium permanganate, benzoyl chloride and silane. These treatments increased the thermophysical properties of the composites.

TABLE 5.4
Density and Thermal Conductivity of PP/Banana Composites

Banana Fibre Treatment	Density, ρ (kg/m³)	Thermal Conductivity, k (W/m K)
PP + untreated fibre	982	0.157
Benzoylated fibre	1012	0.182
KMnO$_4$ treated fibre	1088	0.168
2% alkali-treated fibre	1013	0.163
10% alkali-treated fibre	1041	0.178
Tetraethoxysilane (TEOS)	999	0.160

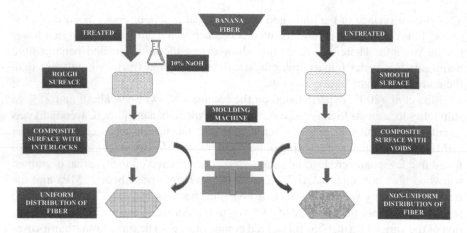

FIGURE 5.2 Variation in the composite texture of the treated and the untreated banana fibres.

The benzoylated fibre composite and the 10% caustic (NaOH)-treated composite had higher thermal conductivity than the untreated fibre composite. This was due to the benzoylation treatment, leading to reduced hydrophilicity of the fibre. This increases the compatibility of the fibre with the hydrophobic polypropylene matrix. Sodium hydroxide pre-treatment removed the hydrogen bonds in the cellulose and activated the fibre surface for benzoylation. The cavities created during the treatment, interlocked with the matrix surface and contributed to the strengthening of the composite materials.

Due to the treatment with alkali, the banana fibre surface became rough, and it created small voids for interlocking with the matrix, which is schematically portrayed in Figure 5.2. Irrespective of the type of treatments, the chemicals affect the thermal conductivity of the composites. As the alkali concentration increased by 8%, there was an increase of 9% in the thermal conductivity and 10% increase in the thermal diffusivity. On treatment with permanganate, the thermal diffusivity increased by 16% when compared to the untreated fibre composite, whereas the silane (tetraethoxysilane)-treated fibre composite had an increase of 8% when compared to the untreated composite. The benzoylated fibre composite had the highest increase of ~25% in the thermal diffusivity and 16% increase in the thermal conductivity when compared to the untreated fibre composites.

Pujari et al. (2017) studied the thermal conductivity of jute fibre composites and banana fibre composites. As the fibre content in the composite increased, its thermal conductivity decreased. Comparing the thermal conductivity of the banana fibre composite, which was 0.228, the jute fibres had a conductivity greater than 0.231 in a temperature range between 20°C and 300°C. Hence, banana fibres can easily find application in insulation items such as furniture, insulation boards and aircraft components.

In the study by Haneefa et al. (2008), while attempting to increase the qualities of the fibre matrix, banana fibre was combined with glass fibre and polystyrene glass. The associated characters of glass fibres and banana fibres led the composite's elasticity and modulus of elasticity to increase in relation to the volumetric expansion of glass fibres with the use of polystyrene as the matrix. The effect of fibre content, fibre stacking and shearing on the mechanical properties of the volume fraction of the glass fibre, such as stiffness, modulus of elasticity, elongation-at-break and flexural properties, was evaluated in terms of the total fibre fraction, which, in addition to the elongation-at-break, extended the values of all the mechanical properties tested. It was discovered that when the stacking of (percent by volume) rose, the deformability and flexural characteristics of the composites improved. The modification of the banana fibre surface further developed the ideal properties of the fibre structure.

Shivankar and Mukhopadhyay (2019) used three types of banana fibre to determine the bending length by using a polypropylene pleat as the matrix. Grain Naine's banana nonwoven texture exhibited a high percentage of elongation for cross-machine support. Fleece structure blend saw was used as a parallel laid (P.L.)-100% banana fleece structure. The shot fastness of the P. nonwoven texture was higher than the Cross-laid (C.) nonwoven texture for three assortments of banana strands, both on the machine and in the cross-pass. The curvature length of the nonwoven structure C. banana was greater than that of nonwoven structure.

Laly et al. (2003) experimented on banana fibre along with unsaturated polyester resin. The tensile strength of the banana fibre was increased when added with the polyester material. The greatest rigidity was attained at 30 mm fibre length, while the sway strength yielded the most extreme incentive for the 40 mm fibre length. The joining of 40% untreated filaments, yielded a 20% increment in the rigidity and a 341% expansion in the sway strength. On treatment with silane coupling agent, the composites showed a 28% expansion in rigidity and a 13% increment in the flexural strength. Studies showed a reduction in the rigidity of the composites. The exploratory elasticity values were contrasted with hypothetical expectations, hence concurring with the Piggot condition. Electron microscopy studies were completed in order to comprehend the morphology of the fibre surface, fibre pull-out and point of interaction holding. Water assimilation studies showed an expansion in water take-up with increases in the fibre content. In conclusion, the authors contrasted the properties of banana fibre in polyester composites and other commonly available natural fibre-supported composites.

Jeyasekaran et al. (2016) investigated the tensile strength of glass fibre and banana glass fibre-reinforced epoxy composites. The value of the experimental stress for the glass fibre-reinforced composite was 567.7 MPa, which was higher than that for banana glass fibre-reinforced composite having a value of 295.38 MPa. This vast difference was due to the fabrication defects, such as improper alignment of fibres, voids, non-uniformly distributed matrix and low inter-bonds between matrix and fibres. There was no major difference between the average load-displacement of these two fibre composites. Hence, they can be used for the manufacturing of household materials, such as partition boards, which require moderate strength.

Sreekumar et al. (2008) investigated banana fibre, used in composites and arranged by the RTM method. All investigations were accomplished by shifting the length of the fibre and its stacking. SEM studies were carried out in order to have an understanding into network cooperation, fibre direction along its breakage or deformation. The idea of the fibre-reinforced composites relies upon few factors such as fibre length, fibre content and void substance. The reliance of the mechanical properties, e.g., elasticity and Young's modulus, is highly dependent on the length of the fibre in the composite. Obviously, as the length of the fibre increases from least 10 to a maximum of 30 mm, the elasticity and Young's modulus show a rising pattern and these two were most extreme at 30 mm. The pressure strain examination was conducted on the banana-supported polyester composites with fibre stacking. Here, additionally at minimum strain esteems, the composites showed a direct trend of behaviour and when the strain % increased, a nonlinear way of behaviour was noticed. The increment in the fibre content builds the pliability, which is apparent from the elongation-at-break values. The elasticity and Young's modulus of the composites were 223% and 226% higher than those of the unreinforced plastic. At lower fibre stacking, scattering of fibre is extremely poor so much so that the pressure transmission across the composite would not happen, as expected. At greater fibre stacking, there was a solid inclination for fibre connection. In those frameworks, the break initiation and its spread will be more straightforward. The SEM for the elastic surface of composites tests was undertaken in order to comprehend its crack component that are tractable via fractography, for the composites containing 40 and 50 vol.% of fibre. Over the cracked surface of the composite, fibre breakage, fibre pull-out and fibrillation can be observed. At the maximum fibre stacking point, there was void development because of fibre pull-out, which was additionally high and was obvious from the SEM results.

Tuan et al. (2014) studied the regular threads from banana tree bark (banana strands), a material that is readily available in Vietnam. These threads were utilised to support the composite material. Banana strands were removed from banana strips, and then pre-treated with 5% NaOH and sliced to a standard length of 30 mm. The banana fibre was made with mass percent of 10, 15, 20 and 25 wt.% for an epoxy pitch Epikote resin 240 matrix. The results were evaluated, based on the underlying morphology, mechanical qualities, fire resistance and thermal properties. The tractable, compressive and sway properties of the biosynthetic materials, up to the 20% by weight, had expanded when compared to epoxy perfect, according to preliminary findings. The fire resistance and warmth were maintained, while the 20 wt.% banana fibre inclusion in the epoxy resin provided a restricting oxygen record of 20.8% and a good warmth consistency. Table 5.5 describes the various composites derived from the application of banana fibres.

5.3 CONCLUSION

Accordingly, it is presumed that because of banana fibre's enticing properties and its varying uses will serve a significant part in the advancement of its anticipated popularity in the fibre industry. The results of different tests conducted on the fibre indicate the fact that it is increasingly being used in a variety of industries. Fibres can be used as a filler in a mix with other strands in order to improve the qualities of basic filaments. Economically effective composites with high elasticity and flexural properties can be developed through the legal use of banana filaments. Banana filaments show a high ability to change along the length and between strands. This is a grade of normal filaments to fully delineate the banana strands and allow

TABLE 5.5

Various Composites of Banana Fibre and Outcomes of Their Analysis

S. No.	Composition of the Composite	Analysis	Inferences	References
1	Fibres of banana and pineapple, epoxy resin, hardener	Wear test using Pin-on-disc type friction and wear monitor	The wear rate was directly proportional to the normal applied load and the sliding distance, whereas the coefficient of friction remains constant. On comparing other hybrids, the efficient bonding between matrix and the fibre in the hybrid composition ratio of 3B:7P, showed the minimum weight loss and wear value at various loads with a sliding velocity of 1 m/s.	Chandu et al. (2018)
2	Unidirectional fibres of glass and banana, epoxy resin and hardener	Tensile strength, surface morphology by using the SEM (scanning electron microscope)	Compared to the glass banana fibre composite, the unidirectional glass fibre composites had a higher tensile strength up to 567.7 MPa on a load of 60 kN. In the application viewpoint, they can be used for household appliances that require moderate strength.	Jeyasekaran et al. (2016)
3	Banana particulate (fibre), polyvinyl chloride, sodium hydroxide, kankara kaolin clay	Mechanical and physical properties, compression analysis	The uniform distribution of the fibres was clearly analysed in the SEM results. XRD (X-ray Diffraction) confirms the compounds of kaolin clay. The tensile strength increased up to 42 MPa and reduced steadily.	Dan Asabe et al. (2016)
4	Banana stem fibre, epoxy resin, hardener	Tensile properties, morphological fractures by using the SEM	The tensile strength is increased by 90% due to the addition of banana fibre to the composite. A 40% increase of impact strength of the fibre-reinforced epoxy material leads to higher toughness properties of the material. The SEM result showed the ductility of the fibres and the high capacity of their strength.	Maleque et al. (2006)
5	Fibres of banana, glass ropes, polyester resin	Bending strength	The single-layer composite had the bending strength, highest of 13.085 N/mm^2 and lowest of 8.957 N/mm^2. The double-layer composite had a bending strength ranging from 16.834 N/mm^2 to 18.196 N/mm^2. These composites can be used for moderate-strength articles of indoor and outdoor.	Sevgi Hoyur and Kerim Cetinkaya (2012)

(Continued)

TABLE 5.5 (Continued)
Various Composites of Banana Fibre and Outcomes of Their Analysis

S. No.	Composition of the Composite	Analysis	Inferences	References
6	Banana fibres, polyurethane matrix form castor oil, NaOH, polyol, prepolymer, resin	Tensile properties, surface morphology using SEM, ATR-FTIR (Attenuated Total Reflectance Infrared Spectroscopy)	The tensile strength increased to a maximum of 268.9 MPa with an increase in treatment time. The dynamic tensile properties are proportional to the quantity and length of banana fibres in the composite. The absorption rate of 2273 cm^{-1} represents the free isocyanate groups in the concentrated polyurethane. The absence of these bands indicates the cross-linking between the isocyanate groups with the hydroxyl groups on the treated banana fibres. This increases the tensile strength and the storage modulus.	Merlini (2011)
7	Banana fibres, epoxy resin, NaOH, hardener, petroleum jelly	Physical and mechanical properties	The increase in length of the fibre in the composite determined the increase of the tensile strength, compressive strength and flexural strength, moisture absorption and void content. The flexural strength increased with the fibre length of 25 mm. The maximum fibre length up to 15 mm increased the tensile strength to its maximum.	Sumaila (2013)
8	Fibres of banana and sisal, polyester resin, methyl ethyl ketone peroxide (curing agent), cobalt naphthenate (catalyst)	Tensile properties, flexural properties, impact properties, surface morphology	As the banana fibre fraction increased, the tensile strength of the hybrid composite also enhanced. As Vf of banana was fixed at 0.19, the flexural modulus attained its maximum. More economical and value-added composites can be produced from this hybrid combination of banana sisal fibres.	Maries Indicula (2004)
9	Banana pseudostem stalk, alkali, 25% sulphidity, portland cement as matrix	Flexural properties, fracture energy, water absorption rate	The fracture toughness depends upon the aspect ratio and the fibre content. A maximum of 16% fibre mass was studied and analysed with the resulting fracture toughness. It was 1.79 kJ m^2, which was 45–50 times higher than that of the matrix material. There was a decrease in the modulus of elasticity from 14 to 6 GPa, with an increase in the fibre content from 2% to 16%.	Zhu and Tobias (1993)
10	Fibres of banana and sisal, wastes of kraft pulp and *Eucalyptus grandis* pulp, alkali, sulphidity, plain cement	Physical and mechanical properties	The composite with 12% fibre weight content had flexural strength of 20 MPa and a fracture toughness of 1–1.5 kJ/m^2.	Shivankar and Mukhopadhyay (2019)

suitable applications in regular fibre composites. A composite frame comprising cross-grip panels with strength and unyielding was created using wear-resistant banana fibre material. The banana strip shows promise as a material for furthering a worldwide bioremediation strategy for effluents containing hazardous combinations. In order to ameliorate the ductile properties of the banana, a synthetic modification to the regular polymer system (antacid treatment, benzoyl chloride treatment and polystyrene maleic anhydride) could be carried out. Composite expansion showed a copious reduction in the hydrophilicity of the banana fibre and improved fibre/structure similarity through mechanical fixation of the body and substance retention. By hybridising sisal fibre and banana yielded an economically intelligent banana standard for the construction of the polymer with adequate strength, and, hence, composites of cushioning characteristics can be created. As a result of the current research, it was determined that banana fibre may be used as an anticipated support in a polyester network.

REFERENCES

Al Rashid A, Khalid MY, Imran R, Ali U, Koc M. Utilization of banana fiber-reinforced hybrid composites in the sports industry. *Materials*. 2020;13(14):3167.

Alavudeen A, Thiruchitrambalam M, Athijayamani A. Clutch plate using woven hybrid composite materials. *Materials Research Innovations* 2011;15(4):229.

Alothman OY, Jawaid M, Senthilkumar K, Chandrasekar M, Alshammari BA, Fouad H, et al. Thermal characterization of date palm/epoxy composites with fillers from different parts of the tree. *Journal of Materials Research and Technology* 2020;9(6): 15537–46.

Bilba K, Arsene MA, Ouensanga A. Study of banana and coconut fibers: Botanical composition, thermal degradation and textural observations. *Bioresource Technology* 2007;98(1):58–68.

Chandrasekar M, Senthilkumar K, Jawaid M, Alamery S, Fouad H, Midani M. Tensile, Thermal and physical properties of Washightonia trunk fibres/pineapple fibre biophenolic hybrid composites. *Journal of Polymer Environment* 2022;30(10): 4427–34.

Chandrasekar M, Siva I, Kumar TSM, Senthilkumar K, Siengchin S, Rajini N. Influence of fibre inter-ply orientation on the mechanical and free vibration properties of banana fibre reinforced polyester composite laminates. *Journal of Polymer Environment* 2020;28(11): 2789–800.

Chandu KV, Krishna SG, Subrahmanyam BV, Rao KV. Experimental analysis on wear behaviour of banana-pineapple hybrid natural fiber composites. *International Journal of Engineering Science Invention (IJESI)* 2018;14–2.

Corbière-Nicollier T, Laban BG, Lundquist L, Leterrier Y, Månson JA, Jolliet O. Life cycle assessment of biofibres replacing glass fibres as reinforcement in plastics. *Resources, Conservation and Recycling* 2001;33(4):267–87.

Dan-Asabe B, Yaro AS, Yawas DS, Aku SY, Samotu IA. Mechanical, spectroscopic and microstructural characterization of banana particulate reinforced PVC composite as piping material. *Tribology in Industry* 2016;38(2):255.

Ghosh R, Ramakrishna A, Reena G. Effect of air bubbling and ultrasonic processing on water absorption property of banana fibre-vinylester composites. *Journal of Composite Materials*. 2014;48(14):1691–7.

Haneefa A, Bindu P, Aravind I, Thomas S. Studies on tensile and flexural properties of short banana/glass hybrid fiber reinforced polystyrene composites. *Journal of Composite Materials* 2008;42(15):1471–89.

Hanifawati IN, Hanim A, Sapuan SM, Zainuddin ES. Tensile and flexural behavior of hybrid banana Pseudostem/glass fibre reinforced polyester composites. *Key Engineering Materials* 2011 Jun 16;471:686–91.

Hariprasad T, Dharmalingam G, Praveen Raj P. Study of mechanical properties of banana-coir hybrid composite using experimental and FEM techniques, *Journal of Mechanical Engineering and Sciences (JMES)* 2013;4:518–531.

Herrera-Estrada L, Pillay S, Vaidya U. Banana fiber composites for automotive and transportation applications. In 8th annual automotive composites conference and exhibition (ACCE) 2008 (pp. 16–18). https://www.researchgate.net/publication/265891790.

Hoyur S. Production of banana/glass fiber bio-composite profile and its bending strength. *Usak University Journal of Material Sciences* 2012;1(1):43–49.

Idicula M, Neelakanta NR, Oommen Z, Joseph K, Thomas S. A study of the mechanical properties of randomly oriented short banana and sisal hybrid fiber reinforced polyester composites. *Journal of Applied Polymer Science* 2005;96(5):1699–1709.

Jannah M, Mariatti M, Abu Bakar A, Abdl Khalil HP. Effect of chemical surface modifications on the properties of woven banana-reinforced unsaturated polyester composites. *Journal of Reinforced Plastics and Composites* 2009;28(12):1519–32.

Jeyasekaran AS, Kumar KP, Rajarajan S. Numerical and experimental analysis on tensile properties of banana and glass fibers reinforced epoxy composites. *Sadhana* 2016;41(11):1357–67.

Kalaprasad G, Joseph K, Thomas S. Influence of short glass fiber addition on the mechanical properties of sisal reinforced low density polyethylene composites. *Journal of Composite Materials* 1997;31(5):509.

Karimi A, Afghahi SS, Shariatmadar H, Ashjaee M. Experimental investigation on thermal conductivity of MFe2O4 (M= Fe and Co) magnetic nanofluids under influence of magnetic field. *Thermochimica Acta* 2014;598:59–67.

Kiruthika AV, Veluraja K. Experimental studies on the physico-chemical properties of banana fibre from various varieties. *Fibers and Polymers* 2009;10(2):193–9.

Kusić D, Božič U, Monzón M, Paz R, Bordón P. Thermal and mechanical characterization of banana fiber reinforced composites for its application in injection molding. *Materials* 2020;13(16): 3581.

Maleque MA, Belal FY, Sapuan SM. Mechanical properties study of pseudostem banana fiber reinforced epoxy composite. *The Arabian Journal for Science and Engineering* 2007;32(2B):359–64.

Merlini C, Soldi V, Barra GM. Influence of fiber surface treatment and length on physico-chemical properties of short random banana fiber-reinforced castor oil polyurethane composites. *Polymer Testing* 2011;30(8):833–40.

Mohan TP, Kanny K. Tribological properties of nanoclay infused banana fiber (NC-BF) reinforced epoxy composites, *Journal of Tribology* 2019;141(5):052003. doi:10.1115/1.4042873

Nasimudeen NA, Karounamourthy S, Selvarathinam J, Kumar Thiagamani SM, Pulikkalparambil H, Krishnasamy S, et al. Mechanical, absorption and swelling properties of vinyl ester based natural fibre hybrid composites. *Applied Science and Engineering Progress* 2021;14(4):680–8.

Paul SA, Boudenne A, Ibos L, Candau Y, Joseph K, Thomas S. Effect of fiber loading and chemical treatments on thermophysical properties of banana fiber/polypropylene commingled composite materials. *Composites Part A: Applied Science and Manufacturing.* 2008;39(9):1582–8.

Pothan LA, Oommen Z, Thomas S. Dynamic mechanical analysis of banana fiber reinforced polyester composites. *Composites Science and Technology* 2003;63(2):283–293.

Pothan LA, Thomas S. Effect of hybridization and chemical modification on the water-absorption behavior of banana fiber–reinforced polyester composites. *Journal of Applied Polymer Science* 2004;91(6):3856–65.

Prabhakar CG, Babu KA, Kataraki PS, Reddy S. A review on natural fibers and mechanical properties of banyan and banana fibers composites. *Materials Today: Proceedings* 2021.

Prabu VA, Manikandan V, Uthayakumar M, Kalirasu S. Investigations on the mechanical properties of red mud filled sisal and banana fiber reinforced polyester composites. *Materials Physics and Mechanics* 2012;15(2):173–9.

Pujari S, Ramakrishna A, Balaram Padal KT. Investigations on thermal conductivities of jute and banana fiber reinforced epoxy composites. *Journal of the Institutions of Engineers (India): Series D* 2017;98(1):79–83.

Pujari S, Venkatesh T, Seeli H. Experimental investigations on thermal conductivity of fenugreek and banana composites. *Journal of the Institution of Engineers (India): Series D* 2018;99(1):51–5.

Ramdhonee A, Jeetah P. Production of wrapping paper from banana fibres. *Journal of Environmental Chemical Engineering* 2017;5(5):4298–306.

Samivel P, Babu AR. Mechanical behavior of stacking sequence in kenaf and banana fiber reinforced-polyester laminate. *International Journal of Mechanical Engineering and Robotics Research.* 2013;2(10): 32–65.

Senthilkumar K, Saba N, Chandrasekar M, Jawaid M, Rajini N, Siengchin S, et al. Compressive, dynamic and thermo-mechanical properties of cellulosic pineapple leaf fibre/polyester composites: Influence of alkali treatment on adhesion. *International Journal of Adhesion and Adhesives* 2021;106: 102823.

Senthilkumar K, Siva I, Rajini N, Jeyaraj P. Effect of fibre length and weight percentage on mechanical properties of short sisal/polyester composite. *International Journal of Computer Aided Engineering and Technology* 2015;7(1):60.

Senthilkumar K, Subramaniam S, Ungtrakul T, Kumar TSM, Chandrasekar M, Rajini N, et al. Dual cantilever creep and recovery behavior of sisal/hemp fibre reinforced hybrid biocomposites: Effects of layering sequence, accelerated weathering and temperature. *Journal of Industrial Textiles* 2022;51(2_suppl):2372S–2390S.

Shahroze RM, Chandrasekar M, Senthilkumar K, Senthil Muthu Kumar T, Ishak MR, Rajini N, et al. Mechanical, interfacial and thermal properties of silica aerogel-infused flax/epoxy composites. *International Polymer Processing* 2021;36(1): 53–9.

Shankar PS, Reddy KT, Sekhar VC, Sekhar VC. Mechanical performance and analysis of banana fiber reinforced epoxy composites. *International Journal of Recent Trends in Mechanical Engineering* 2013;1(4):1–0.

Shivankar VS, Mukhopadhyay S. Some studies on 100% banana parallel laid and 60:40% banana: Polypropylene cross laid non-woven fabrics. *Fashion and Textiles* 2019;6:7.

Sreekumar PA, Albert P, Unnikrishnan G, Joseph K, Thomas S. Mechanical and water sorption studies of ecofriendly banana fiber-reinforced polyester composites fabricated by RTM. *Journal of Applied Polymer Science* 2008;109(3):1547–55.

Srinivasababu N, Rao KM, Kumar JS. Experimental determination of tensile properties of okra, sisal and banana fiber reinforced polyester composites. *Indian Journal of Science and Technology* 2009;2(7):35–8.

Srinivasan VS, Boopathy SR, Sangeetha D, Ramnath BV. Evaluation of mechanical and thermal properties of banana–flax based natural fibre composite. *Materials & Design* 2014;60:620–7.

Subramanya R, Reddy DS, Sathyanarayana PS. Tensile, impact and fracture toughness properties of banana fibre-reinforced polymer composites. *Advances in Materials and Processing Technologies* 2020;6(4):661–8.

Sumaila M, Amber I, Bawa M. Effect of fiber length on the physical and mechanical properties of random oriented, non-woven short banana (*Musa balbisiana*) fiber/epoxy composite. *Cellulose* 2013;62:64.

Teli MD, Valia SP. Acetylation of banana fibre to improve oil absorbency. *Carbohydrate Polymers* 2013;92(1):328–33.

Thanushan K, Sathiparan N. Mechanical performance and durability of banana fibre and coconut coir reinforced cement stabilized soil blocks. *Materialia* 2022;21:101309.

Thiagamani SMK, Pulikkalparambil H, Siengchin S, Ilyas RA, Krishnasamy S, Muthukumar C, et al. Mechanical, absorption, and swelling properties of jute/kenaf/banana reinforced epoxy hybrid composites: Influence of various stacking sequences. *Polymer Composites* 2022;43(11):8297–307.

Thomas SK, Parameswaranpillai J, Krishnasamy S, Begum PMS, Nandi D, Siengchin S, et al. A comprehensive review on cellulose, chitin, and starch as fillers in natural rubber biocomposites. *Carbohydrate Polymer Technologies and Applications* 2021;2:100095.

Tuan HM, Quan TM, Dang LH. Characterization of biology and morphological traits of native Bac Kan banana (Musaxparadisiaca). *Science and Technology Journal of Agriculture & Rural Development (Ha Noi, Vietnam)* 2014:49–55.

Venkateshwaran N, Elayaperumal A. Banana fiber reinforced polymer composites-a review. *Journal of Reinforced Plastics and Composites* 2010;29(15):2387–96.

Vigneshkumar S, Rajasekaran T. Experimental analysis on tribological behavior of fiber reinforced composites. In *IOP Conference Series: Materials Science and Engineering* 2018 Aug 1 (Vol. 402, No. 1, p. 012198). IOP Publishing.

Wang J, Hu Y. Novel particleboard composites made from coir fiber and waste banana stem fiber. *Waste and Biomass Valorization* 2016;7:1447–58. doi:10.1007/s12649-016-9523-3

Zeleke Y, Rotich GK. Design and development of false ceiling board using polyvinyl acetate (PVAc) composite reinforced with false banana fibres and filled with sawdust. *International Journal of Polymer Science* 2021;2021:5542329.

Zhu WH, Tobias BC, Coutts RS, Langfors G. Air-cured banana fiber-reinforced cement composites. *Cement and Concrete Composites* 1994;16(1):3–8.

6 A Review on Palm Fibre-Reinforced Polyester Composites

Mohd Nor Faiz Norrrahim
Universiti Pertahanan Nasional Malaysia

Muhammad Syukri Mohamad Misenan
Yildiz Technical University

Nurjahirah Janudin
Universiti Pertahanan Nasional Malaysia

Nur Sharmila Sharip
Nextgreen Pulp & Paper Sdn Bhd

Mohd Azwan Jenol
Universiti Putra Malaysia

Syed Umar Faruq Syed Najmuddin
Universiti Malaysia Sabah

Siti Shazra Shazleen
Universiti Putra Malaysia

Norizan Mohd Nurazzi
Universiti Sains Malaysia

R. A. Ilyas
Universiti Teknologi Malaysia

M. R. M. Asyraf
Universiti Teknologi Malaysia

DOI: 10.1201/9781003270980-6

CONTENTS

6.1   INTRODUCTION

Polyesters, which feature an ester functional group in their backbone, are particularly significant polymers. Polyesters can be found in nature or synthesized [1]. Synthetic polyesters are typically nonbiodegradable but natural polyesters are biodegradable. The polyester can be classified into two classes which are thermoplastic and thermoset, according to their chemical structure [2]. Similarly, depending on the primary chain, polyesters can be aliphatic, aromatic, or semi-aromatic. Aromaticity improves thermal stability, mechanical strength, glass transition temperature, and chemical stability in polyesters. Polycaprolactone, polylactic acid, polyhydroxyalkanoate, polyethylene adipate, polyethylene terephthalate, polybutylene terephthalate, and polyethylene naphthalate are some of the examples (Figure 6.1).

Moreover, moisture resistance, fire resistance, mechanical and thermal properties, liquid crystal qualities, and environmental stability are all advantages of synthetic polyesters. Films, capacitor components, liquid crystal displays, car interiors, holograms, wire insulation, jet engine seals, tapes, and textiles are all made from polyesters [3]. In addition, car body parts and watercraft have both used polyester resin and glass fibre (GF) composites [4].

Since the 1980s, the utilization of natural fibres in polyester composites has piqued interest. Hemp, jute, banana, sisal, coir, kenaf, bamboo, rice, oil palm, and flax were among the plant fibres utilized to reinforce the polyester matrix. These fibres have attracted renewed interest as replacements for artificial fibres such as carbon, aramid, and glass in the automotive industry. In contrast, natural fibres provide a number of advantages over synthetic fibres, including low cost, low density, acceptable specific strength qualities, ease of separation, carbon dioxide sequestration, and

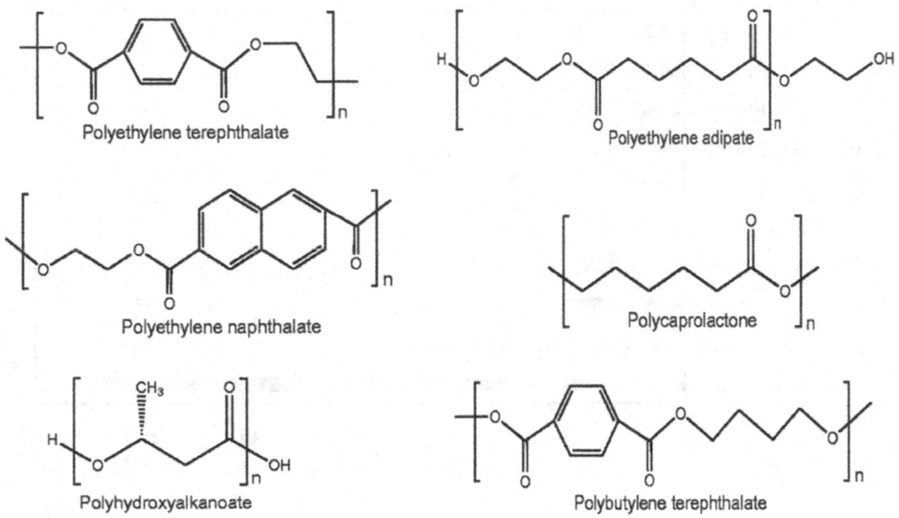

FIGURE 6.1 Examples of polyester.

biodegradability. The fibres in fibre-reinforced composites operate as reinforcement, providing strength and stiffness to the composite structure [5–11]. In addition, these natural fibre composites are found to have a better electrical resistance, good thermal and acoustic insulating properties, and higher resistance to fracture. Moreover, they are also renewable, have high strength and stiffness, and cause no skin irritation. Nonetheless, in a humid environment, these fibres are very vulnerable to water absorption. Chemical treatment of these fibres' surfaces improved their interfacial characteristics, lowering the composite's water absorption [12–14].

Among the natural fibres, the usage palm fibre has been increased. In recent, palm tree extracts have been used by a number of researchers in a variety of matrices [15–19]. On the other hand, leaf fibres, meshes, Ijuk, and empty fruit bunches have demonstrated distinct properties [20]. Every year, palm harvesting generates tons of biomass fibrous wastes around the world, which can be recycled and used to make palm natural fibre composite. The majority of palm fibre research has been done in Malaysia and India, as illustrated in Figure 6.2, due to the high availability of these resources in those countries.

The ability of palm natural fibre composite to be used in a variety of applications has been demonstrated by the results of many investigations. Nonetheless, the physical, mechanical, and morphological properties of these composites, as well as matrix/fibre bonding, production methods, and dispersion characteristics, must all be thoroughly investigated before they are used in industrial applications to ensure dependable and repeatable results. However, significant drawbacks exist, such as fibre microstructure modification when subjected to pressures and water absorption. In addition, natural fibres are known for having limited heat stability, as most fibres begin to degrade at temperatures exceeding 200°F. To avoid the decomposition of lignocellulose components, which would prevent the composite from performing

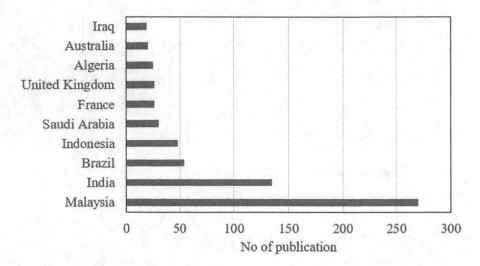

FIGURE 6.2 Statistics from Scopus database. (Search Keywords: "Natural Fibers Composites" and "Palm"; Date: 26 February 2022. Documents by country or territory.)

optimally, the processing temperature and operation parameters must be carefully chosen. Some of these flaws are resolved through fibre treatments and matrix modification [21].

Therefore, the goal of this review is to classify and summarize recent research on the main categories of palm natural fibre composites and their mechanical, thermal, and wettability properties. Additionally, the factors that affect the palm natural fibre-reinforced polyester are discussed in this review.

6.2 PALM FIBRE

Elaeis guineensis (oil palm tree) is one of the most valuable plants in Malaysia, Indonesia, and Thailand. Palm trees are planted for oil on around 15 million hectares, and each hectare produces about 11 tonnes of dry biomass. Currently, oil palm production has increased by double in the last decades, and it has been the most productive fruit crop on the globe for over two decades. Due to this reason, biomass waste from the oil palm sector has been estimated in millions of tonnes due to the expansion of the palm industry. If we effectively utilize oil palm biomass, we will not only solve the disposal problem, but we will also be able to add value to this biomass product. One of the agricultural by-products is fronds (leaves), with an estimated global production of 164 million tonnes of fronds per year. The palm trees also give two additional sources of fibre in addition to the fronds. The fibrous empty fruit bunches or known as oil palm empty fruit bunch (OPEFB) have been explored as possible fibre sources after the seeds have been harvested. Approximately 22 g of palm oil is produced every kilogram of fruit bunch, while roughly a kilogram of OPEFB is produced. After the oil is extracted, the mesocarp remains in the seed, which can be also used as a source of fibre. Essentially, natural retting method can

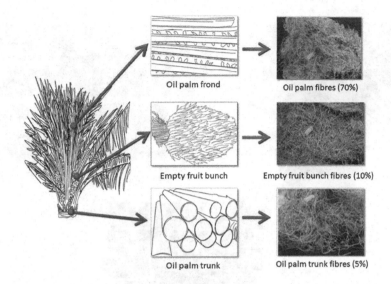

Oil palm frond

Oil palm fibres (70%)

Empty fruit bunch

Empty fruit bunch fibres (10%)

Oil palm trunk

Oil palm trunk fibres (5%)

FIGURE 6.3 Oil palm biomass and oil palm fibres. (Reproduced from Ref. [22].)

provide roughly 400 g of fibres from each OPEFB on average. Typical OPEFB fibre composition is classified as lignocellulosic residues containing about 63% cellulose, 18% hemicellulose, and 18% lignin. Due to the high content of cellulose, natural cellulose fibres have been isolated from a variety of palm tree species and sections. Figure 6.3 depicts the oil palm tree and its biomass.

There are several other types of palm fibres such as sugar palm and date palm. For example, *Hyphaene thebaica* (doum palm), a plant native to upper Africa, was used to remove fibres from the folioles and leaf stalks. The plant portions were split into fibres for mechanical extraction by beating and grating to free the fibres, which were then dried in the air. Another example is sugar palm fibre (SPF) extracted from *Arenga Pinnata* plant, a Palmae family plant originated from Indonesia. Fibre from sugar palm tree could be extracted from four different morphological parts, i.e. fronds, trunk, bunch, and *Ijuk* (surface of trunk). The extraction of SPFs is started by soaking them in a water tank to remover dirt before the palm stalks are occasionally stirred to accelerate the separation process. The SPFs were dried at room temperature before use [23]. According to Ishak et al. [24], the content of cellulose, hemicellulose, and lignin in sugar palm trees is determined by the palm tree's height. In comparison to fibres of 5–15 m height, the composition of 1 m height fibres contained more impurities such as silica. Indication of presence silica is related to ash content. Ash content for 5–15 m height of fibres is less than 10%, as compared to 1 m fibres which contained more than 30%. In general, the composition of cellulose, hemicelluloses, and lignin in 1 m fibres was lower (37.3%, 4.71%, and 17.93%, respectively), compared to the composition of cellulose, hemicellulose, and lignin in fibres at 5 m and above (53.41%–56.8%, 7.45%–7.93%, 20.45%–24.92%, respectively).

In Middle Eastern countries, the date palm (*Phoenix dactylifera*) is an important source of economic and social life. They are common in Qatar, which has abundance

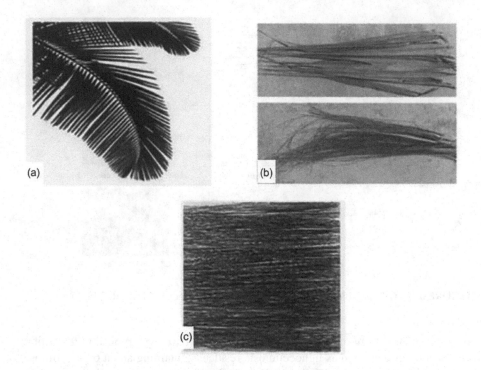

FIGURE 6.4   (a) Green date palm, (b) dried date palm, and (c) date palm leaflet fibres. (Reproduced from Ref. [25].)

of palm trees. The date palm's fibres can be harvested and used in a variety of applications. Khalasa date palm fibres, which were taken from stems, are one of the examples. The Khalasa leaflet was submerged in water, shredded into fibre bundles, and lastly dried at room temperature. Long strands of fibres were painstakingly extracted using a specially designed needle brush. The green date palm, dried date palm leaflet, and continuous date palm leaflet fibres are depicted in Figure 6.4. In the Khalasa date palm, cellulose contributes for 47.14% of its weight, followed by lignin 36.73%, and hemicellulose 16.13% [25].

Palm fibres are promising materials for numerous applications such as absorbent material for different oils [28,29], thermal and acoustic insulation [30,31], reinforcement after acid and alkali treatment [32], cement composites [33,34], thermoplastic composites [35,36], and nanocomposites [37–40]. For example, Neoh et al. [41] investigated the composition of diacylglycerol (DAG) and lauric acid extracted using different hexane extraction procedures (cold, soxhlet, and reflux). The goal is to find the best palm pressed fibre (PPF) oil for premium hard stock manufacture. PPF is a by-product of oil palm fruit extraction. PPF has unique properties as a result of the mixture of palm mesocarp fibre, kernel shell, and crushed kernel. As a result, PPF oil inherited all of the products' properties, such as high triglycerides, palmitic and oleic content from the mesocarp, and gaining high DAG and lauric acid from the palm kernel. Data revealed that as compared to soxhlet extraction, reflux and

TABLE 6.1

Specification of Different Types of Palm Fibre [22,24,26,27]

Properties	Type		
	Oil Palm Fibre	Sugar Palm Fibre	Date Palm Fibre
Moisture content (%)	15–18	5.36–8.7	24
Diameter (μm)	45–65	99–311	100–1000
Length (mm)	5–15	10–15	20–250
Tensile strength (MPa)	50–4000	15.5–270	58–203
Young's modulus	0.57–9	0.49–2.68	2–7.5
Strain (%)	19.6	18.8	10
Density (kg/m³)	114–231	1.26	754
Elongation (%)	2.5–18	12.8	5–10
Thermal conductivity (W/mk)	0.049–0.054	0.032–0.07	0.083

cold extraction yielded a higher recovered oil yield. Cold extraction is the preferred approach for acylglyceride extraction since it yields more DAG. Therefore, the author proposed that cold extraction delivers a higher amount of DAG and a reasonable amount of oil with minimal effort, which is ideal for premium hardstock raw material manufacturing.

On the other hand, Chaktouna et al. [42] used date palm fibres as novel photocatalyst for efficient dye removal. The authors used date palm fibres from the leaves of date palm trees grown in Morocco and combined with titanium dioxide as green photocatalyst of cationic and anionic dyes photodegradation. X-ray diffraction, Fourier transform infrared spectrometry (FTIR), scanning electron microscopy coupled to energy-dispersive X-ray spectroscopy, thermogravimetric analysis, and nitrogen adsorption–desorption were used to evaluate the green photocatalyst. The results reveal that it has a high capacity for adsorption of Congo red and methylene blue dyes under optimum parameters, with excellent regeneration after a few cycles and simple separation. Taban et al. [43] created sound-absorbing panels from date palm fibres. The absorption coefficient of panels of varying thicknesses was examined using an impedance tube and modelled using both the Johnson–Champoux–Allard and the Attenborough models. The findings indicate that samples with a thickness of 55 mm and a density of 175 kg/m³ have the best sound absorption performance and thus can be implemented to improve room acoustics characteristics.

Modification of palm fibre with other materials could enhance the performance of palm fibres. For instance, Darryle et al. [44] employed hydrogen peroxide to modify date palm fibre and evaluated its adsorption capacity against sodium diclofenac in an aqueous solution. The surface functionalization was tailored using a plasma device. The adsorption investigation was conducted with a few variables such as pH, adsorbent size, and temperature. The results showed that plasma-modified palm fibres had excellent diclofenac adsorption performance, which was strongly dependent on the pH solution. As an adsorption mechanism, non-covalent interactions such as electrostatic interaction, hydrogen bonding, and van der Waals forces are involved.

TABLE 6.2

Other Applications of Palm Fibre in Various Industries

Type of Palm Fibre	Application	Performance	Ref.
Oil palm fibre	Bioplastic composites	- Melting temperature of composites decreased - Better improvement in tensile strength and Young's modulus - Water sensitivity and permeability improved due to presence of oil	[45]
Oil palm fibre	Biochar for sulphur dioxide adsorption	- Oil palm fibre-activated biochar displayed better SO_2 adsorption than oil palm fibre biochar - Regeneration of biochar was conducted using thermal treatment	[46]
Oil palm fibre	Solid-state supercapacitor	- Viscous nature of the gel electrolyte reduced the ionic diffusion and capacitance of the device - Reduced size promotes ionic mobility and effective diffusion - Stable electrochemical double layer	[47]
Sugar palm fibre	Thermoplastic polyurethane composites	- Combination of alkaline silane-treated hybrid composites promotes lowest density, thickness swelling, and great water absorption as compared to other composites - Enhanced thermal stability - Possibility for automotive industry	[48]
Sugar palm fibre	Hybrid composites	- Benzoylation of sugar palm fibres (SPF) and its hybridization in glass fibres (GF)-reinforced epoxy composites successfully conducted through a traditional hand lay-up technique - Flexural and compressive properties of hybrid composites improved with increase in GF content	[49]
Sugar palm fibre	Propylene-reinforced composites	- Better thermal stability as compared to untreated reinforced composites - The benzoylation treatment gave a good interfacial bonding, where the polymer acted as a barrier to prevent the degradation of the natural fibres	[50]
Sugar palm fibre	Polyvinylidene fluoride reinforced composites	- Improved hardness, Young's modulus, stiffness, and fracture phenomena - Excellent resistance to water absorption, moisture content, and thickness swelling	[51]
Date palm fibre	Epoxy composite-hybridization with bamboo fibre	- Highest mechanical properties compared to date palm fibre composite without hybridization - Excellent tensile strength, flexural strength, and impact toughness values - Reduced thickness swelling and water absorption	[52]
Date palm fibre	Poly(butylene succinate) as bio-composite	- Biodegradable composites at low cost and high rigidity - Enzyme treatment contributed to more fibrillose structures and proper surface	[53]
Date palm fibre	Fibre–cement composite	- Treatment of date palm fibres with linseed oil cover reduces water absorption and improves mortar workability - Bending strength of mortars improved by fibres treated with sodium hydroxide	[54]

Therefore, palm fibres may be promoted as a low-cost adsorbent material for controlling water pollution caused by pharmaceuticals. Esterified OPEFB fibre was explored as an oil absorption material in the marine sector by Asadpour et al. [29]. Their kinetic and isotherm studies, as well as their physical and chemical characterization, are well discussed in this study. The presence of hydroxyl and ester groups in the FTIR spectra proved the esterification of oil palm fibres. The pseudo-second-order model was better fitted by R^2, which represents the correlation coefficient, and q_e, which represents the adsorption quantity. The crude oil was adsorbed as a monolayer onto the surface of the treated oil palm fibre, according to batch isotherm data. Hence, the use of fatty acid oil palm fibres as an economical sorbent for crude oil clean-up could be widely considered.

6.3 PRETREATMENT OF PALM FIBRE

6.3.1 THE IMPORTANT OF PRETREATMENT OF FIBRE FOR COMPOSITES PRODUCTION

Pretreatment of fibre is a very important step in composites production, as it would remove the lignin present (i.e. delignification) by altering or modifying the composition of lignocellulosic macrostructure and microstructure, which would therefore result in the increase of its surface area and improve the porosity of lignocellulosic material [9,55–60]. Apart from that, pretreatment methods could also reduce the crystallinity of cellulose [57]. It is understandable that each operating facility for composites production would aim for a more resourceful pretreatment method (i.e. cheap, easy to operate, and form a substantial percentage of yield), and it is important to note that the pretreatment methods need to be selected thoroughly based on the lignin profile as the amount of lignin present varies in each type of oil palm biomass (i.e. oil palm mesocarp fibre (OPMF), oil palm empty fruit bunch (OPEFB), oil palm trunk (OPT), oil palm frond (OPF), and oil palm leaves (OPL)) [5,58,61]. For instance, oil palm mesocarp (OPMF) recorded the highest amount of lignin (24.88%), whereas OPEFB had the lowest composition of lignin (14.25%) [58]. On a related note, selection of pretreatment method should also be made upon consideration of the effects on the final product and its composition [58].

6.3.2 SEVERAL TYPES OF PRETREATMENT THAT HAVE BEEN APPLIED TO THE PALM FIBRE

There are several types of pretreatment methods, which are grouped into three main categories: (1) physical pretreatment, (2) chemical pretreatment, and (3) biological pretreatment [58]. Both physical and chemical pretreatment methods are widely used due to their low cost and user-friendliness but the biological pretreatment method could offer more in terms of its reduced energy consumption, better productivity, greater specificity, and also forming non-toxic components [62]. Physical pretreatment method includes milling, pyrolysis, and hydrothermal treatment, while acid, alkaline, and organic solvents are used in the chemical pretreatment, and biological pretreatment involves the use of microorganisms and enzymatic reaction [58].

6.3.1.1 Physical Pretreatment

6.3.1.1.1 Milling

This green approach (i.e. it does not produce any toxic or inhibitory compounds and does not require any harmful chemicals) can reduce the particle size to 0.2 mm and indirectly leads to improved enzymatic hydrolysis in later steps as milling can loosen the cellulose structure and retain the micro-fibre structure in the hydrogen bond fracture [63]. Additionally, it can also increase the degree of polymerization of the biomass and decrease crystalline structure. Examples of milling pretreatment methods are ball milling, two-roll milling, colloid milling, rod milling, vibratory milling, hammer milling, and wet disc milling. Despite all the advantages it possesses, this approach consumes high energy for the operation of machinery (i.e. expensive) which exceeds the cost of particle size reduction as well as requires the capital cost of mechanical equipment [64].

6.3.1.1.2 Extrusion

Similar to milling technique, extrusion machines (single-screw extruders and twin-screw extruders) increase accessibility of cellulose, does not produce inhibitors such as furfural and HMF, and are eco-friendly but demand high energy consumption [65]. The mechanism of extrusion starts when the raw materials are passing through the rotating screws positioned in a tight barrel of the extruders resulting in the breaking down of the lignocellulose biomass. The high temperature (>300°C) of the barrel also causes chemical changes in the biomass [65]. It is interesting to note that extruders can be modified according to needs as the design of the screws, the rotational speed, and the temperature within the barrel are the main parameters controlling the extrusion pretreatment method [66].

6.3.1.2 Chemical Pretreatment

6.3.1.2.1 Acid pretreatment

Inorganic acids (i.e. H_2SO_4, H_3PO_4, HNO_3, and HCl) or organic acids (i.e. formic acid, maleic acid, oxalic acid, acetic acid, and citric acid) are employed as a catalyst to hydrolyze carbohydrates (i.e. hemicellulose), thereby weakening the lignin–hemicellulose barrier [67]. In general, the use of organic acids holds several advantages over the inorganic acids as they do not generate lignocellulose-derived compounds during fermentation process, less toxic, and produce comparatively fewer inhibitory compounds in the hydrolysate [68]. On a related note, pretreatment can be carried out using either diluted acids (0.1% to 10% at 120°C–200°C) or concentrated acids (30%–70% at lower than 100°C) [69]. Diluted acids are often preferred over concentrated acids for lignocellulose pretreatment due to their high reaction rate and ability to support continuous production processes without the need for recycling. In addition, diluted acids create a more gradual decrease in pH, allowing microorganisms used in subsequent fermentation steps to better adapt to the low-pH environment. This can lead to more efficient conversion of the lignocellulosic material to biofuels or other value-added products [70]. On the other hand, there are several drawbacks when using the acid pretreatment such as the need of using individual reactors to withstand corrosion and requirements to perform subsequent detoxification, washing, and neutralizing before further processing, which would impact the operational and maintenance costs and be time-consuming [71]. Other than that, Eom et al. [72] observed that acid pretreatment causes starch to degrade to form HMF under severe conditions.

6.3.1.2.2 Alkaline pretreatment

Hydroxides such as NaOH, H_2O_2, NH_4OH, KOH, and $Ca(OH)_2$ are used to break down the intermolecular ester linkage between hemicellulose and lignin through a saponification reaction [58]. This results in the solubilization of lignin and changes to the lignin structure, which exposes the cellulose and enlarges its internal surface area. This enhanced surface area facilitates enzymatic hydrolysis in subsequent steps, leading to higher sugar yields that can be converted into valuable products [73]. Previous studies showed that it can also decrease the polymerization and degree of crystallinity of cellulose as well as remove acetyl and various uronic acid in hemicellulose [58]. Alkali pretreatment has several advantages as it can be operated at low temperature and pressure, making it a less-energy consumption process, degrades less sugar, and only solubilizes a limited amount of hemicellulose, but it is important to note that this method is only suitable for biomass with low lignin content and requires extensive washing and neutralizing before proceeding to further enzymatic hydrolysis [64]. In addition, Noorshamsiana et al. [71] stated that alkali salts can be converted to irrecoverable salts and may get absorbed by the biomass, making them difficult to recover.

6.3.1.2.3 Organic solvent pretreatment

Ethanol, methanol, acetone, glycerol, aqueous phenol, ethylene glycol, triethylene glycol, and aqueous n-butanol are some of the organic solvents used in this type of chemical pretreatment [58]. This inexpensive and less toxic method reaction is performed at temperatures between 150°C and 200°C (i.e. low boiling points), whereby the organic solvent would hydrolyze the biomass internal bonds (i.e. unbounding the linkage between lignin and hemicellulose) and increase the pore size, thus increasing its accessibility for further reactions [64]. In addition, Kapoor et al. [64] outlined that the reaction rate would also increase with the addition of acids or bases as catalysts. Other than that, increasing the concentration of organic solvent such as ethanol would enhance the delignification process outcome, thus promoting cellulose and hemicellulose to be converted into sugars through acid hydrolysis. Although the organic solvent is considered less toxic towards environment, it may be hazardous when handled in large quantities and is highly flammable [58].

6.3.1.3 Biological Pretreatment

Biological pretreatment approaches can be performed either by using complete/whole organisms such as bacteria or fungi (brown-rot fungi that can break down cellulose and have hydrolytic role, white-rot fungi which can selectively degrade lignin, and soft-rot fungi) or enzymes extracted from these organisms or synthesized artificially such as lignin peroxidase, versatile peroxidase, manganese peroxidase, and laccases [71,74]. Of note, enzymes such as lignin peroxidase and manganese peroxidase play a role as oxidizing agent to oxidize non-phenolic lignin and phenolic ring lignin, respectively, whereas the versatile peroxidase can oxidize both types of lignin compounds [73]. Nevertheless, despite the fact it is a green and eco-friendly process, this fermentation approach has certain drawbacks as it has slower rate of conversion in comparison to physical and chemical pretreatment methods and is costly (i.e. dependency on enzymes).

6.4 PALM FIBRE-REINFORCED POLYESTER COMPOSITE PROPERTIES

6.4.1 MECHANICAL PROPERTIES

Utilization of palm fibre as reinforcement materials in polyester composites generally imparted better mechanical properties. As reported by Hussein [75], a 20 vol.% date palm fibre reinforcement increased the ultimate compression strength and flexural strength of unsaturated polyester by 24% (from 72 to 89 MPa) and 49% (from 40.5 to 60 MPa), respectively. The presence of date palm fibre was reported to prevent the cracks propagation and rapid failure of the polyester, through the formation of good bonding between the matrix and the filler. The date fibres also restricted the chain mobility of the unsaturated polyester and hence remarkably improved its Young's modulus by 77% (from 540 to 955 MPa). The enhancement of flexural strength and modulus by date palm fibre inclusion was also reported in another study [76].

Similarly, a study by Norizan et al. [77] revealed remarkable increment of the tensile modulus and flexural strength by 13% and 98% when incorporated with palm fibre despite the decrement in the tensile strength. A similar finding on tensile strength decrement was reported earlier by de Farias et al. [78]. The randomly oriented 10 wt.% palm peach fibre in unsaturated polyester imparted reduction to the composite tensile strength by more than half of those by neat polymer. The tensile strength was better for sample with weaved oriented fibre, where the value was similar to the neat. Reports on the palm fibre polyester tensile strength were, however, varied in trend. In an earlier report by Wazzan [79], ultimate tensile strength, elongation at break, and Young's modulus of date palm polyester composites were all higher than those of the neat polyester regardless of the fibre volume fraction and structure (unidirectional and woven). The incorporation of oil palm fibre also successfully increased the tensile strength for neat polyester from 25 to 35 MPa [80]. Additionally, Raghavendra and Lokesh [81] who investigated the effect of different fibre origin revealed that almost all polyester containing 50 vol.% date palm fibre showed higher tensile strength (8%–24%), flexural strength (64%–92%), and impact strength (63%–83%).

The impact strength of polyester composites generally were lower than the neat itself, and reduced with increase in fibre content [75,79,82]. However, de Farias et al. [78] reported an exception to this trend. They found that incorporating 10% peach palm fibres into an unsaturated polyester matrix, with fibres oriented in different directions, led to higher impact strength values ranging from approximately 0.54 to 0.85 J/m, compared to the neat matrix value of approximately 0.33 J/m. Higher Young's modulus through palm peach fibre reinforcement was also reported in a study, suggesting that the fibre provides toughness to the polymer matrix.

6.4.2 THERMAL PROPERTIES

Improvement in thermal conductivity behaviour of unsaturated polyester through reinforcement of date palm fibre from 0.4 to 0.8 W/m.C° and reduction in dielectric strength from 33.7 to 32.6 KV/mm were also reported [75]. Meanwhile, Norizan et al. [83] who studied the effect of palm sugar fibre on the thermal stability of unsaturated polyester/glass hybrid composite reported that the presence of fibre which are less thermally stable caused decrement in the thermal stability in terms of degradation onset and the maximum thermal decomposition. Alteration of sugar palm and glass fraction in the unsaturated polyester hybrid composites also influence the thermal stability; for example, composites with 50 wt.% sugar palm/50 wt.% glass exhibited much higher thermal stability than those with 70 wt.% sugar palm/30 wt.% glass.

6.4.3 WETTABILITY PROPERTIES

Palm fibre as natural fibre is generally hydrophilic and hence often imparts changes in the physical properties of polymer composites by means of wettability or water absorption and swelling behaviour. A study [84] showed relatively four times higher water absorption and five times higher thickness swelling percentage of composites with SPF as compared to the neat matrix which was the vinyl ester. This was in agreement with earlier studies whereby palm peach fibre in unsaturated polyester led to increasing weight gain due to water absorption as compared to the neat matrix [78].

6.5 FACTORS AFFECTING THE PROPERTIES OF PALM FIBRE-REINFORCED POLYESTER

The palm fibre volume fraction in composites mixture plays an important role especially in their mechanical properties. Increasing the date fibre content in polyester up to 60% led to continuous increment of ultimate tensile strength, elongation at break, and Young's modulus, whereas the impact strength showed vice versa [79]. In other report, the Young's modulus, compression and flexural strength of unsaturated polyester steadily increased with increasing date palm fibre up to 20% volume fraction before prominent decrement through addition of 30% volume fraction [75]. Similar trends were observed in the study by Norizan et al. [77], where inclusion of SPF up to 30 wt.% increased the tensile modulus of unsaturated polyester from 3.54 to 4.43 GPa. However, with further increases in fibre content, the tensile modulus gradually decreased to 4.01 and 3.90 GPa at 40 and 50 wt.%, respectively. The oil palm fibre polyester composites prepared by authors of Refs. [85,86] showed increasing trend; for example, increasing fibre content up to 45 wt.% steadily increased the tensile strength.

Apart from fibre content, the orientation also implied a great influence on the mechanical properties. Fibre aligned to the direction of force evidently contributed to better stress transfer, which was translated through highest tensile strength, elongation at break, tensile modulus, impact and compressive strength for 0° orientation, followed by 45° and 90° as the lowest [77]. This was in agreement with earlier report, which showed higher mechanical properties by unidirectional fibre alignment as compared to the woven pattern [79].

The tensile strength of composite also relies on the interfacial interaction between the matrix and filler. Therefore, surface modification or functionalization using coupling agents was often opted to enhance the compatibility, and eventually the tensile strength of composites. For example, the use of silane agents such as chloropropyltriethoxysilane (CPTS) and tetraethylorthosilicate (TEOS) imparts Si and Cl elements to the materials and hence improve the interfacial interaction between the filler and the matrix. Findings by Azmami et al. [87] proved the efficiency of these two silane agents in improving the interactions between the palm fibre and the polyester as evident through prominent increment of tensile strength by almost two-folds (from 157 N to approximately 300 N) for both CPTS- and TEOS-treated samples. Additionally, these treated palm fibre/polyester composites exhibited lower water absorption capacity by almost half of that untreated sample due to their increased in hydrophobicity, besides higher thermal stability and flame retardancy.

Anggawan and Mohamad [86] conducted a study on alkali treatment to enhance the matrix–filler adhesion and improve the tensile strength of oil palm fibre/polyester composites. Their results showed that treating oil palm fibre with 2% NaOH for 2 hours gave highest tensile strength, and further increase in NaOH concentration to 10% caused lowest tensile strength. Earlier investigations by Hill and Abdul Khalil [85] revealed better compatibility of acetylated oil palm fibre translated through its highest tensile strength (4.8 MPa), while the other two treatments silane and titanate gave 38.3 and 38.8 MPa tensile strength, respectively. More hydrophobic properties of the cellulosic oil palm fibre provide stronger interfacial bond with the polyester, and hence better stress transfer upon applying load.

Several manufacturing methods can be used in preparing palm polyester composites by which the method used may influence the properties of the fabricated composites. The methods include hand lay-up, resin transfer moulding, filament winding and compression moulding. Hand lay-up technique is a process in which fibre is impregnated with matrix by hand. The process could be done using roller or brushes and nip-roller type impregnator was usually used to press the materials before left to cure [88]. Despite the fact that it is a simple and widely used technique, hand lay-up is highly dependent on personal skills by which complete impregnation can be difficult to be ensured. A similar concept to hand lay-up technique is by spray layup, which can be more easily automated and suitable for large size pieces. The limitation is, however, due to its incapability of loading high fibre volume fraction, and utilization of long fibres, as well and high viscosity resins can be unsuitable. In resin transfer moulding method, a single or multi-stack preforms are laid up and could be prepressed into mould by using binder. Resin will be injected into the cavity of the mould and left to cure. Similar to spray layup, this method depends on viscosity of the resin, and therefore is more suitable for less viscous resin like unsaturated polyester, vinyl ester, and epoxy [89]. This method allows incorporation of high fibre volume with low void content. Nevertheless, despite its capability in producing excellent surface quality, a fibre wash in which fibre in the mould is moved due to the pressure (when resin was injected in) can happen and might influence the mechanical properties of the composites [88,90,91].

6.6 CHALLENGES AND FUTURE RECOMMENDATION

The length, diameter, thickness, mechanical properties (tensile, Young's modulus, flexibility), chemical properties (stability, water permeability, reactivity), thermal conductivity, availability, cost-effectiveness, and cellulose content are all key aspects to consider when contemplating the use of palm fibre in a variety of applications. Essentially, palm fibres have some disadvantages over synthetic fibres, such as low thermal stability, sensitivity to the environment, poor interfacial adhesion, and low durability. Palm fibres will be a versatile material for reinforced composites in a variety of applications once all of these drawbacks are addressed. Oil palm fibres for example, are not suited for high-performance applications such as automotive and supercapacitor because of their low strength, sensitivity to environment, and poor moisture resistance, which causes reinforced composites to lose strength and stiffness. As a result, combining oil palm fibres with other fibres (glass, jute, sisal) will improve their mechanical and thermal stability [22].

Palm fibres also offer a lot of potential as reinforcement in cement-insulating composites. Date palm fibre, for example, will improve the thermal degradation and flammability of phenolic resins. Phenolic resins are a type of polymer that is highly cross-linked. The outer layer of phenolic resin transforms into a char layer during a fire, making it fire-resistant. The reinforced composites contained date palm fibres lower phenolic composites' brittleness, thus improving their impact and thermal properties [92].

On the other hand, SPF has a unique set of properties that make it not only robust, but also resistant to seawater. It is easily available in woven fibres, making it simple to prepare. However, the use of SPF as a composite material is not yet completely investigated. Therefore, more research is needed to unveil its relevance and capabilities. There are lot of advantages which could be offered by SPFs such as cost-effectiveness, eco-friendly, and great mechanical properties. Employing the SPFs as composites not only increased the productivity but could also influence environmental as waste management can be efficiently practiced [24].

To date, low interfacial adhesion in palm fibre has been a major problem. Matrix plays a function in composites by transferring load via shear stresses at the matrix–fibre contact. The bonding between the matrix and the fibre determines the load transfer capacity. If the interfacial adhesion is weak, it will have a negative impact on the composite's mechanical properties. Finding an appropriate matrix for fibre-reinforced composites is one possible answer to this problem. Fibre modification, such as introducing a new functional group or treating it with alkali or acid, can also be a solution. The introduction of a functional group may improve the bonding between the matrix and the fibre [93].

Eventually, the potential of palm fibre as a nanomaterial is yet to be fully explored. The mechanical and chemical properties of nano-sized palm fibres have undoubtedly improved, resulting in a wider range of applications. Reinforcing composites in various thermoset and thermoplastic materials to make nanocomposites using nano-sized palm fibre could have a great prospect. We also believe that expanding the usage of palm fibres by developing bio-composites and hybrid composites based on them is still a major issue. This technique could resolve the issue of palm fibre compatibility with other natural or synthetic fibres.

ACKNOWLEDGEMENT

The authors would like to highly acknowledge the financial support (UPNM/2018/CHEMDEF/ST/4) from Ministry of Education Malaysia. The authors also gratefully acknowledge Universiti Pertahanan Nasional Malaysia (National Defence University of Malaysia).

CONFLICT OF INTEREST

There are no conflicts of interest to declare.

REFERENCES

1. Rehm, B.H.A. Polyester synthases: Natural catalysts for plastics. *The Biochemical Journal* **2003**, *376*, 15–33, doi:10.1042/BJ20031254.
2. Aziz, S.H.; Ansell, M.P. The effect of alkalization and fibre alignment on the mechanical and thermal properties of kenaf and hemp bast fibre composites: Part 1 – polyester resin matrix. *Composites Science and Technology* **2004**, *64*, 1219–1230, doi:10.1016/j.compscitech.2003.10.001.
3. Kausar, A. Review of fundamentals and applications of polyester nanocomposites filled with carbonaceous nanofillers. *Journal of Plastic Film and Sheeting* **2019**, *35*, 22–44, doi:10.1177/8756087918783827.
4. Rachchh, N.V.; Trivedi, D.N. Mechanical characterization and vibration analysis of hybrid E-glass/bagasse fiber polyester composites. *Materials Today: Proceedings* **2018**, *5*, 7692–7700, doi:10.1016/j.matpr.2017.11.445.
5. Asyraf, M.R.M.; Ishak, M.R.; Norrrahim, M.N.F.; Nurazzi, N.M.; Shazleen, S.S.; Ilyas, R.A.; Rafidah, M.; Razman, M.R. Recent advances of thermal properties of sugar palm lignocellulosic fibre reinforced polymer composites. *International Journal of Biological Macromolecules* **2021**, *193*, 1587–1599.
6. Lee, C.H.; Lee, S.H.; Padzil, F.N.M.; Ainun, Z.M.A.; Norrrahim, M.N.F.; Chin, K.L. Biocomposites and nanocomposites. In *Composite Materials*; Amit Sachdeva, Pramod K. Singh, Hee Woo Rhee Eds.; CRC Press: Boca Raton, 2021; pp. 29–60.
7. Ariffin, H.; Norrrahim, M.N.F.; Yasim-Anuar, T.A.T.; Nishida, H.; Hassan, M.A.; Ibrahim, N.A.; Yunus, W.M.Z.W. Oil palm biomass cellulose-fabricated polylactic acid composites for packaging applications. In *Bionanocomposites for Packaging Applications*; Springer: Cham, Switzerland, 2018; pp. 95–105.
8. Bhat, A.; Naveen, J.; Jawaid, M.; Norrrahim, M.N.F.; Rashedi, A.; Khan, A. Advancement in fiber reinforced polymer, metal alloys and multi-layered armour systems for ballistic applications–A Review. *Journal of Materials Research and Technology* **2021**, doi:10.1016/j.jmrt.2021.08.150.
9. Norrrahim, M.N.F. *Superheated Steam Pretreatment of Oil Palm Biomass for Improving Nanofibrillation of Cellulose and Performance of Polypropylene/Cellulose Nanofiber Composites*; Universiti Putra Malaysia: Selangor, Malaysia, 2018.
10. Ilyas, R.A.; Zuhri, M.Y.M.; Norrrahim, M.N.F.; Misenan, M.S.M.; Jenol, M.A.; Samsudin, S.A.; Nurazzi, N.M.; Asyraf, M.R.M.; Supian, A.B.M.; Bangar, S.P.; et al. Natural fiber-reinforced polycaprolactone green and hybrid biocomposites for various advanced applications. *Polymers* **2022**, *14*, doi:10.3390/polym14010182.
11. Norrrahim, M.N.F.; Ariffin, H.; Yasim-Anuar, T.A.T.; Hassan, M.A.; Ibrahim, N.A.; Yunus, W.M.Z.W.; Nishida, H. Performance evaluation of cellulose nanofiber with residual hemicellulose as a nanofiller in polypropylene-based nanocomposite. *Polymers* **2021**, *13*, 1064, doi:10.3390/polym13071064.

12. Haghdan, S.; Smith, G.D. Natural fiber reinforced polyester composites: A litera-ture review. *Journal of Reinforced Plastics and Composites* **2015**, *34*, 1179–1190, doi:10.1177/0731684415588938.

13. Zuhri, M.Y.M.; Sapuan, S.M.; Ismail, N. Oil palm fibre reinforced polymer composites: A review. *Progress in Rubber, Plastics and Recycling Technology* **2009**, *25*, 233–246, doi:10.1177/147776060902500403.

14. Norrrahim, M.N.F.; Ilyas, R.A.; Nurazzi, N.M.; Rani, M.S.A.; Atikah, M.S.N.; Shazleen, S.S. Chemical pretreatment of lignocellulosic biomass for the production of bioprod-ucts: An overview. *Applied Science and Engineering Progress* **2021**, *14*, 588–605.

15. Zakaria, M.R.; Norrrahim, M.N.F.; Hirata, S.; Hassan, M.A. Hydrothermal and wet disk milling pretreatment for high conversion of biosugars from oil palm mesocarp fiber. *Bioresource Technology* **2015**, *181*, 263–269, doi:10.1016/j.biortech.2015.01.072.

16. Norizan, M.N.; Alias, A.H.; Sabaruddin, F.A.; Asyraf, M.R.M.; Shazleen, S.S.; Mohidem, N.A.; Kamarudin, S.H.; Norrrahim, M.N.F.; Rushdan, A.I.; Ishak, M.R.; et al. Effect of silane treatments on mechanical performance of kenaf fibre rein-forced polymer composites: A review. *Functional Composites and Structures* **2021**, *3*, 045003.

17. Norrrahim, M.N.F.; Ariffin, H.; Yasim-Anuar, T.A.T.; Ghaemi, F.; Hassan, M.A.; Ibrahim, N.A.; Ngee, J.L.H.; Yunus, W.M.Z.W. Superheated steam pretreatment of cel-lulose affects its electrospinnability for microfibrillated cellulose production. *Cellulose* **2018**, *25*, 3853–3859, doi:10.1007/s10570-018-1859-3.

18. Norrrahim, M.N.F.; Ariffin, H.; Hassan, M.A.; Ibrahim, N.A.; Yunus, W.M.Z.W.; Nishida, H. Utilisation of superheated steam in oil palm biomass pretreatment process for reduced chemical use and enhanced cellulose nanofibre production. *International Journal of Nanotechnology* **2019**, *16*, 668–679.

19. Yasim-anuar, T.A.T.; Ariffin, H.; Norrrahim, M.N.F.; Hassan, M.A. Factors affecting spinnability of oil palm mesocarp fiber cellulose solution for the production of microfi-ber. *BioResources* **2017**, *12*, 715–734.

20. Alhijazi, M.; Zeeshan, Q.; Safaei, B.; Asmael, M.; Qin, Z. Recent developments in palm fibers composites: A review. *Journal of Polymers and the Environment* **2020**, *28*, 3029–3054, doi:10.1007/s10924-020-01842-4.

21. Mukhtar, I.; Leman, Z.; Ishak, M.R.; Zainudin, E.S. Sugar palm fibre and its compos-ites: A review of recent developments. *Bioresources* **2016**, *11*, 10756–10782.

22. Khalil, H.P.S.A.; Jawaid, M.; Hassan, A.; Paridah, M.T.; Zaidon, A. *Oil Palm Biomass Fibres and Recent Advancement in Oil Palm Biomass Fibres Based Hybrid Biocomposites*; Intech: Rijeka, 2012; pp. 187–220.

23. Bachtiar, D.; Sapuan, S.M.; Zainudin, E.S.; Khalina, A.; Dahlan, K.Z.M. The tensile properties of single sugar palm (*Arenga pinnata*) fibre. *IOP Conference Series: Materials Science and Engineering* **2010**, *11*, 012012, doi:10.1088/1757-899x/11/1/012012.

24. Ishak, M.R.; Sapuan, S.M.; Leman, Z.; Rahman, M.Z.A.; Anwar, U.M.K.; Siregar, J.P. Sugar palm (*Arenga pinnata*): Its fibres, polymers and composites. *Carbohydrate Polymers* **2013**, *91*, 699–710, doi:10.1016/j.carbpol.2012.07.073.

25. Mahdi, E.; Ochoa, D.R.H.; Vaziri, A.; Dean, A.; Kucukvar, M. Khalasa date palm leaf fiber as a potential reinforcement for polymeric composite materials. *Composite Structures* **2021**, *265*, 113501, doi:10.1016/j.compstruct.2020.113501.

26. Ghori, W.; Saba, N.; Jawaid, M.; Asim, M. A review on date palm (*phoenix dacty-lifera*) fibers and its polymer composites. *IOP Conference Series: Materials Science and Engineering* **2018**, *368*, doi:10.1088/1757-899X/368/1/012009.

27. Faris M. AL-Oqla, Othman Y. Alothman, M. Jawaid, S. M. Sapuan, and M.H.E.-S. Processing and properties of date palm fibers and its composites. In *Biomass and Bioenergy: Processing and Properties*; 2014; Vol. 9783319076, pp. 1–367.

28. Martins, L.S.; Monticelli, F.M.; Mulinari, D.R. Influence of the granulometry and fiber content of palm residues on the diesel S-10 oil sorption in polyurethane /palm fiber biocomposites. *Results in Materials* **2020**, *8*, 100143, doi:10.1016/j.rinma.2020.100143.

29. Asadpour, R.; Yavari, S.; Kamyab, H.; Ashokkumar, V.; Chelliapan, S.; Yuzir, A. Study of oil sorption behaviour of esterified oil palm empty fruit bunch (OPEFB) fibre and its kinetics and isotherm studies. *Environmental Technology and Innovation* **2021**, *22*, 101397, doi:10.1016/j.eti.2021.101397.

30. Ramlee, N.A.; Naveen, J.; Jawaid, M. Potential of oil palm empty fruit bunch (OPEFB) and sugarcane bagasse fibers for thermal insulation application – A review. *Construction and Building Materials* **2021**, *271*, 121519, doi:10.1016/j.conbuildmat.2020.121519.

31. Agrebi, F.; Ghorbel, N.; Rashid, B.; Kallel, A.; Jawaid, M. Influence of treatments on the dielectric properties of sugar palm fiber reinforced phenolic composites. *Journal of Molecular Liquids* **2018**, *263*, 342–348, doi:10.1016/j.molliq.2018.04.130.

32. Echaroj, S.; Pannucharoenwong, N.; Rattanadecho, P.; Benjapiyaporn, C.; Benjapiyaporn, J. Investigation of palm fibre pyrolysis over acidic catalyst for bio-fuel production. *Energy Reports* **2021**, *7*, 599–607, doi:10.1016/j.egyr.2021.07.093.

33. Yaro, N.S.A.; Bin Napiah, M.; Sutanto, M.H.; Usman, A.; Saeed, S.M. Performance evaluation of waste palm oil fiber reinforced stone matrix asphalt mixtures using traditional and sequential mixing processes. *Case Studies in Construction Materials* **2021**, *15*, e00783, doi:10.1016/j.cscm.2021.e00783.

34. AlShuhail, K.; Aldawoud, A.; Syarif, J.; Abdoun, I.A. Enhancing the performance of compressed soil bricks with natural additives: Wood chips and date palm fibers. *Construction and Building Materials* **2021**, *295*, 123611, doi:10.1016/j.conbuildmat.2021.123611.

35. Nurazzi, N.M.; Asyraf, M.R.M.; Fatimah Athiyah, S.; Shazleen, S.S.; Rafiqah, S.; Harussani, M.M.; Kamarudin, S.H.; Razman, M.R.; Rahmah, M.; Zainudin, E.S.; et al. A review on mechanical performance of hybrid natural fiber polymer composites for structural applications. *Polymers* **2021**, *13*, 2170.

36. Maou, S.; Meghezzi, A.; Grohens, Y.; Meftah, Y.; Kervoelen, A.; Magueresse, A. Effect of various chemical modifications of date palm fibers (DPFs) on the thermo-physical properties of polyvinyl chloride (PVC)–high-density polyethylene (HDPE) composites. *Industrial Crops and Products* **2021**, *171*, 113974, doi:10.1016/j.indcrop.2021.113974.

37. Yasim-Anuar, T.A.T.; Ariffin, H.; Norrrahim, M.N.F.; Hassan, M.A.; Andou, Y.; Tsukegi, T.; Nishida, H. Well-dispersed cellulose nanofiber in low density polyethylene nanocomposite by liquid-assisted extrusion. *Polymers* **2020**, *12*, 1–17.

38. Norrrahim, M.N.F.; Ariffin, H.; Yasim-Anuar, T.A.T.; Hassan, M.A.; Nishida, H.; Tsukegi, T. One-pot nanofibrillation of cellulose and nanocomposite production in a twin-screw extruder. *IOP Conference Series: Materials Science and Engineering* **2018**, *368*, 1–9.

39. Norrrahim, M.N.F.; Kasim, N.A.M.; Knight, V.F.; Ujang, F.A.; Janudin, N.; Razak, M.A.I.A.; Shah, N.A.A.; Noor, S.A.M.; Jamal, S.H.; Ong, K.K.; et al. Nanocellulose: The next super versatile material for the military. *Materials Advances* **2021**, doi:10.1039/D0MA01011A.

40. Ilyas, R.A.; Sapuan, M.S.; Norizan, M.N.; Norrrahim, M.N.F.; Ibrahim, R.; Atikah, M.S.N.; Huzaifah, M.R.M.; Radzi, A.M.; Izwan, S.; Azammi, A.M.N.; et al. Macro to nanoscale natural fiber composites for automotive components: Research, development, and application. In *Biocomposite and Synthetic Composites for Automotive Applications*; Sapuan, M.S., Ilyas, R.A., Eds.; Woodhead Publishing: Amsterdam, Netherland, 2020.

41. Neoh, B.K.; Thang, Y.M.; Zain, M.Z.M.; Junaidi, A. Palm pressed fibre oil: A new opportunity for premium hardstock? *International Food Research Journal* **2011**, *18*, 769–773.

42. Chakhtouna, H.; Zari, N.; Bouhfid, R.; Benzeid, H. Novel photocatalyst based on date palm fibers for efficient dyes removal. *Journal of Water Process Engineering* **2021**, *43*, 102167, doi:10.1016/j.jwpe.2021.102167.

43. Taban, E.; Amininasab, S.; Soltani, P.; Berardi, U.; Abdi, D.D.; Samaei, S.E. Use of date palm waste fibers as sound absorption material. *Journal of Building Engineering* **2021**, *41*, 102752, doi:10.1016/j.jobe.2021.102752.

44. Darryle, C.M.; Acayanka, E.; Takam, B.; Line, L.N.; Kamgang, G.Y.; Laminsi, S.; Sellaoui, L.; Bonilla-Petriciolet, A. Influence of plasma-based surface functionalization of palm fibers on the adsorption of diclofenac from water: Experiments, thermodynamics and removal mechanism. *Journal of Water Process Engineering* **2021**, *43*, doi:10.1016/j.jwpe.2021.102254.

45. Yang, J.; Ching, Y.C.; Chuah, C.H.; Nguyen, D.H.; Liou, N.S. Synthesis and characterization of starch/fiber-based bioplastic composites modified by citric acid-epoxidized palm oil oligomer with reactive blending. *Industrial Crops and Products* **2021**, *170*, 113797, doi:10.1016/j.indcrop.2021.113797.

46. Iberahim, N.; Sethupathi, S.; Bashir, M.J.K.; Kanthasamy, R.; Ahmad, T. Evaluation of oil palm fiber biochar and activated biochar for sulphur dioxide adsorption. *Science of the Total Environment* **2022**, *805*, 150421, doi:10.1016/j.scitotenv.2021.150421.

47. Tobi, A.R.; Dennis, J.O.; Zaid, H.M. Evaluation of low-cost high performance solid-state supercapacitor derived from physically and chemically activated oil palm fiber. *Materials Letters* **2021**, *285*, 129127, doi:10.1016/j.matlet.2020.129127.

48. Atiqah, A.; Jawaid, M.; Sapuan, S.M.; Ishak, M.R.; Ansari, M.N.M.; Ilyas, R.A. Physical and thermal properties of treated sugar palm/glass fibre reinforced thermoplastic polyurethane hybrid composites. *Journal of Materials Research and Technology* **2019**, doi:10.1016/j.jmrt.2019.06.032.

49. Safri, S.N.A.; Sultan, M.T.H.; Saba, N.; Jawaid, M. Effect of benzoyl treatment on flexural and compressive properties of sugar palm/glass fibres/epoxy hybrid composites. *Polymer Testing* **2018**, *71*, 362–369, doi:10.1016/j.polymertesting.2018.09.017.

50. Izwan, S.M.; Sapuan, S.M.; Zuhri, M.Y.M.; Mohamed, A.R. Thermal stability and dynamic mechanical analysis of benzoylation treated sugar palm/kenaf fiber reinforced polypropylene hybrid composites. *Polymers* **2021**, *13*, doi:10.3390/polym13172961.

51. Alaaeddin, M.H.; Sapuan, S.M.; Zuhri, M.Y.M.; Zainudin, E.S.; AL-Oqla, F.M. Physical and mechanical properties of polyvinylidene fluoride - Short sugar palm fiber nanocomposites. *Journal of Cleaner Production* **2019**, *235*, 473–482, doi:10.1016/j.jclepro.2019.06.341.

52. Supian, A.B.M.; Jawaid, M.; Rashid, B.; Fouad, H.; Saba, N.; Dhakal, H.N.; Khiari, R. Mechanical and physical performance of date palm/bamboo fibre reinforced epoxy hybrid composites. *Journal of Materials Research and Technology* **2021**, *15*, 1330–1341, doi:10.1016/j.jmrt.2021.08.115.

53. Chaari, R.; Khlif, M.; Mallek, H.; Bradai, C.; Lacoste, C.; Belguith, H.; Tounsi, H.; Dony, P. Enzymatic treatments effect on the poly (butylene succinate)/date palm fibers properties for bio-composite applications. *Industrial Crops and Products* **2020**, *148*, 112270, doi:10.1016/j.indcrop.2020.112270.

54. Ali-Boucetta, T.; Ayat, A.; Laifa, W.; Behim, M. Treatment of date palm fibres mesh: Influence on the rheological and mechanical properties of fibre-cement composites. *Construction and Building Materials* **2021**, *273*, 121056, doi:10.1016/j.conbuildmat.2020.121056.

55. Nurazzi, N.M.; Harussani, M.M.; Aisyah, H.A.; Ilyas, R.A.; Norrrahim, M.N.F.; Khalina, A.; Abdullah, N. Treatments of natural fiber as reinforcement in polymer composites—A short review. *Functional Composites and Structures* **2021**, *3*, 024002.

56. Nurazzi, N.M.; Asyraf, M.R.M.; Rayung, M.; Norrrahim, M.N.F.; Shazleen, S.S.; Rani, M.S.A.; Shafi, A.R.; Aisyah, H.A.; Radzi, M.H.M.; Sabaruddin, F.A.; et al. Thermogravimetric analysis properties of cellulosic natural fiber polymer composites: A review on influence of chemical treatments. *Polymers* **2021**, *13*, 2710.

57. Zulkiple, N.; Maskat, M.Y.; Hassan, O. Pretreatment of oil palm empty fruit fiber (OPEFB) with aqueous ammonia for high production of sugar. *Procedia Chemistry* **2016**, *18*, 155–161.

58. Diyanilla, R.; Hamidon, T.S.; Suryanegara, L.; Hussin, M.H. Overview of pretreatment methods employed on oil palm biomass in producing value-added products: A review. *BioResources* **2020**, *15*, 9935.

59. Lawal, A.A.; Hassan, M.A.; Zakaria, M.R.; Yusoff, M.Z.M.; Norrrahim, M.N.F.; Mokhtar, M.N.; Shirai, Y. Effect of oil palm biomass cellulosic content on nanopore structure and adsorption capacity of biochar. *Bioresource Technology* **2021**, doi:10.1016/j.biortech.2021.125070.

60. Farid, M.A.A.; Hassan, M.A.; Roslan, A.M.; Ariffin, H.; Norrrahim, M.N.F.; Othman, M.R.; Yoshihito, S. Improving the decolorization of glycerol by adsorption using activated carbon derived from oil palm biomass. *Environmental Science and Pollution Research* **2021**, *28*, 27976–27987.

61. Asyraf, M.R.M.; Ishak, M.R.; Syamsir, A.; Nurazzi, N.M.; Sabaruddin, F.A.; Shazleen, S.S.; Norrrahim, M.N.F.; Rafidah, M.; Ilyas, R.A.; Abd Rashid, M.Z.; et al. Mechanical properties of oil palm fibre-reinforced polymer composites: A review. *Journal of Materials Research and Technology* **2022**, *17*, 33–65.

62. Taylor, M.J.; Alabdrabalameer, H.A.; Skoulou, V. Choosing physical, physicochemical and chemical methods of pre-treating lignocellulosic wastes to repurpose into solid fuels. *Sustainability* **2019**, *11*, 3604, doi:10.3390/su11133604.

63. Nabilah-Jansar, K.; Roslan, A.M.; Hassan, M.A. Appropriate hydrothermal pretreatment of oil palm biomass in palm oil mill. *Pertanika Journal of Scholarly Research Reviews* **2018**, *4*, 31–40.

64. Kapoor, M.; Semwal, S.; Gaur, R.; Kumar, R.; Gupta, R.P.; Puri, S.K. The pretreatment technologies for deconstruction of lignocellulosic biomass. In *Waste to Wealth*; Springer: Singapore, 2018; pp. 395–421.

65. Peinemann, J.C.; Pleissner, D. Continuous pretreatment, hydrolysis, and fermentation of organic residues for the production of biochemicals. *Bioresource Technology* **2020**, *295*, 122256.

66. Gatt, E.; Rigal, L.; Vandenbossche, V. Biomass pretreatment with reactive extrusion using enzymes: A review. *Industrial Crops and Products* **2018**, *122*, 329–339.

67. Risanto, L.; Fajriutami, T.; Hermiati, E. Enzymatic saccharification of liquid hot water and dilute sulfuric acid pretreated oil palm empty fruit bunch and sugarcane bagasse. In *Proceedings of the IOP Conference Series: Earth and Environmental Science*; 2018; Vol. 14, p. 012025.

68. Oktaviani, M.; Hermiati, E.; Thontowi, A.; Laksana, R.P.B.; Kholida, L.N.; Andriani, A.; Mangunwardoyo, W. Production of xylose, glucose, and other products from tropical lignocellulose biomass by using maleic acid pretreatment. In *Proceedings of the In IOP Conference Series: Earth and Environmental Science*; 2019; Vol. 251, p. 012013.

69. Sari, E.; Effendy, M.; Kanani, N.; Rusdi, W. Utilization of empty fruit bunch fiber of palm oil industry for bio-hydrogen production. *International Journal on Advanced Science, Engineering and Information Technology* **2018**, *8*, 842–848.

70. Singh, J.K.; Vyas, P.; Dubey, A.; Upadhyaya, C.P.; Kothari, R.; Tyagi, V. V.; Kumar, A. Assessment of different pretreatment technologies for efficient bioconversion of lignocellulose to ethanol. *Frontiers in Bioscience - Scholar* **2018**, *10*, 350–371.

71. Noorshamsiana, A.W.; Nur Eliyanti, A.O.; Fatiha, I.; Astimar, A.A. A review on extraction processes of lignocellulosic chemicals from oil palm biomass. *Journal of Oil Palm Research* **2017**, *29*, 512–527.

72. Eom, I.Y.; Yu, J.H.; Jung, C.D.; Hong, K.S. Efficient ethanol production from dried oil palm trunk treated by hydrothermolysis and subsequent enzymatic hydrolysis. *Biotechnology for Biofuels* **2015**, *8*, 1–11.

73. Baruah, J.; Nath, B.K.; Sharma, R.; Kumar, S.; Deka, R.C.; Baruah, D.C.; Kalita, E. Recent trends in the pretreatment of lignocellulosic biomass for value-added products. *Frontiers in Energy Research* **2018**, *6*, 141.

74. Norrrahim, M.N.F.; Huzaifah, M.R.M.; Farid, M.A.A.; Shazleen, S.S.; Misenan, M.S.M.; Yasim-Anuar, T.A.T.; Naveen, J.; Nurazzi, N.M.; Rani, M.S.A.; Hakimi, M.I.; et al. Greener pretreatment approaches for the valorisation of natural fibre biomass into bioproducts. *Polymers* **2021**, *13*, 2971.

75. Hussein, S.M. A Study of some mechanical and physical properties for palm fiber/polyester composite. *Engineering and Technology Journal* **2020**, *38*, 104–114, doi:10.30684/etj.v38i3B.598.

76. Al-Kaabi, K.; Al-Khanbashi, A.; Hammami, A. Natural fiber reinforced composites from date palm fibers. In *Proceedings of the 11th European Conference on Composite Materials*; Rhodes: Greece, 2004.

77. Nurazzi, N.M.; Khalina, A.; Chandrasekar, M.; Aisyah, H.A.; Rafiqah, S.A.; Ilyas, R.A.; Hanafee, Z.M. Effect of fiber orientation and fiber loading on the mechanical and thermal properties of sugar palm yarn fiber reinforced unsaturated polyester resin composites. *Polimery* **2020**, *65*, 115–124, doi:10.14314/polimery.2020.2.5.

78. de Farias, M.A.; Farina, M.Z.; Pezzin, A.P.T.; Silva, D.A.K. Unsaturated polyester composites reinforced with fiber and powder of peach palm: Mechanical characterization and water absorption profile. *Materials Science and Engineering C* **2009**, *29*, 510–513, doi:10.1016/J.MSEC.2008.09.020.

79. Wazzan, A.A. Effect of fiber orientation on the mechanical properties and fracture characteristics of date palm fiber reinforced composites. *International Journal of Polymeric Materials and Polymeric Biomaterials* **2005**, *54*, 213–225, doi:10.1080/00914030390246379.

80. Yousif, B.F.; El-Tayeb, N.S.M. High-stress three-body abrasive wear of treated and untreated oil palm fibre-reinforced polyester composites. *Proceedings of the Institution of Mechanical Engineers, Part J: Journal of Engineering Tribology* **2008**, *222*, 637–646, doi:10.1243/13506501JET412.

81. Raghavendra, S.; Lokesh, G.N. Evaluation of mechanical properties in date palm fronds polymer composites. *AIP Conference Proceedings* **2019**, *2057*, doi:10.1063/1.5085592.

82. Anyakora, A.N. Investigation of impact strength properties of oil and date palm frond fiber reinforced polyester composites. *Research Article International Journal of Current Engineering and Technology Accepted* **2013**, *3*, 493–497.

83. Mohd Nurazzi, N.; Khalina, A.; Sapuan, S.M.; Ilyas, R.A.; Ayu Rafiqah, S.; Hanafee, Z.M. Thermal properties of treated sugar palm yarn/glass fiber reinforced unsaturated polyester hybrid composites. *Journal of Materials Research and Technology* **2019**, *9*, 1606–1618, doi:10.1016/j.jmrt.2019.11.086.

84. Huzaifah, M.R.M.; Sapuan, S.M.; Leman, Z.; Ishakc, M.R. Comparative study of physical, mechanical, and thermal properties on sugar palm fiber (*Arenga pinnata* (Wurmb) Merr.) reinforced vinyl ester composites obtained from different geographical locations. *BioResources* **2019**, *14*, 619–637, doi:10.15376/biores.14.1.619-637.

85. Hill, C.A.S.; Abdul Khalil, H.P.S. Effect of fiber treatments on mechanical properties of coir or oil palm fiber reinforced polyester composites. *Journal of Applied Polymer Science* **2000**, *78*, 1685–1697, doi: 10.1002/1097-4628(20001128)78:9<1685::AID-APP150>3.0.CO;2-U.

86. Anggawan, A.D.; Mohamad, N. Effect of different fibre treatment of oil palm fibre reinforced polyester composite. *AIP Conference Proceedings* **2020**, *2233*, 020002, doi:10.1063/5.0003796.

87. Azmami, O.; Sajid, L.; Boukhriss, A.; Majid, S.; El Ahmadi, Z.; Benayada, A.; Gmouh, S. Sol-gel and polyurethane based flame retardant and water repellent coating for Palm/ PES nonwovens composite. *Journal of Sol-Gel Science and Technology* **2021**, *97*, 92–105, doi: 10.1007/s10971-020-05429-2.

88. Hassanin, A.H.; Elseify, L.A.; Hamouda, T. Date palm fiber composites fabrication techniques. In *Date Palm Fiber Composite ; Processing, Properties and Applications*; Midani, M., Saba, N., Alothman, O.Y., Eds.; Springer: Singapore, 2020; pp. 161–183.

89. Sbiai, A.; Maazouz, A.; Fleury, E.; Sautereau, H.; Kaddami, H.; Einstein, A. Short date palm tree fibers/polyepoxy composites prepared using RTM process: Effect of tempo mediated oxydation of the fibers. *BioResources* **2010**, *5*, 672–689.

90. Advani, S.G.; Hsiao, K.-T. *Manufacturing Techniques for Polymer Matrix Composites (PMCs)*; Woodhead Publishing Limited: Cambridge, UK, 2012.

91. Strong, A.B. *Fundamentals of Composites Manufacturing: Materials, Methods and Applications*; Society of Manufacturing Engineers: Dearborn, MI, 2008.

92. Asim, M.; Jawaid, M.; Khan, A.; Asiri, A.M.; Malik, M.A. Effects of date palm fibres loading on mechanical, and thermal properties of date palm reinforced phenolic composites. *Journal of Materials Research and Technology* **2020**, *9*, 3614–3621, doi:10.1016/j.jmrt.2020.01.099.

93. Huzaifah, M.R.M.; Sapuan, S.M.; Leman, Z.; Ishak, M.R.; Maleque, M.A. A review of sugar palm (*Arenga pinnata*): Application, fibre characterisation and composites. *Multidiscipline Modeling in Materials and Structures* **2017**, *13*, 678–698, doi:10.1108/ MMMS-12-2016-0064.

7 Coir Fiber–Polyester Composites

Carlo Santulli
Università di Camerino

CONTENTS

7.1 INTRODUCTION

Coir fibers are obtained from the hair of the coconut fruit and therefore constitute a product from a multifaceted cropping system. The most added value products are obtained from the exploitation of coconuts as for flesh and milk, also marketed and prepared to serve as additives in food, e.g., for confectionery products. Furthermore, coconut (*Cocos nucifera*) palms offer other products such as coconut palm sheath leaves (Obi Reddy et al., 2010). Most recently, the use of coconut palm wood in buildings has been promoted as a strategy that would be beneficial for the sustainability of the construction industry. However, there are a number of impediments to be removed in the process, starting from the cost of processing and low perception and image of this type of wood (Sodangi et al., 2020). It is important to notice that coir, according to the different maturation phases of the fruit, may be distinguished as white and brown coir, and the latter is typically more frequently used in composites (Figure 7.1), where it offers better performance in terms of mechanical and wear resistance (Valášek et al., 2018).

Dealing specifically with the coconut shell, coir fiber is not the only product it offers. Coir pith is produced from coconut shell, which is more powder-like than fiber, and has a number of uses. These include, for example, the possibility of application as a peat substitute (Meerow 1994), which also highlighted some potential in terms of bioremediation from heavy metals (Namasivayam and Sangeetha 2006) and wastewater treatment (Kavitha and Namasivayam 2007), and the use as filler

DOI: 10.1201/9781003270980-7

121

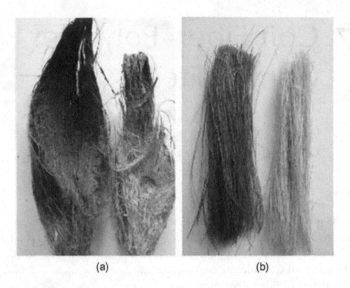

(a)   (b)

FIGURE 7.1 Brown coir (a) and white coir (b), as received and prepared for experimentation. (Source: Valášek et al. 2018.)

in cementitious composites (Brasileiro et al., 2013). Also, coir pith offers a huge possibility to be compressed up to 8–10 times its initial volume and can be shaped into blocks, such as those represented in Figure 7.2. In this case the locking effect between pith particles is able to offer a solid material, capable of withstanding transportation, only with the addition of a limited amount of thermosetting resin, such as unsaturated polyester, although of course a higher amount of resin would enhance its resistance (Panamgama and Peramune 2018).

On the other side, the nutshells present a value in themselves, specifically in their combustion or in the production of active carbons (Iqbaldin et al., 2013). However, it is important to note that, the complete exploitation of all products from the coconut system makes it desirable to use also the shell waste powder as the filler of thermoplastic matrices, such as polyethylene and polypropylene (Bradley and Conroy 2019). Coconut shells have also found some application in the construction industry, typically in road lower layers, in combination with compacted soil (Ting et al., 2015). Coir has also been proposed for structural upgradation and compared for performance and ease of positioning and setting up with other well-known natural fibers, in particular bamboo, jute and sisal (Sen and Reddy 2011).

During the past decades, coir fiber has often been considered as a suitable reinforcement for composites. A large variety of matrices has been used for this purpose, starting from cementitious ones (Asasutjarit et al., 2007; Andiç-Çakir et al., 2014; Yan et al., 2016), also in hybrid combination e.g., with flax (Yan et al., 2015), to a considerable variety of polymer ones, an aspect which is also elucidated by Reddy et al. (2019).

In contrast with other traditional fibers, such as jute and sisal, sometimes coir proved to be inferior, especially when applied with matrices of interest for the automotive industry, such as polypropylene, yet was considered more reliable for

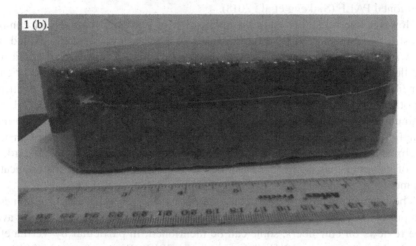

FIGURE 7.2 Coir pith blocks (a) top view and (b) side view. (Source: Panamgama and Peramune 2018.)

its reduced water absorption (Sudhakara et al., 2015). These fibers spawned some interest also for the production of composites, on which a review encompassing all the routes from the fiber to the production and characterization of composites was recently produced by Oladele et al. (2022).

The production of composites is principally carried out using some traditional thermosetting matrices, such as phenol-formaldehyde (Bharath et al., 2019), vinyl-ester (Velumani et al., 2014), or epoxy (Biswas et al., 2011; Kumar et al., 2014), and unsaturated polyester, which is the specific focus of this work. With thermosetting matrices, the profile of application of coir fibers as their reinforcement has been often considered very high and perfectly comparable with most common fibers, such as glass and even carbon. In practice, also the application of coir–epoxy composites as ballistic armors has been attempted in the understanding that the complexity of their damage absorption modes would much exceed the expectations that can be raised from their quite modest tensile strength (Luz et al., 2017). In the case of a comparative study between pineapple leaf fibers (PALF)–glass and coir–glass hybrids with epoxy matrix, the potential offered by coir was particularly concentrated on the improvement of impact properties, which were able to exceed 10 kJ/m^2 for a 15% glass and 15% coir reinforcement (Mittal and Chaudhary 2018).

In view of the application for ballistic purposes, such as for helmet shells, the fabrication of carbon/coir–epoxy composites has been investigated (Singh et al., 2020). On the other side, the creation of hybrid composites including carbon and natural fibers also obtained some success, so that a specific review has been dedicated to these (Santulli 2019). More recently, also thermoplastic matrices, including petrochemical ones, such as polypropylene (Islam et al. 2010; Sudakhara et al., 2013) and starch-derived ones, such as wheat gluten (Muensri et al., 2011), poly(lactic acid) (PLA) (Yusoff et al., 2016; Rigolin et al., 2019) have been introduced in the production of coir fiber reinforced composites. Further studies on PLA were also carried out in comparison with other fibers, such as the previously mentioned PALF (Siakeng et al., 2018).

However, the reliability and easy application of polyester matrices for composites still make them preferable for a number of uses, which especially extend over the wider construction sector (Scheirs and Long 2005). Their ease of application and flexibility of use allowed for a facile laying of randomly discontinuous oriented coir fibers (Priya et al., 2014) and also the production of hybrid composites including other fibers with coir. Over time, these included jute (Saw et al., 2012), oil palm (Zainudin et al., 2014) and glass, which in particular was studied considering intermingling with coir (Pavithran et al., 1991). Most commonly, and in view of their machinability, a skin (glass)–core (coir) approach was selected: in this regard, the use of polyester matrix was demonstrated to be highly suitable for the application (Hamouda et al., 2017).

The chapter partially draws upon a number of reviews that exist on coir fiber composites, which of course are usually more general. Among the previously available reviews on coir fibers, some can be mentioned, in particular by Verma et al. (2013), Jayavani et al. (2016), Adeniyi et al. (2019), Bongarde et al. (2019) and Hasan et al. (2021).

Characterization of obtained composites from mechanical and thermal points of view also needs to be investigated. On the other side, to improve the resistance of coir fibers and hence the performance of composites, treatments may be necessary, on which a specific review exists by Dugvekar and Dixit (2022). Treatment is not the only way to improve fiber–matrix compatibility: another method is the addition of compatibilizer, such has been the case for lignin in the case of polypropylene/coir composites (Morandim-Giannetti et al., 2012), as an alternative to the usual maleic anhydride and silanization/compatibilizations (Santos et al. 2009). Further reviews concern other applications of coir fibers, which do not specifically concern the production of composites. This is the case, for example, of the review on soil stabilization applications of coir fibers by Maurya et al. (2015), on the improvement of sand properties, as reported by Sridhar (2017), and in road construction by Ting et al. (2015).

7.2 CHARACTERISTICS OF COIR FIBERS

To start investigating the potential of coir fibers in polyester composites, a number of fiber characteristics are essential to be known, as they are extracted from fruit hair: these include chemical composition, thermal degradation properties and mechanical performance of the fibers. Some data are reported in Table 7.1 on untreated fibers, which are able to offer an idea of the scattering in properties observed.

The amount of lignin present in coir fibers is higher than what is obtained from fibers extracted from other parts of the coconut plant: for example, a study on leaf sheath fibers reported values of 27%, 40% and 26% for lignin, cellulose and hemicellulose, respectively (Suresh Kumar et al., 2014). The relatively high amount of lignin in coir fibers, together with their comparatively high microfibrillar angle, does affect the tensile and flexural properties of the obtained composites. On the other hand, it has been suggested that they are particularly suitable for impact-resistant applications, especially by making use of thermosetting matrices, such as unsaturated polyesters (Wambua et al., 2003).

The elongation of coir fibers, as it is apparent from Table 7.2, may be very variable, and actually strongly depends on the geometrical characteristics of fiber, in particular on their diameter and length. Some data are offered in Sindhu et al. (2007): in this study, increasing the length of fibers employed from 10 to 30 mm long fibers, resulted in an enhancement of tensile strength from 15 to 24.2 MPa, and in Young's modulus growing from 831 to 952 MPa. However, a further consequence was also a loss in elongation, which was that elongation was reduced from 9% to 5%.

TABLE 7.1
Chemical Composition of Untreated Coir Fibers

Lignin	Cellulose	Hemicellulose	Ref.
37	42	n.c.	Verma et al. (2013)
42.10	32.69	22.56	Muensri et al. (2011)
49.2	39.3	2	Abraham et al. (2013)
45	43	n.c.	Jústiz-Smith et al. (2008)
33	44	12	Khalil et al. (2006)
40–45	32–43	0.150.25	Mohanty et al. (2005)

TABLE 7.2
Mechanical Properties of Untreated Coir Fibers

Diameter (μm)	Strength (MPa)	Elongation at Break (%)	Young's Modulus (GPa)	Ref.
100–450	106–175	4.7	3–6	Verma et al. (2013)
100–460	131–175	15–40	4–6	Silva et al. (2000)
92–314	92	16.5	0.55	Setyanto et al. (2013)
–	593	30	4–6	Ku et al. (2011)
100–460	108–252	15–40	4–6	Sanjay et al. (2018)

In general terms, it can be suggested that the most repeatable and consistent data for coir originate from fiber diameter. This has been suggested also from comparative studies, for example, with banana and bagasse fibers, where coir performed quite at the same level than the former and showed to be much superior to the latter (Jústiz-Smith et al., 2008).

Early studies on thermal characterization of untreated coir fibers, after a period for moisture evaporation, which terminates at around 120°C, with a weight loss usually not exceeding 8%, indicated two main transition intervals peaking at 334°C and 476°C for cellulose and lignin degradation, respectively (Silva et al., 2000). The thermal stability of cellulose can be affected even at higher temperatures, such as 357°C, although the transition interval may be very wide, such as from 223°C to 395°C, as reported by Dos Santos et al., 2018.

7.3   COIR FIBER–POLYESTER COMPOSITES

7.3.1   GENERAL CONSIDERATIONS

Apart from the use of coconut shell powder, which showed some promise as the filler of polyester matrices (Islam et al., 2017) and has been recently repurposed for this aim in the form of micro- or nanofiller (Karthik Babu et al. 2020), coir fibers have been used for fabrication of composites using polyester matrices from the very early experiments, according to Prasad et al. (1983). In particular, it was suggested even from the 1980s that this fiber–matrix combination would be suitable for a number of applications, such as covers for voltage stabilizers or projectors, mirror casing and roofing (Satyanarayana et al., 1986). More recent experiments also suggested some potential for coir–polyester composites as material for liquid storage tanks, with some possible concerns to be attributed to fiber strength deteriorating during the aging process (Yousif and Ku 2012).

First evidence was that the interface between the hydrophobic polyester matrix and the relatively hydrophilic coir fibers was unsatisfactory. These initial studies revealed that treatment with a 5% sodium hydroxide solution almost doubled the debonding stress of coir fibers so that a volume fraction equal to 0.3 was successfully introduced. Other types of alkaline treatments, such as with sodium bicarbonate, do not affect the mechanical strength of the fibers: also, coir/polyester composites, though inferior to coir/epoxy ones in terms of tensile or flexural strength, outperform these as far as stiffness is concerned (Dos Santos et al., 2019).

Over time, composite fabrication with higher tenors of fibers has been progressively experimented, to try to lead to substantial advantages in terms of mechanical performance. An important study was carried out by Monteiro et al. (2008), where comparison between two different modes for the introduction of coir fibers in a polyester matrix, namely, as either pressed mat or tangled mass, allowed pushing the tenor of reinforcement up to 80 wt.%. With growing amounts of fibers, the strength, especially the flexural one, of the composite decreases: however, this can be justified by obtaining two different types of products, rigid composites with less than 50 wt.% of fibers, and agglomerates with more than this.

7.3.2 EFFECT OF FIBER TREATMENTS

Apart from the typical alkali treatment for ligno-cellulosic fibers with sodium hydroxide, other possible surface modifications have been attempted. Rout et al. (2001) and (2003) compared a few of these in their investigation of tensile, flexural and impact performance of coir–polyester composites. In particular, alkali treatment with concentrations of 2 to 10%, bleaching defatted fibers with sodium chlorite, then neutralized at pH 7, were applied. In comparison, cyanoethylation and vinyl grafting applied on 5% alkali-treated fibers were experimented, therefore limiting possible damage produced by excessively concentrated solutions onto the fiber surface. The results indicated that every surface modification is able to improve the mechanical properties of 17 wt.% coir–polyester composites. Silane treatment did generally considerably and consistently improve tensile strength and modulus of coir–polyester with respect to the application of untreated fibers, up to 70 wt.% of reinforcement added (Anyakora 2013). However, in general terms, tensile strength decreases with the addition of a higher amount of coir fibers while the opposite occurs for stiffness.

Other types of treatment have also been experimented, such as acetylation at high temperature (Khalil et al., 2000). This treatment was performed in the understanding that the improvement of temperature would result in a possible removal of loose matter, apart from moisture, with the consequence of enhancing the mechanical properties of coir–polyester composites. This proved true up to 100°C, while the results of treatment with a further increased temperature were contrasting.

7.3.3 PRODUCTION OF HYBRID COMPOSITES

Coir–cotton hybrid polyester with fiber content of up to 20 wt.% were studied: the introduction of coir compares unfavorably against that of cotton in tensile, flexural and Charpy impact testing with frequent occurrences of pullout of coir fibers (Balaji and Senthil Vadivu 2017). The introduction of coir in hybrids with other fibers with superior tensile performance, namely, PALF and bamboo, however, proved to be successful to compensate with positive interlocking effect between the three fibers (Rihayat et al., 2018). More complex hybrids were also proposed more recently, where in an unsaturated polyester matrix, sisal (50%–90%), sorghum (5%–25%) and coir (5%–25%) were introduced (Kumar 2020). An attempt also of self-reinforced hybrid composites was developed including poly(ethylene terephtalate) (PET) fibers from textile waste, adding to this olive root waste and coir pith microparticles, as reported in Figure 7.3 (Tufail et al., 2021).

Another possible trend was the addition of particulate ceramics, such as eggshell powder (ESP) and rice husk (around 70% in silica), in large amounts into a polyester–coir composite with fibers having different lengths, from 10 to 50 mm. The amount of polyester was kept constant at 60 wt.%, while coir fibers and particulate reinforcements were balanced in a compromised manner, with a maximum of particulate equal to 25% and subsequently with a minimum amount of coir equal to 15% (Bharathiraja et al., 2016). An optimization study was carried out in a similar occurrence of addition of coir and particulates, in this case rice husk and red mud from aluminum production, with the aim of identifying the optimal fabrication conditions

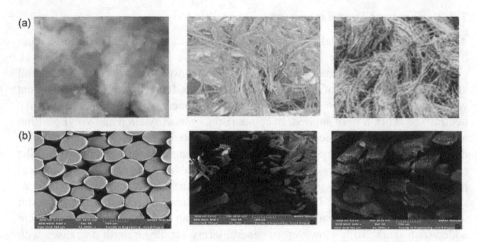

FIGURE 7.3   Images (a) of SEM micrograph and (b) of PET fibers, olive root waste and coir pith fibers. (Source: Tufail et al. 2021.)

about the amount of bioparticles + coir (total 40%) and coir fiber length (between 26 and 29 mm) (Bharathiraja et al., 2017).

The impregnation of short coir fibers in polyester matrix using calcium carbonate was also studied by full factorial design (Sathiyamurthy et al., 2012). This method was also followed in a complete study by Rajamuneeswaran et al. (2016), where calcium carbonate filler was extracted from crab carapace: in this case, polyester matrix was found to offer properties inferior to vinylester and epoxy to the composite. In particular, with 25% of fibers, 50 mm long and with 0.25 mm diameter and filler content of 2%, a tensile strength equal to 17.8 MPa, a flexural strength equal to 22.8 MPa and impact strength equal to 38.4 kJ/m² were obtained.

A complex hybrid with jute was also realized recently including two types of ceramic fillers, ESP and montmorillonite nanoclay (MMT-NC) under dry and wet (freshwater and seawater) conditions. This was performed with the idea to lead to its optimization with respect to the contents of the fillers and the different mechanical properties, i.e., tensile, flexural and impact strength, examined (Ganesan et al., 2021). The results indicated that the addition of equal amounts of 1.5 wt.% of MMT-NC and ESP offered, as a whole, optimum properties.

Coir–unsaturated polyester composites (50/50) demonstrated to be also suitable for mixing with other types of waste materials, such as bio-particulate, for example, red mud from aluminum production and termite mound soil. The best mechanical (tensile, flexural and impact) performance was obtained with 30 mm long fibers and 20% bioparticulate (10 red mud + 10 termite mound) (Bharathiraja et al., 2014). Another study concerned the filling of coir–polyester laminates using alumina evaluating the combined effect of fiber length and diameter and filler content (Sathiyamurthy et al., 2013).

7.3.4   MECHANICAL CHARACTERIZATION

A recent review is available on mechanical properties of coir-reinforced composites, which offered some data also on coir–polyester composites, namely, on tensile and flexural strength, Young's modulus and elongation (Guyat et al., 2021).

A significant characteristic of coir fiber has also been revealed in a drilling study against glass fiber when inserted into polyester matrix, and they indicated a lower proneness to delamination than the latter (Balaji et al., 2014). The general issue of machinability of coir/polyester composites has also been modeled, since it depends on a number of different variables during the drilling process, such as thrust force, torque and tool wear. The effect of these can be globally treated into a design of experiments (DOE) procedure (Jayabal et al., 2010, 2011a,b). In terms of wear properties, it was noticed that coir fibers did break in the polyester matrix before pulling out, therefore confirming the remarkable level of adhesion, as reported by Herrera-Franco and Valadez-Gonzalez (2004). However, the fabrication of two types of coir–polyester composites with three and four layers, respectively, resulted in a limited but not fully consistent reduction of friction, hence with a large variation among different samples, resulting in a higher specific wear rate (Yousif et al., 2009). This was attributed to the presence of micro-plowing in the resin-rich regions, which is a typical occurrence in coir fiber–reinforced composites, possibly for the irregular surface profile of these fibers (Siva et al., 2012). Another tribological study compared E-glass, coir and their hybrid polyester composites, all with 48 (±1) wt. % total reinforcement content.

Impact strength can be considered to be the most consistently behaving mechanical data: in particular, the increase in volume fraction of fibers (10%–30%) and of laminate thickness (from 2 to 6 mm) proved advantageous, allowing Charpy impact strength to grow with a rather predictable trend. To have an idea of the possible improvement obtainable, it can be noticed that a 10 vol% 2 mm thick laminate offered an impact strength of around 0.2 J/m, while a 30 vol% 6 mm thick laminate yielded an impact strength approximating 1.6 J/m (Prasad et al., 2017). A comparative study between impact strengths of different polyester matrix composites, reinforced, respectively, with E-glass, coir or oil palm fibers, and two hybrids, including E-glass/coir and E-glass/oil palm, and the addition of sand filler, was carried out by Wong et al. (2010). In this context, different fiber volume fractions (40, 50 and 60%) and fiber lengths (3, 7, 10 mm) were employed to study coir composites with fibers disposed either longitudinally or transversely to load application, with up to four layers and different spacing among them. A larger number of layers enhance the impact strength, but this is depressed by higher amounts of fibers. In an attempt to use longer coir fibers, yet untreated, of up to 35 mm and aligned on the longitudinal direction of tensile samples, it was noticed that the effect on the strength of the resin was only positive for 10 wt.% fiber content. On the other side, stiffness was improved up to 40 wt.% fiber (Júnior et al., 2010).

A study topic that is not very usual among natural fibers is the attempt to predict mechanical properties, using artificial neural network (ANN) methods: in the case of a study on coir/polyester composites, the idea was to elucidate for which amount of coir fibers, the peak of flexural strength was reached for laminates with different thicknesses. The result obtained was for 21 wt.% and 5 mm, with a flexural strength of 144.82 MPa, while the experimental testing, carried out at every 5% of further fiber addition, indicated at 20% fibers, 142.04 MPa. The idea, which was successful, was that ANN was able to predict the trend of the flexural strength vs. fiber weight behavior with higher accuracy than what is realizable with experimental testing (Prasad et al., 2016).

A preliminary fatigue evaluation has been carried out on 10 wt.% coir–polyester composites to try to elucidate possible S-N curves at 6 Hz frequency and load ratio R equal to 0.1, which indicated a more than halving of ultimate stress between 10^3 and 10^5 cycles (Mulinari et al., 2011).

7.3.5 THERMAL CHARACTERIZATION

The treatment of coir fibers has been indicated to improve the thermal stability of the fibers, to a very different extent, depending on the type of chemical used. In particular, it was highlighted that the most successful one is still mercerization, while less effective are oxidation by hydrogen peroxide and then acetylated coir, although all of them provide some positive effect (Akintayo et al., 2016).

A thermogravimetric study on coir–polyester composites was performed by Santafé et al. (2011) and subsequently reported and compared in Monteiro et al. (2012). In particular, for a 30% coir–polyester composite, two peaks were observed at 285°C and 348°C, attributed to hemicellulose and cellulose degradation, respectively. Comparative study carried out by the same authors on neat polyester suggested that its decomposition occurred at higher temperatures, such as at 385°C, with a tail as high as 414°C.

Balaji et al. (2020) observed that the addition of coir fibers (15%) to zea fibers (25%) in a polyester matrix offered some increase of the crystalline temperature, precisely from 92°C to 98°C, while zea fibers already gave rise to a considerable shift from 64°C to 92°C.

7.4 CONCLUSIONS

Coir–polyester composites are a very diffuse matrix–fiber combination for coconut fruit fibers, second possibly only to epoxy. It allowed in recent years the fabrication of composites with up to 70% of coir fibers, normally treated to obtain an increase of crystallinity and some improvement of thermal stability, other than the removal of looser material, hence a higher hardness and stiffness. It needs to be considered that in mechanical terms this is convenient for obtaining a higher stiffness and a better impact strength, for a consistent interlocking effect, less so in the case of tensile and flexural strength. However, the fabrication of hybrid composites, both with other natural fibers, such as jute, PALF, sisal, etc., and/or with other lignocellulosic material and ceramic waste, especially silica or calcium carbonate-based material, also proved to be successful. This suggests the flexibility of unsaturated polyester matrix to obtain a sufficient compatibility and interface with coir fibers, also for the experience accumulated over a few decades of experimentation on the manufacturing and characterization of coir–polyester composites.

REFERENCES

Abraham, E., Deepa, B., Pothen, L. A., Cintil, J., Thomas, S., John, M. J., ... & Narine, S. S. (2013). Environmental friendly method for the extraction of coir fibre and isolation of nanofibre. *Carbohydrate Polymers*, 92(2), 1477–1483.
Adeniyi, A. G., Onifade, D. V., Ighalo, J. O., & Adeoye, A. S. (2019). A review of coir fiber reinforced polymer composites. *Composites Part B: Engineering*, 176, 107305.

Akintayo, C. O., Azeez, M. A., Beuerman, S., & Akintayo, E. T. (2016). Spectroscopic, mechanical, and thermal characterization of native and modified Nigerian coir fibers. *Journal of Natural Fibers*, 13(5), 520–531.

Andiç-Çakir, Ö., Sarikanat, M., Tüfekçi, H. B., Demirci, C., & Erdoğan, Ü. H. (2014). Physical and mechanical properties of randomly oriented coir fiber–cementitious composites. *Composites Part B: Engineering*, 61, 49–54.

Anyakora, A. N. (2013). Correlation between fiber treatment and ash content on the tensile behavior of coir reinforced polyester composite. *World Journal of Engineering and Pure & Applied Sciences*, 3(1), 1.

Asasutjarit, C., Hirunlabh, J., Khedari, J., Charoenvai, S., Zeghmati, B., & Shin, U. C. (2007). Development of coconut coir-based lightweight cement board. *Construction and Building Materials*, 21(2), 277–288.

Balaji, N. S., Chockalingam, S., Ashokraj, S., Simson, D., & Jayabal, S. (2020). Study of mechanical and thermal behaviours of zea-coir hybrid polyester composites. *Materials Today: Proceedings*, 27, 2048–2051.

Balaji, N. S., Jayabal, S., Kalyana Sundaram, S., Rajamuneeswaran, S., & Suresh, P. (2014). Delamination analysis in drilling of coir-polyester composites using design of experiments. *Advanced Materials Research* 984, 185–193.

Balaji, V., Senthil Vadivu K (2017). Mechanical characterization of coir fiber and cotton fiber reinforced unsaturated polyester composites for packaging applications. *Journal of Applied Packaging Research*, 9(2), 2.

Bharath, K. N., Sanjay, M. R., Jawaid, M., Harisha, Basavarajappa, S., & Siengchin, S. (2019). Effect of stacking sequence on properties of coconut leaf sheath/jute/E-glass reinforced phenol formaldehyde hybrid composites. *Journal of Industrial Textiles*, 49(1), 3–32.

Bharathiraja, G., Jayabal, S., & Kalyana Sundaram, S. (2017). Gradient-based intuitive search intelligence for the optimization of mechanical behaviors in hybrid bioparticle-impregnated coir-polyester composites. *Journal of Vinyl and Additive Technology*, 23(4), 275–283.

Bharathiraja, G., Jayabal, S., Kalyana Sundaram, S., Rajamuneeswaran, S., & Manjunath, B. H. (2016). Mechanical behaviors of rice husk and boiled egg shell particles impregnated coir-polyester composites. *Macromolecular Symposia*, 361(1), 136–140.

Bharathiraja, G., Jayabal, S., Prithivirajan, R., & Sathiyamurthy, S. (2014). Optimization of mechanical behaviors of bio particulates filled coir-polyester composites using simulated annealing. *ARPN Journal of Engineering and Applied Sciences*, 9(4), 789.

Biswas, S., Kindo, S., & Patnaik, A. (2011). Effect of fiber length on mechanical behavior of coir fiber reinforced epoxy composites. *Fibers and Polymers*, 12(1), 73–78.

Bongarde, U. S., & Khot, B. K. (2019). A review on coir fiber reinforced polymer composite. *International Research Journal of Engineering and Technology*, 6(4), 793–95.

Bradley, W. L., & Conroy, S. (2019). Using agricultural waste to create more environmentally friendly and affordable products and help poor coconut farmers. In *E3S Web of Conferences* (Vol. 130, p. 01034). EDP Sciences.

Brasileiro, G. A. M., Vieira, J. A. R., & Barreto, L. S. (2013). Use of coir pith particles in composites with Portland cement. *Journal of Environmental Management*, 131, 228–238.

dos Santos, J. C., de Oliveira, L. Á., Vieira, L. M. G., Mano, V., Freire, R. T., & Panzera, T. H. (2019). Eco-friendly sodium bicarbonate treatment and its effect on epoxy and polyester coir fibre composites. *Construction and Building Materials*, 211, 427–436.

dos Santos, J. C., Siqueira, R. L., Vieira, L. M. G., Freire, R. T. S., Mano, V., & Panzera, T. H. (2018). Effects of sodium carbonate on the performance of epoxy and polyester coir-reinforced composites. *Polymer Testing*, 67, 533–544.

Dugvekar, M., & Dixit, S. (2022). Chemical treatments for modification of the surface morphology of coir fiber: A review. *Journal of Natural Fibers*, 19(15), 11940–11961.

Ganesan, K., Kailasanathan, C., Rajini, N., Ismail, S. O., Ayrilmis, N., Mohammad, F., ... & Aldhayan, D. M. (2021). Assessment on hybrid jute/coir fibers reinforced polyester composite with hybrid fillers under different environmental conditions. *Construction and Building Materials*, 301, 124117.

Goyat, V., Ghangas, G., Sirohi, S., Kumar, A., & Nain, J. (2021). A review on mechanical properties of coir based composites. *Materials Today: Proceedings*, 62, 1738–1745.

Hamouda, T., Hassanin, A. H., Kilic, A., Candan, Z., & Safa Bodur, M. (2017). Hybrid composites from coir fibers reinforced with woven glass fabrics: Physical and mechanical evaluation. *Polymer Composites*, 38(10), 2212–2220.

Hasan, K. F., Horváth, P. G., Bak, M., & Alpár, T. (2021). A state-of-the-art review on coir fiber-reinforced biocomposites. *RSC Advances*, 11(18), 10548–10571.

Herrera-Franco, P. J., & Valadez-Gonzalez, A. (2004). Mechanical properties of continuous natural fibre-reinforced polymer composites. *Composites Part A: Applied Science and Manufacturing*, 35(3), 339–345.

Hill, C. A. S., & Abdul Khalil, H. P. S. (2000). Effect of fiber treatments on mechanical properties of coir or oil palm fiber reinforced polyester composites. *Journal of Applied Polymer Science*, 78(9), 1685–1697.

Iqbaldin, M. M., Khudzir, I., Azlan, M. M., Zaidi, A. G., Surani, B., & Zubri, Z. (2013). Properties of coconut shell activated carbon. *Journal of Tropical Forest Science*, 25, 497–503.

Islam, M. N., Rahman, M. R., Haque, M. M., & Huque, M. M. (2010). Physico-mechanical properties of chemically treated coir reinforced polypropylene composites. *Composites Part A: Applied Science and Manufacturing*, 41(2), 192–198.

Islam, M. T., Das, S. C., Saha, J., Paul, D., Islam, M. T., Rahman, M., & Khan, M. A. (2017). Effect of coconut shell powder as filler on the mechanical properties of coir-polyester composites. *Chemical and Materials Engineering*, 5(4), 75–82.

Jayabal, S., & Natarajan, U. (2011). Drilling analysis of coir-fibre-reinforced polyester composites. *Bulletin of Materials Science*, 34(7), 1563–1567.

Jayabal, S., Natarajan, U., & Sathiyamurthy, S. (2011c). Effect of glass hybridization and stacking sequence on mechanical behaviour of interply coir–glass hybrid laminate. *Bulletin of Materials Science*, 34(2), 293–298.

Jayabal, S., Natarajan, U., & Sekar, U. (2011a). Regression modeling and optimization of machinability behavior of glass-coir-polyester hybrid composite using factorial design methodology. *The International Journal of Advanced Manufacturing Technology*, 55(1), 263–273.

Jayabal, S., Natarajan, U., & Sekar, U. (2011b). Regression modeling and optimization of machinability behavior of glass-coir-polyester hybrid composite using factorial design methodology. *The International Journal of Advanced Manufacturing Technology*, 55(1), 263–273.

Jayabal, S., Sathiyamurthy, S., Loganathan, K. T., & Kalyanasundaram, S. (2012). Effect of soaking time and concentration of NaOH solution on mechanical properties of coir–polyester composites. *Bulletin of Materials Science*, 35(4), 567–574.

Jayabal, S., Velumani, S., Navaneethakrishnan, P., & Palanikumar, K. (2013). Mechanical and machinability behaviors of woven coir fiber-reinforced polyester composite. *Fibers and Polymers*, 14(9), 1505–1514.

Jayavani, S., Deka, H., Varghese, T. O., & Nayak, S. K. (2016). Recent development and future trends in coir fiber-reinforced green polymer composites: Review and evaluation. *Polymer Composites*, 11(37), 3296–3309.

Júnior, H. S., Lopes, F. P. D., Costa, L. L., & Monteiro, S. N. (2010). Mechanical properties of tensile tested coir fiber reinforced polyester composites. *Revista Materia*, 15(2), 113–118.

Jústiz-Smith, N. G., Virgo, G. J., & Buchanan, V. E. (2008). Potential of Jamaican banana, coconut coir and bagasse fibres as composite materials. *Materials Characterization*, 59(9), 1273–1278.

Kanagaraj, C., Velu, P. S., Durkaieswaran, P., & Ravikumar, K. (2013) Numerical simulation of car interior door panel using coir-polyester composites. *Journal of Huazhong University of Science and Technology* 50 (6), 1–8.

Karthik Babu, N. B., Muthukumaran, S., Arokiasamy, S., & Ramesh, T. (2020). Thermal and mechanical behavior of the coir powder filled polyester micro-composites. *Journal of Natural Fibers*, 17(7), 1058–1068.

Kavitha, D., & Namasivayam, C. (2007). Experimental and kinetic studies on methylene blue adsorption by coir pith carbon. *Bioresource Technology*, 98(1), 14–21.

Khalil, H. A., Rozman, H. D., Ahmad, M. N., & Ismail, H. (2000). Acetylated plant-fiber-reinforced polyester composites: A study of mechanical, hygrothermal, and aging characteristics. *Polymer-Plastics Technology and Engineering*, 39(4), 757–781.

Khalil, H. S. A., Alwani, M. S., & Omar, A. K. M. (2006). Chemical composition, anatomy, lignin distribution, and cell wall structure of Malaysian plant waste fibers. *BioResources*, 1(2), 220–232.

Ku, H., Wang, H., Pattarachaiyakoop, N., & Trada, M. (2011). A review on the tensile properties of natural fiber reinforced polymer composites. *Composites Part B: Engineering*, 42(4), 856–873.

Kumar, S. S. (2020). Dataset on mechanical properties of natural fiber reinforced polyester composites for engineering applications. *Data in Brief*, 28, 105054.

Kumar, S. S., Duraibabu, D. A., & Subramanian, K. (2014). Studies on mechanical, thermal and dynamic mechanical properties of untreated (raw) and treated coconut sheath fiber reinforced epoxy composites. *Materials & Design*, 59, 63–69.

Li, Z., Wang, L., & Ai Wang, X. (2007). Cement composites reinforced with surface modified coir fibers. *Journal of Composite Materials*, 41(12), 1445–1457.

Luz, F. S. D., Monteiro, S. N., Lima, E. S., & Lima, É. P. (2017). Ballistic application of coir fiber reinforced epoxy composite in multilayered armor. *Materials Research*, 20, 23–28.

Maurya, S., Sharma, A. K., Jain, P. K., & Kumar, R. (2015). Review on stabilization of soil using coir fiber. *International Journal of Engineering Research*, 4(6), 296–299.

Meerow, A. W. (1994). Growth of two subtropical ornamentals using coir (coconut mesocarp pith) as a peat substitute. *HortScience*, 29(12), 1484–1486.

Mittal, M., & Chaudhary, R. (2018). Development of PALF/Glass and COIR/Glass fiber reinforced hybrid epoxy composites. *Journal of Materials Science and Surface Engineering*, 6(5), 851–861.

Mohanty, A. K., Misra, M., & Drzal, L. T. (Eds.). (2005). *Natural Fibers, Biopolymers, and Biocomposites*. CRC Press, Boca Raton, FL.

Monteiro, S. N., Calado, V., Rodriguez, R. J. S., Margem, F. M. (2012), Thermogravimetric behavior of natural fibers reinforced polymer composites—An overview. *Materials Science and Engineering: A*, 557, 17–28.

Monteiro, S. N., Terrones, L. A. H., & D'Almeida, J. R. M. (2008). Mechanical performance of coir fiber/polyester composites. *Polymer Testing*, 27(5), 591–595.

Morandim-Giannetti, A. A., Agnelli, J. A. M., Lanças, B. Z., Magnabosco, R., Casarin, S. A., & Bettini, S. H. (2012). Lignin as additive in polypropylene/coir composites: Thermal, mechanical and morphological properties. *Carbohydrate Polymers*, 87(4), 2563–2568.

Muensri, P., Kunanopparat, T., Menut, P., & Siriwattanayotin, S. (2011). Effect of lignin removal on the properties of coconut coir fiber/wheat gluten biocomposite. *Composites Part A: Applied Science and Manufacturing*, 42(2), 173–179.

Mulinari, D. R., Baptista, C. A. R. P., Souza, J. V. C., & Voorwald, H. J. C. (2011). Mechanical properties of coconut fibers reinforced polyester composites. *Procedia Engineering*, 10, 2074–2079.

Namasivayam, C., & Sangeetha, D. (2006). Recycling of agricultural solid waste, coir pith: Removal of anions, heavy metals, organics and dyes from water by adsorption onto $ZnCl_2$ activated coir pith carbon. *Journal of Hazardous Materials*, 135(1–3), 449–452.

Obi Reddy, K., Sivamohan Reddy, G., Uma Maheswari, C., Varada Rajulu, A., & Madhusudhana Rao, K. (2010). Structural characterization of coconut tree leaf sheath fiber reinforcement. *Journal of Forestry Research*, 21(1), 53–58.

Oladele, I. O., Adelani, S. O., Makinde-Isola, B. A., & Omotosho, T. F. (2022). Coconut/coir fibers, their composites and applications. In Sanjay Mavinkere Rangappa, Jyotishkumar Parameswaranpillai, Suchart Siengchin, Togay Ozbakkaloglu, & Hao Wang (Eds.). *Plant Fibers, their Composites, and Applications* (pp. 181–208). Woodhead Publishing, Sawston, Cambridge.

Panamgama, L. A., & Peramune, P. R. U. S. K. (2018). Coconut coir pith lignin: A physicochemical and thermal characterization. *International Journal of Biological Macromolecules, 113*, 1149–1157.

Pavithran, C., Mukherjee, P. S., & Brahmakumar, M. (1991). Coir-glass intermingled fibre hybrid composites. *Journal of Reinforced Plastics and Composites*, 10(1), 91–101.

Prasad, G. E., Gowda, B. K., & Velmurugan, R. (2017). A study on impact strength characteristics of coir polyester composites. *Procedia Engineering*, 173, 771–777.

Prasad, G. L., Gowda, B. S., & Velmurugan, R. (2016). Prediction of flexural properties of coir polyester composites by ANN. In Yong Zhu, Raman Singh, Piyush R. Thakre, Carter Ralph, Gyaneshwar Tandon, & Pablo Zavattieri (Eds.). *Mechanics of Composite and Multi-functional Materials, Volume 7* (pp. 173–180). Springer, Cham.

Prasad, S. V., Pavithran, C., & Rohatgi, P. K. (1983). Alkali treatment of coir fibres for coir-polyester composites. *Journal of Materials Science*, 18(5), 1443–1454.

Priya, N. A. S., Raju, P. V., & Naveen, P. N. E. (2014). Experimental testing of polymer reinforced with coconut coir fiber composites. *International Journal of Emerging Technology and Advanced Engineering*, 4(12), 453–460.

Rajamuneeswaran, S., & Jayabal, S. (2016). A lexicographic multiobjective genetic algorithm for optimization of mechanical properties of crab carapace impregnated coir–polyester composites. *Polymer Composites*, 37(3), 844–853.

Reddy, N. (2019). Composites from coir fibers. In Reddy, N. (Ed.). *Sustainable Applications of Coir and Other Coconut By-products* (pp. 141–185). Springer, Cham.

Rigolin, T. R., Takahashi, M. C., Kondo, D. L., & Bettini, S. H. P. (2019). Compatibilizer acidity in coir-reinforced PLA composites: matrix degradation and composite properties. *Journal of Polymers and the Environment*, 27(5), 1096–1104.

Rihayat, T., Suryani, S., Fauzi, T., Agusnar, H., Wirjosentono, B., Alam, P. N., & Sami, M. (2018, March). Mechanical properties evaluation of single and hybrid composites polyester reinforced bamboo, PALF and coir fiber. In *IOP Conference Series: Materials Science and Engineering* (Vol. 334, No. 1, p. 012081). IOP Publishing.

Rout, J., Misra, M., Mohanty, A. K., Nayak, S. K., & Tripathy, S. S. (2003). SEM observations of the fractured surfaces of coir composites. *Journal of Reinforced Plastics and Composites*, 22(12), 1083–1100.

Rout, J., Misra, M., Tripathy, S. S., Nayak, S. K., & Mohanty, A. K. (2001). The influence of fibre treatment on the performance of coir-polyester composites. *Composites Science and Technology*, 61(9), 1303–1310.

Sanjay, M. R., Madhu, P., Jawaid, M., Senthamaraikannan, P., Senthil, S., & Pradeep, S. (2018). Characterization and properties of natural fiber polymer composites: A comprehensive review. *Journal of Cleaner Production*, 172, 566–581.

Santafé Jr, H. P., Rodriguez, R. J., Monteiro, S. N., & Castillo, T. E. (2011), Characterization of thermogravimetric behavior of polyester composites reinforced with coir fiber. In S. N. Monteiro, D. E. Verhulst, V. I. Lakshmanan, P. N. Anyalebechi, S. Bontha, J. A. Pomykala (eds.), *EPD Congress 2011* (pp. 305–310). Hoboken, NJ, USA: John Wiley & Sons, Inc.

Santos, E. F., Mauler, R. S., & Nachtigall, S. M. (2009). Effectiveness of maleated-and silanized-PP for coir fiber-filled composites. *Journal of Reinforced Plastics and Composites*, 28(17), 2119–2129.

Santulli, C. (2019). Mechanical and impact damage analysis on carbon/natural fibers hybrid composites: A review. *Materials, 12*(3), 517.

Sathiyamurthy, S., Thaheer, A. S. A., & Jayabal, S. (2012). Mechanical behaviours of calcium carbonate-impregnated short coir fibre-reinforced polyester composites. *Proceedings of the Institution of Mechanical Engineers, Part L: Journal of Materials: Design and Applications*, 226(1), 52–60.

Sathiyamurthy, S., Thaheer, A., & Jayabal, S. (2013). Modelling and optimization of mechanical behaviors of Al_2O_3-coir-polyester composites using response surface methodology. *Indian Journal of Engineering and Materials Sciences* 20, 59–67.

Satyanarayana, K. G., Sukumaran, K., Kulkarni, A. G., Pillai, S. G. K., & Rohatgi, P. K. (1986). Fabrication and properties of natural fibre-reinforced polyester composites. *Composites*, 17(4), 329–333.

Saw, S. K., Sarkhel, G., & Choudhury, A. (2012). Preparation and characterization of chemically modified Jute–Coir hybrid fiber reinforced epoxy novolac composites. *Journal of Applied Polymer Science*, 125(4), 3038–3049.

Scheirs, J., & Long, T. E. (Eds.). (2005). *Modern Polyesters: Chemistry and Technology of Polyesters and Copolyesters*. John Wiley & Sons, Hoboken, NJ.

Sen, T., & Reddy, H. J. (2011). Application of sisal, bamboo, coir and jute natural composites in structural upgradation. *International Journal of Innovation, Management and Technology*, 2(3), 186.

Setyanto, R. H., Diharjo, K., Miasa, I. M., & Setyono, P. (2013). A preliminary study: The influence of alkali treatment on physical and mechanical properties of coir fiber. *Journal of Materials Science Research*, 2(4), 80.

Siakeng, R., Jawaid, M., Ariffin, H., & Salit, M. S. (2018). Effects of surface treatments on tensile, thermal and fibre-matrix bond strength of coir and pineapple leaf fibres with poly lactic acid. *Journal of Bionic Engineering*, 15(6), 1035–1046.

Silva, G. G., De Souza, D. A., Machado, J. C., & Hourston, D. J. (2000). Mechanical and thermal characterization of native Brazilian coir fiber. *Journal of Applied Polymer Science*, 76(7), 1197–1206.

Sindhu, K., Joseph, K., Joseph, J. M., & Mathew, T. V. (2007). Degradation studies of coir fiber/polyester and glass fiber/polyester composites under different conditions. *Journal of Reinforced Plastics and Composites*, 26(15), 1571–1585.

Singh, Y., Singh, J., Sharma, S., Lam, T. D., & Nguyen, D. N. (2020). Fabrication and characterization of coir/carbon-fiber reinforced epoxy based hybrid composite for helmet shells and sports-good applications: Influence of fiber surface modifications on the mechanical, thermal and morphological properties. *Journal of Materials Research and Technology*, 9(6), 15593–15603.

Siva, I., Jappes, J. T. W., & Suresha, B. (2012). Investigation on mechanical and tribological behavior of naturally woven coconut sheath-reinforced polymer composites. *Polymer Composites*, 33(5), 723–732.

Sridhar, R. (2017). A review on performance of coir fiber reinforced sand. *International Journal of Engineering and Technology*, 9(1), 249–256.

Sudhakara, P., Jagadeesh, D., Wang, Y., Prasad, C. V., Devi, A. K., Balakrishnan, G., ... & Song, J. I. (2013). Fabrication of Borassus fruit lignocellulose fiber/PP composites and comparison with jute, sisal and coir fibers. *Carbohydrate Polymers*, 98(1), 1002–1010.

Ting, T. L., Jaya, R. P., Hassan, N. A., Yaacob, H., & Jayanti, D. S. (2015). A review of utilization of coconut shell and coconut fiber in road construction. *Jurnal Teknologi*, 76(14).

Tufail, M. R., Jamshaid, H., Mishra, R., Hussain, U., Tichy, M., & Muller, M. (2021). Characterization of hybrid composites with polyester waste fibers, olive root fibers and coir pith micro-particles using mixture design analysis for structural applications. *Polymers*, 13(14), 2291.

Valášek, P., D'Amato, R., Müller, M., & Ruggiero, A. (2018). Mechanical properties and abrasive wear of white/brown coir epoxy composites. *Composites Part B: Engineering, 146*, 88–97.

Velumani, S., Navaneetha Krishnan, P., & Jayabal, S. (2014). Mathematical modeling and optimization of mechanical properties of short coir fiber-reinforced vinyl ester composite using genetic algorithm method. *Mechanics of Advanced Materials and Structures*, 21(7), 559–565.

Verma, D., Gope, P. C., Shandilya, A., Gupta, A., & Maheshwari, M. K. (2013). Coir fibre reinforcement and application in polymer composites: A review. *Journal of Materials and Environmental Science*, 4(2), 263–276.

Wambua, P., Ivens, J., & Verpoest, I. (2003). Natural fibres: Can they replace glass in fibre reinforced plastics? *Composites Science and Technology*, 63(9), 1259–1264.

Wong, K. J., Nirmal, U., & Lim, B. K. (2010). Impact behavior of short and continuous fiber-reinforced polyester composites. *Journal of Reinforced Plastics and Composites*, 29(23), 3463–3474.

Yan, L., Chouw, N., Huang, L., & Kasal, B. (2016). Effect of alkali treatment on microstructure and mechanical properties of coir fibres, coir fibre reinforced-polymer composites and reinforced-cementitious composites. *Construction and Building Materials*, 112, 168–182.

Yan, L., Su, S., & Chouw, N. (2015). Microstructure, flexural properties and durability of coir fibre reinforced concrete beams externally strengthened with flax FRP composites. *Composites Part B: Engineering*, 80, 343–354.

Yousif, B. F. (2009). Frictional and wear performance of polyester composites based on coir fibres. *Proceedings of the Institution of Mechanical Engineers, Part J: Journal of Engineering Tribology*, 223(1), 51–59.

Yousif, B. F., & Ku, H. (2012). Suitability of using coir fiber/polymeric composite for the design of liquid storage tanks. *Materials & Design (1980–2015)*, 36, 847–853.

Yusoff, R. B., Takagi, H., & Nakagaito, A. N. (2016). Tensile and flexural properties of polylactic acid-based hybrid green composites reinforced by kenaf, bamboo and coir fibers. *Industrial Crops and Products*, 94, 562–573.

Zainudin, E. S., Yan, L. H., Haniffah, W. H., Jawaid, M., & Alothman, O. Y. (2014). Effect of coir fiber loading on mechanical and morphological properties of oil palm fibers reinforced polypropylene composites. *Polymer Composites*, 35(7), 1418–1425.

8 Wood Fiber-Reinforced Polyester Composite

Anthony Chidi Ezika
Tshwane University of Technology, Pretoria, South Africa

Emmanuel Rotimi Sadiku
Tshwane University of Technology, Pretoria, South Africa

Suprakas Sinha Ray
DST-CSIR National Center for Nanostructured
Industrial Research, Pretoria, South Africa
University of Johannesburg, South Africa

*Henry Chukwuka Oyeoka, Martin Emeka
Ibenta, and Victor Ugochukwu Okpechi*
Nnamdi Azikiwe University, Awka, Nigeria

CONTENTS

8.1 Introduction 138
8.2 Wood Fiber 138
8.3 Chemical Treatment of Wood Fibers... 139
    8.3.1 Polyester.. 141
        8.3.1.1 Unsaturated Polyester Resin 142
        8.3.1.2 Saturated Polyester... 144
8.4 Method of Preparation of Wood Fiber-Reinforced Polyester Composites ... 145
8.5 Properties of Wood–Polyester Composites 147
    8.5.1 Mechanical Properties... 147
    8.5.2 Water Properties 150
    8.5.3 Thermal Properties 151
    8.5.4 Physico-Chemical Properties 152
    8.5.5 Morphological Properties 153
8.6 Applications of Wood–Polyester Composites.. 154
8.7 Conclusion 154
References... 155

137

8.1 INTRODUCTION

The rise in ecological awareness has led to the use of natural resources as technological materials for different applications, and government modern ecological rules and regulations have tasked various engineering industries to find new materials [1]. Presently, synthetic fiber-reinforced composites are extensively used in different engineering applications. However, the difficulties in the handling of synthetic fibers such as their non-degradability and skin irritations have been overemphasized [2]. Composites are materials which consist of two or more suitably ordered or distributed physically distinct phases. The matrix is referred to as the continuous phase while the fillers or reinforcements are the distributed phase [3].

Over the past couple of decades, considerable interest has been channeled toward natural fibers as suitable substitute for man-made (synthetic) fiber reinforcements in polymer composites [4–6]. These interests are driven by low cost, low density, simple handling, relatively good mechanical properties, renewability, and eco-friendliness of the natural fiber-reinforced composites [7,8].

Wood is one of the commonly used natural fibers in engineering applications, because of their low cost, low weight, and better recyclability [9,10]. It is also the most preferred material for structural applications; this is because of its inexpensive, aesthetically attractive quality, and good physical strength. Wood fiber polymer composites are produced either by incorporating wood fiber as an inclusion in polymer matrix, and molding or pressing under high pressure and temperature [3]. In various industries such as transportation, aerospace, biomedical equipment, automobiles, and mining industries, conventional materials are being widely replaced by fiber-reinforced composites [4,11–13].

Furthermore, the eco-friendliness of waste products from wood industry is one of the positive effects of using them in polymer composites. However, its major drawback is the degradation of the physical and mechanical properties due to environmental variation that leads to bond failure [14]. Another disadvantage is their hygroscopicity, which makes it difficult to achieve a good level of dispersion in the polymer matrix. Hygroscopicity can be minimized, however, by esterifying the wood components. Chemically modifying wood enhances its mechanical characteristics, as well as its stability while reducing water absorption [15,16]. The degradation of wood plastic composites is a function of proportion of wood dust in the composite. A faster degradability in both water and the earth was ensured with higher wood proportion [17].

Unsaturated polyester (UP) is the most popular in preparing composites and accounts for 80% excess of all thermoset resins, and the global UP market has been forecast to grow at a steady rate of 5.3% compound annual growth rate between 2019 and 2029 [18,19]. Finally, to address the increasing demand for the use of wood-based polyester composite, an up-to-date review is required to better understand their properties.

8.2 WOOD FIBER

Wood is a lignocellulosic material composed mostly of lignin, hemicellulose, and cellulose, with extractives and ash as minute constituents [20]. In woods, there are several kinds of specialty cells; the wood cells are elongated, pointed at the end, and oriented along the stem axis. Latewood and earlywood have different cell sizes. When compared to the latewood, which is generated later in the growth season and has darker rings, the earlywood is softer, weaker, and lighter in weight. [21].

Wood fiber is a finely ground wood that can be made from a variety of sources, including chips, wood planar shaving, saw dust, and other neat wood waste from wood processing industries. Particle size distribution, moisture, and density are all observed during processing to ensure constant physical and chemical properties [22]. Wood flour (WF) and long wood fibers are two types of wood fiber. However, long wood fibers are more difficult to handle than WF. Long wood fibers tend to clump together, causing dispersion problems. However, longer wood fibers tend to provide greater reinforcing effects in composites due to their high aspect ratio [22].

The properties of wood are determined by the tree from which it was extracted, while the source of the wood determines its density. Ever since the evolution of man, wood has contributed enormously in human civilization and to the socioeco-nomic lives of the global human population. Due to the over-exploitation of wood throughout history, as a result of large-scale multiple applications, there have been a depletion of forest and this poses a danger to the survival of the ecological system. However, wood polymer composites (WPCs) can proffer solution to this problem. To accomplish this, wood fibers or WF derived from wood waste are utilized to create WPCs with strength and other qualities comparable to, if not superior to, wood.

Woods are classified into two classes: hard woods and soft woods. Gymnosperms (primarily conifers) produce soft woods, while hard woods come from angiosperms (deciduous trees). The woods also differ in terms of their constituents' cells. The structure of hardwood is contained of fibrous elements, axial parenchyma cells, and vessel elements with more complex structure [24]. Figure 8.1 depicts the morphology of softwood and hardwood. Cell walls are layered structures that are stacked con-centrically and have different chemical compositions and orientations (Figure 8.2). The middle lamella is a cellulose-free lamella that binds the separate cells together. The primary wall (P) is distinguished from secondary walls (S) by the transitional arrangement of the cellulose fibrils in the wall (S1, S2, S3) [25]. The cellulose fibrils in the primary wall are oriented in a modest slope perpendicular to the fiber direc-tion, whereas the fibrils are unevenly arranged with the fiber orientation in the sec-ondary wall (S2) while non-structural substances are present in high concentration in secondary wall 3 (S3) layer [26]. There is more cellulose in the secondary walls 1 and 2 (S1 and S2) than in the secondary wall and primary wall 3. Lumen is the cell wall-bound cavity, and the difference in wall thickness between latewood and early-wood is due to differences in S2 thickness; therefore, the bulk cellulose is found [21].

8.3 CHEMICAL TREATMENT OF WOOD FIBERS

The utilization of natural fibers as an inclusion in polyester matrix, to give polyester composite is often limited by the incompatibility nature of the cellulose fibers with the polymer interface [27]. Each of the repeating units of cellulose has three hydroxyl groups which bond with each other to form inter-cohesive hydrogen bond [28]. The hydrogen bond results in high flexural strength between the repeating units,

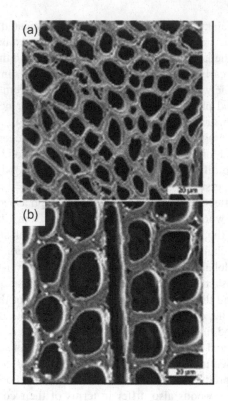

FIGURE 8.1 SEM image of (a) hardwood (*Eucalytus globulus*) and (b) softwood (*Cupressus macrocarpa*) [23].

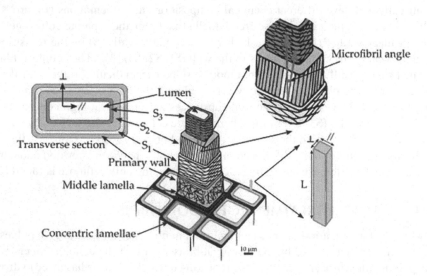

FIGURE 8.2 Diagram of the transversal structure of the tracheid cell wall in black spruce wood [25].

with solvent affinity attributes on the surface of the fiber [27]. A method is required to enhance the homogeneity between wood fibers and polyester matrix to enhance composite materials' physical properties. Materials with coupling ability have been employed to establish a chemical bond at the interface between fiber and polymer to improve the fiber polymer interfacial bonding. The hydroxyl group on the fiber surface reacts with the bonding sites of the coupling agent. Depending on the type of the polyester, the other end is coupled to the polyester matrix's functional group through various bonding such as hydrogen bond, covalent bonding, acid–base interaction, and also through chain entanglement [29].

Natural fibers are treated to improve their water resistance, increase the wettability of their surfaces with polymers (mostly non-polar polymers), and boost interfacial adhesion [30]. Fiber chemical treatments which include peroxide treatment, alkali treatment, acrylation, benzoyl treatment, silane treatment, and isocyanate treatment have been reported to enhance fiber strength and fiber–matrix bonding in wood fiber composites to varying degrees, and are thus used for final composite applications [27,31,32]. Thus, the chemical treatments are utilized to prepare the wood fiber prior to its usage as a filler in polyester matrix, which will result in the formation of wood fiber composite with enhanced properties. The effect of chemical treatment on both the pristine wood fiber and the resulted wood fiber composite is evident on the surface morphology of the treated wood fiber and wood fiber composites, which assumes a soft and curly nature as a result of mechanical entanglement which will enhance fiber adhesion when used as a filler [31]. More so, chemical treatment of wood fibers-filled polymer matrix results in a decline in the voids and irregular matrix cracking, showing outstanding laminar linkages between the polymer matrix and the WF [31].

8.3.1 Polyester

Polyesters are classified into different classes based on the bonding affinity of monomer (e.g., aromatic or aliphatic). This division yields two classes: UPs and saturated polyesters. Poly(L-lactic acid), poly(caprolactone), poly(glycolic acid), and their copolymers are examples of saturated polyesters. Because of their biodegradability and biocompatibility, polyesters have various applications in their biocomposites form [33].

Polyester resins are one of the most common man-made resin categories, which are widely utilized in plastic reinforcement, as well as powder coatings, water-based paints, lacquers, glues, and automotive cements [34]. Polyesters are polymeric large molecules composed of monomers joined by ester linkages. Ester linkages/bonds are covalent chemical bonds formed of both a –COOH carboxylic acid group and a –OH alcohol group [35]. Polyesters are among the most valuable polymer categories in use currently. Polyesters are created in their most basic form by the polycondensation reaction of a glycol with a di-functional carboxylic acid [36].

The most commonly used thermosetting resins are polyesters. Because of their low strength and brittleness, polyesters could not be used for technology-specific purpose without reinforcement, but they are widely used in composite matrices. UPs have high polydispersity value and low molecular weight. Because of the unsaturation in their chain, UPs can be utilized in cure chemical interactions through radical polymerization reactions in the presence of a reactive monomer, say unsaturated styrene, resulting in unsaturated polyester resins (UPRs) which belong to the class of thermosetting polyesters [33].

8.3.1.1 Unsaturated Polyester Resin

UP resin, also known as polyester resin, is a viable and widely known resin, as well as a multifunctional synthetic copolymer used in composites production. They are a type of high-performance engineering polymer that can be found in a variety of applications [37]. UP resins are linear polycondensation products derived from saturated and unsaturated anhydrides/acids as well as oxides or diols. UP resins are typically with a low degree of polymerization, resulting in pale yellow oligomers. These oligomers can be splintery solids or sticky liquids depending on their molecular weight (1200–3000 g/mol) and chemical composition. They were mostly used in compression molding, injection molding, filament winding, resin transfer molding, and hand lay-up. Polyesters are found to be used in the manufacture of 85% of fiber-filled polymeric products, which include car and aircraft components, boats body parts and chairs [38].

The curing effect of UPR is commonly initiated by the co-polymerization of the ester functional group on its main chain with styrene monomer. The unsaturated effect of the main chain provides sites for peroxide initiators to react with double bonds in styrene monomer, resulting in the formation of a 3D network [39]. UP can form very strong bonds with other materials and have high toughness and crack resistance [37]. This type of polyester's unsaturation offers a site for successive crosslinking. These resins are mixed with various reinforcements and inclusions, and cured with radical initiators to produce thermosetting entities with a variety of mechanical and chemical properties depending on the diols, initiators, crosslinking agents, diacids, and other inclusions used. The diversity of the resulted thermoset product's properties has sparked interest in these polyester resins as an essential engineering material for a broad array of material applications [40].

The most important low cost and commonly used monomers for introducing unsaturation in the polyester chain are fumaric acid and maleic anhydride. The UP characteristics are easily modified by introducing varying concentrations of diacids and diols, as well as altering the constituent's ratio [33,40,41]. The characteristics of UPs are determined not only by the degree of the cure reaction, but also by their molecular composition. Hence, it is critical to comprehend the reaction's thermodynamics that occur during curing and their inferences in the process [41].

The conversion of UPs into an engineering material necessitates the inclusion of another material, the unsaturated monomer (UM), also known as diluent and typically added in wt.% ranging from 30% to 40%. The general procedure for obtaining UPRs is classified into two major stages: UP synthesis and curing reaction (Figure 8.3).

The UM is incorporated into the UP for two reasons: (1) to increase the fluidity of the system (which aids resin preparation) and (2) to develop an effective double bond, which initiates a crosslink network in the UP main chain. Styrene is the most common UM used, but other UMs such as alkyl methacrylate vinyl esters, dimethacrylates, and divinylbenzene can also be utilized. Also, due to the radical kinetics of the curing process, an initiator of radical nature must be added to the formulation.

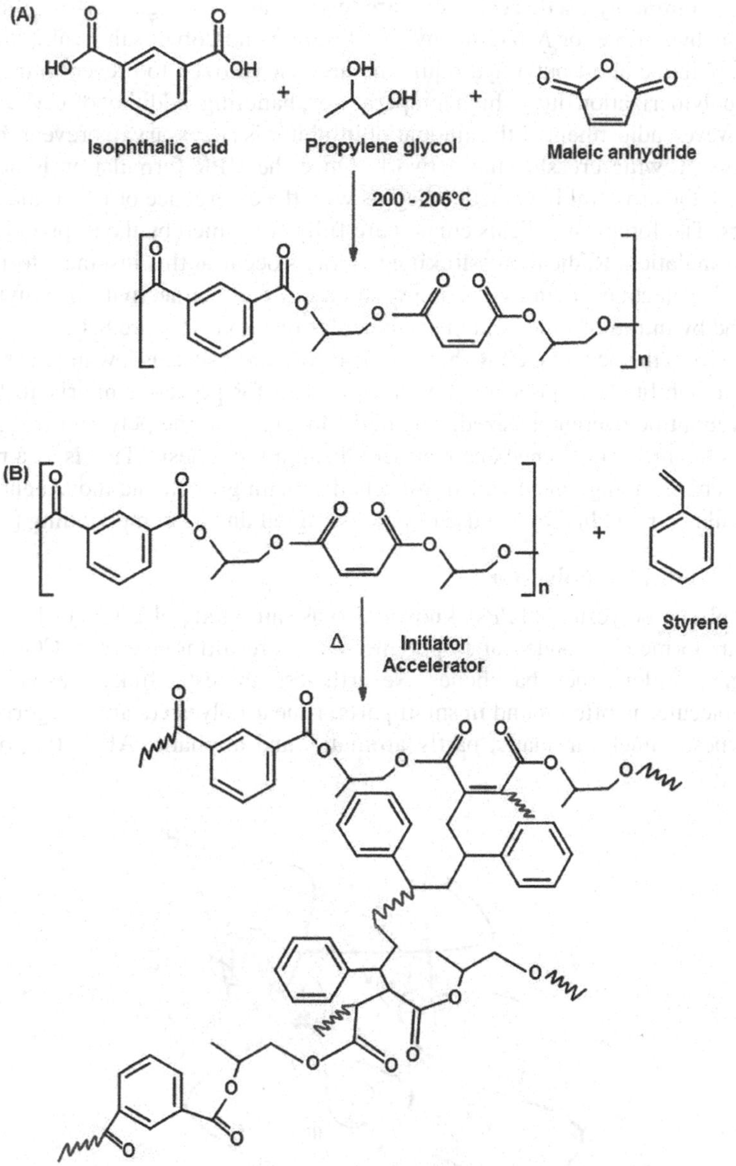

FIGURE 8.3 Path for development of UPRs: (a) polycondensation reaction and (b) cure reaction [33].

Organic peroxides, such as alkyl hydro peroxides, ketone peroxides, dialkyl, or dia-cyl peroxides, are the most commonly used initiators [33,40,42].

A further critical additive for the compounding is the accelerator, which is a chemical component capable of lowering the activation energy and, thereby, low-ering the temperature and enthalpy required for network formation (crosslinking).

The most commonly used accelerators are tertiary amines (e.g., *N,N*-diethylaniline, *N,N*-dimethylaniline, or *N,N*-dimethyl-*p* toluidine) and cobalt salts. Inhibitors such as hydroquinone and t-butyl hydroquinone are also utilized to prevent unnecessary radical polymerization due to high temperatures, handling, addition of UM, and storage. However, adjustment of the amount of inhibitor is necessary to prevent the slowing of the growing crosslinking network. Once the UPR formulation is heated or irradiated, the crosslinking reaction begins with the emergence of free radicals from initiators. The foremost radicals created are fully consumed by the trapped inhibitor in the formulation. Radical crosslinking reactions occur at this instance, forming an elongated molecular chain by the stepwise end-to-end connectivity of vinyl monomers, and by intramolecular and intermolecular reactions (Figure 8.4).

Aside from the benefits, UPRs have some drawbacks, such as low impact strength, high flammability, and poor toughness. The need for polyester matrix to be reinforced cannot be overemphasized. The high shrinkage of the polyester resin during curing is the major challenge encountered during its synthesis. This is as a result of the over chain entanglement which restricts the chain growth and movement, and as such, could be remedied by the use of processing oil during compounding [33,39].

8.3.1.2 Saturated Polyester

Thermoplastic polyesters (TPPs), known also as saturated polyesters or linear polyesters, are formed by condensation polymerization, resulting in ester (–COO–) functional groups along their backbones. Nevertheless, the ester linkage established in TPP' molecules is often found in small parts. Linear polyesters are categorized into three types, namely aromatic, partly aromatic, and aliphatic. Aliphatic polyesters

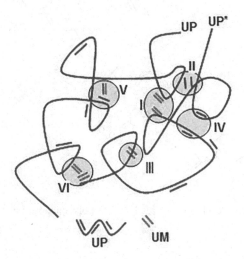

FIGURE 8.4 Schematic representation of the possible reactions in the UM-UP cure reaction: (I) intermolecular crosslinking UP-UM-UP*; (II) intermolecular crosslinking UP-UP*; (III) intramolecular crosslinking UP-UM-UP (or UP*-UM-UP*); (IV) intramolecular crosslinking UP-UP (or UP*-UP*); (V) branching growth UP-UM (or UP*-UM); (VI) UM homopolymerization [33].

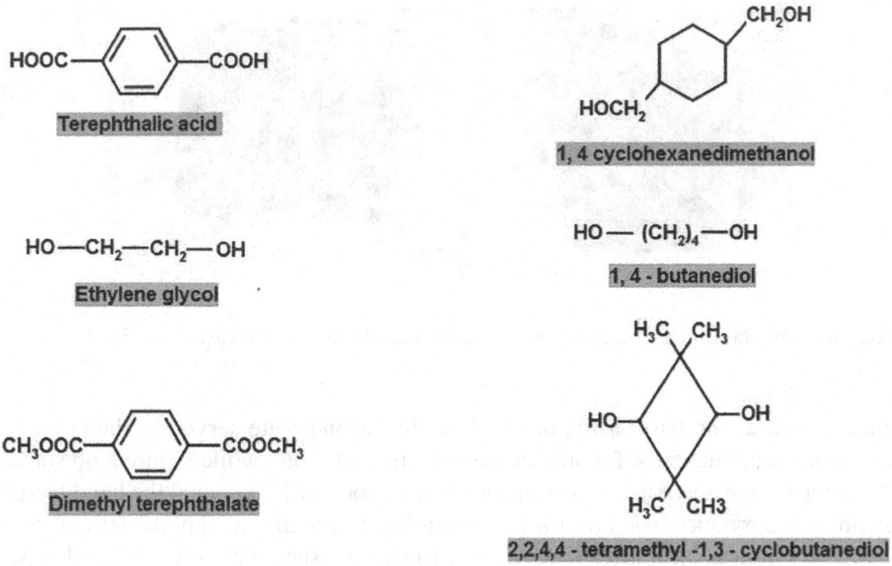

FIGURE 8.5 Fewer examples of co-monomers and monomers utilized for the preparation of partial co-polyesters and aromatic homo-polyesters [43].

are created by combining aliphatic diols with dicarboxylic acids (or esters). Linear aliphatic polyesters have lower melting points, and are susceptible to hydrolysis due to the high mobility nature of the C–O interaction that exists in their main chain. Aliphatic diols and aromatic dicarboxylic acids (or esters) are used to make partially aromatic polyesters [43] (Figure 8.5).

Also, polyesters, however, exist in liquid crystalline forms, with most ester linkages anchored to aromatic rings. The most common TPP among others is poly(ethylene terephthalate) (PET). PET is a clear polymer with good mechanical properties and dimensional stability under varying loads. Furthermore, PET has good gas barrier properties as well as chemical resistance, and also has good mechanical properties when reinforced with both synthetic and bio-fillers. Among so many bio-fillers for the reinforcement of polyesters, wood fiber presents an excellent choice for the preparation of polyester composites, due to its soft, curl nature, and improved mechanical properties when chemically treated.

8.4 METHOD OF PREPARATION OF WOOD FIBER-REINFORCED POLYESTER COMPOSITES

Among the various methods for the production of composites, the hand lay-up technique is the basic approach of composite fabrication. It is the most commonly used method in the wood fiber-reinforced polyester composite fabrication [3,44,45]. Cast mold was employed in the hand lay-up fabrication of composites for the study of the effect of water on the behaviors of saw dust and chopped reeds-filled UP composites [3]. The iron used for the mold consisted of dual plates. The polished defect-free first

FIGURE 8.6   (a) Image of a compression apparatus; (b) image of a cast mold [3].

plate served as the base of the mold while the second plate served as the cover to attain uniform thickness. Figure 8.6a depicts the cast mold, while Figure 8.6b shows the compression apparatus. Ganesan and Kaliyamoorthy [1] adopted the hand lay-up technique to produce wood dust-filled woven jute fiber-reinforced polyester composites. The stainless-steel mold (300×300 mm) was polished with a wax-dipped paper cloth. The thoroughly mixed, accelerated, and catalyzed UPR was poured in the stainless-steel mold for the fabrication of the composite samples.

A lot of curing practices are available in composite fabrication. However, allowing the composite to cure at room temperature is the most basic method. Curing can also be sped up via application of heat and pressure. Coconut and wood fiber-reinforced composites were produced by mixing coconut and wood fiber rigorously in polyester resin to obtain a uniform mixture, which was conveyed into the mold plate and pressure of 2 tons applied for 60 minutes. The samples were then allowed to cure under the sunlight for 3 days [46]. Ahmed and Mei [47] used aluminum mold castings to develop rubber saw dust-reinforced UP composites. The composites were cured after a day at room temperature, post-cured for 6 hours at 60°C and an additional 5 hours at 150°C. Nunez et al. [48] produced wood particle fiber-reinforced composites via compression molding while hand lay-up and compression molding were adopted in the manufacturing of the hybrid composites. After 6 hours at room temperature and under pressure, the plagues were cured and subsequently post-cured at 150°C for 1.5 hours.

Other methods of wood fiber-reinforced composites fabrication have been developed based on the materials, the part design, application, and end use [49]. Glass fiber/UPR/poplar wood composites were produced via prepreg/press method using laboratory instruments and machines [50]. The glass fiber was dipped in a container of hardener and UPR for impregnation of the resin on the fabric. The fabric was extruded with a roller and accelerator sprayed on the fabric. The fabric was finally sandwiched fabric between the poplar wood and cold pressed to produce the composite (Figure 8.7).

Needling mat polyester/wood composite fiber was produced using carded polyester web with wood sprinkled uniformly on the polyester web [51]. Multilayer wood fiber composite web and polyester fiber were finally needled by needle punching to form the polyester/wood composite fiber [51].

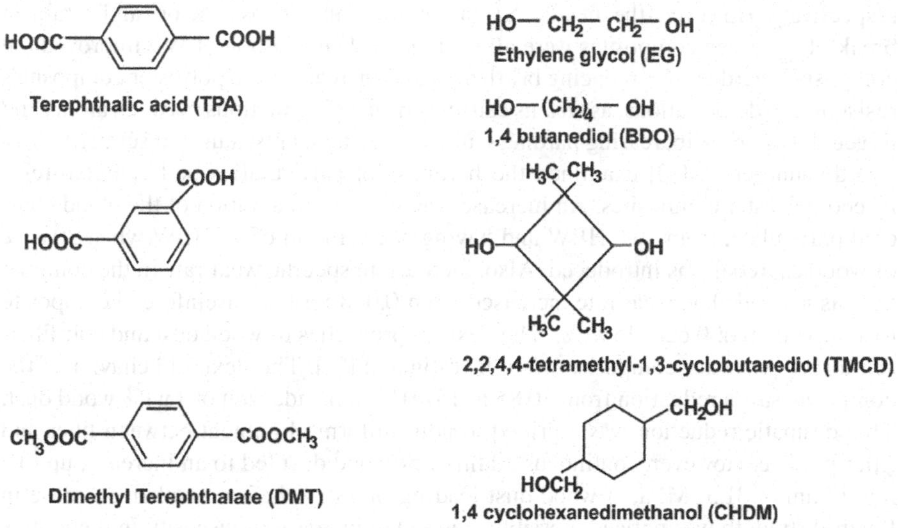

HOOC—⬡—COOH

Terephthalic acid (TPA)

HO—CH_2—CH_2—OH
Ethylene glycol (EG)

HO—$(CH_2)_4$—OH
1,4 butanediol (BDO)

COOH / HOOC—⬡

H_3C CH_3
HO—◇—OH
H_3C CH_3
2,2,4,4-tetramethyl-1,3-cyclobutanediol (TMCD)

CH_3OOC—⬡—$COOCH_3$

Dimethyl Terephthalate (DMT)

CH_2OH / $HOCH_2$
1,4 cyclohexanedimethanol (CHDM)

FIGURE 8.7 The prepreg/press fabrication method of glass fiber/UPR/poplar wood composites [50].

8.5 PROPERTIES OF WOOD–POLYESTER COMPOSITES

8.5.1 MECHANICAL PROPERTIES

The mechanical characteristics of natural fiber composites play a big role in their structural applications. On fiber-filled composites, a variety of mechanical tests are performed, some of which are extremely specialized to a particular application. These tests can be used to define the mechanical behavior of any composite to a significant extent, as well as provide insight into other potential application areas [52].

Farhan [3] studied and compared the Young's modulus and flexural strength of saw dust and chopped reed-filled UP composite. They observed that saw dust-filled polyester composite recorded lower flexural strength and modulus than the chopped reed composites. This decrease was attributed to size of saw dust being lesser than the chopped reeds, thereby decreasing the adhesion between matrix and saw dust. The effect of WF on styrene-diluted UP thermoset resin was also studied [48]. The study recorded an increase in tensile modulus from 3.4 GPa for the unreinforced composites to a maximum of 4.94 GPa when WF was introduced. However, a decrease in the tensile strength was observed.

In another studies, the mechanical behavior of WF-reinforced UPR was evaluated [18]. Before and after accelerated aging, the mechanical attributes of the composites were investigated. The flexural strength was observed to reduce both before (108–74 MPa) and after aging (69–39 MPa). However, the hardness was seen to increase (36–45 MPa) before aging but a reduction was noticed after accelerated aging. In a previous study, Paczkowski et al. [19] synthesized and characterized WF green composite-filled UPR from recycled PET. Inclusion of WF fiber into the resin increased the flexural modulus and Barcol hardness from 3.59 to 3.81 GPa and 36–45 HBa, respectively. However, the flexural strength, mechanical loss factor, and strain at break of the green composites were all decreased. The addition of WF improved the composite's hardness by reducing brittleness and increasing the polyester composite's resistance to deformations, as well as causing stronger interactions between atoms and molecules, thereby increasing hardness and improving its resistance to scratch.

Edoziuno et al. [53] examined the hardness of particulate wood charcoal-reinforced polyester composites. An increase was noticed on addition of the wood charcoal particulate, from 16.6 HBW and having a maximum of 43 HBW when 20 wt% of wood charcoal was introduced. Also, increase in specific wear rate in the composite was noticed. The wear rate increased from 0.048 for the unreinforced composite to a maximum of 0.628 (15 wt%). The flexural properties of wood dust and coir fiber-reinforced polyester composites were investigated [54]. The flexural behavior of the composite saw a reduction from 20.85 to 15.40 MPa on addition of 1 wt% wood dust. This dramatic reduction was ascribed to non-uniform alignment between fiber and filler particle. However, continuous addition of wood dust led to an increase, up to a maximum of 31.32 MPa, at wood dust loading of 7 wt% before another decrease in flexural strength was noticed, possibly, due to the increase in viscosity in the matrix and in turn increase in porosity and decrease in the wettability of the composite.

Barbosa et al. [55] investigated the toughness behavior of eucalyptus wood fiber-reinforced polyester composite by means of Charpy impact test. From their observations, incorporation of wood fiber into polyester matrix considerably improved the impact strength of the composite. Improvement from 23.8 J/m (unreinforced) to 251.1 J/m (30 wt% wood fiber) was recorded.

Apparently, as a result of the minimal eucalyptus fiber/polyester matrix boundary shear stress, an increase in toughness of the composite was achieved. This resulted in large rupture areas following a higher absorbed energy due to a longitudinal propagation of the cracks all through the interface. The long-term behavior of wood–polyester composite material, reinforced with glass fiber, was reported [56]. The purpose of the study was to evaluate the long-term performance of the materials by the Goldenblat–Kopnov criterion. The studied model was a sandwich type, where the upper and lower flanges are glass/polyester composite material while the intermediate layer is pine wood. The observations of the calculated results show that the conditions of strength to the dangerous points of the plate were met for this type of material.

Laksono, Ernawati, and Maryanti [44] recorded the modulus of elasticity of Bangkirai wood-reinforced polyester composite. Increase was observed in the result as the wood fibers were introduced into the polyester composite (1.46–2.22 GPa) (Figure 8.8). The improvements in the elastic modulus were attributed to the effective distribution of pressure between fillers and matrix [57]. Also, the mechanical behavior of hybrid wood dust and coir-reinforced polyester composites was evaluated [45]. The tensile performance of the composites was raised by nearly 10% on addition of 7.5% particles of Teak wood dust to the composite. Further increase up to 24% was also reported on increasing the wood particles to 20%. Jose et al. also reported that addition of wood dust in the composite increased the flexural characteristic. The flexural performance of the specimen, which has 7.5% wood dust with 15% of fiber content, was 12% greater than that of composites with no wood dust.

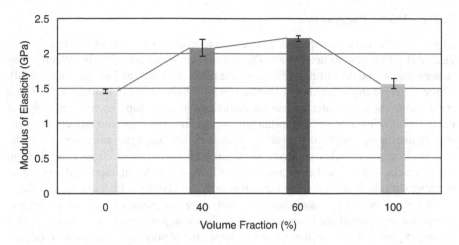

FIGURE 8.8 Modulus of elasticity of various volume fractions of waste Bangkira wood–polyester matrices [44].

The hardness, braking time, and wear properties of wood powder and coconut fiber polyester composite were investigated [46]. The composition with 40% wood powder and 0% coconut fiber had the maximum hardness value (59 VHN). The hardness of the UP composites reduced with growing composition of coconut fiber. However, inclusion of coconut fiber led to a reduced braking time and wear properties (Figure 8.9). Marcovich et al. [58] studied sisal and WF composites made from UP thermoset. The use of sisal and wood fibers led to an improvement in flexural strength. The flexural strength obtained was higher than that observed for WF composites and sisal composites. In a previous study [59] the reliance of mechanical performance of WF–polyester composite on moisture content was examined. The results noticed that the compressive and flexural strength determined on the wet samples were poor. These properties were worse as the humidity of the environment increases. This was ascribed to the detriment of wood fiber strength in wet conditions.

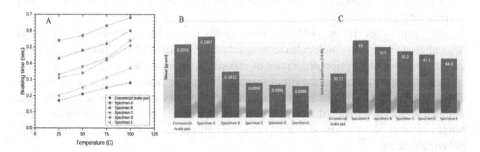

FIGURE 8.9 (a) Braking time, (b) wear, and (c) and Vicker's hardness of commercial brake pad and wood–coconut hybrid polyester composites [46].

8.5.2 WATER PROPERTIES

Moisture diffusion in composites can cause other qualities, such as mechanical and electrical properties, to deteriorate. The amount of water absorbed is influenced by factors such as the type of material, exposure time, additives, and temperature [60,61]. Three different mechanisms are involved in diffusion of water molecules. The first mechanism involves water molecules diffusing via tiny gaps in between polymer chains. Following that is the capillary transport into the gaps and defects at fiber–matrix interfaces, and finally, swelling processes that propagate micro-cracks in the matrix. Water molecules penetrate in natural fiber-reinforced composites through micro-cracks, when in wet environment. This leads to a reduced interfacial bonding of the fiber with the matrix, thereby leading to rebinding between fiber and matrix [62].

The effect of water in saw dust-filled polyester composites were studied [3] and compared to chopped reed composites. Saw dust composite had the minimum value of weight gain of water due to the bonding between the polyester and saw dust reduced the penetration of water to the material. Ahmad and Mei [47] showed the water absorption of rubber wood saw dust-filled UPR composite from recycled PET. The effect of both sodium hydroxide (NaOH)-treated and untreated wood was investigated. It was observed that raising the filler content led to increased water absorption properties. However, the value considerably decreased with surface change of the wood. The untreated rubber wood saw dust/UPR composite had deficient wettability and their bonding toward the UPR is ascribed to the hydrophilic property of the filler.

Ansari, Skifvars, and Berglund [63] revealed the moisture sorption properties of wood cellulose nanofiber-reinforced UPR. The moisture sorption of the composites was studied in different relative humidity (RH50, RH90). The sorption properties of the composite were seen to increase on the introduction of wood cellulose nanofiber into the UPR. The percentage water uptake of Garjon wood and Gamari wood particles-reinforced composites was also studied [64]. From their research findings, pristine polyester had minimal water absorption inclination. However, the water absorption of the composite increased with the rise in the wood particle content (Figure 8.10). Water absorption in composites leads to swelling of fibers till the cell walls are completely filled with water. This saturation leads to void formation or composite delamination as water exists as free water in the void structure. Another effect of water absorption is the deterioration of the interfacial bonding and hydrolytic degradation of both fiber and matrix [65].

High water absorption could be caused by fiber cracking and blistering, whereas degradation causes tiny molecules to be leached [66]. Abdul Mun'aim et al. [67] evaluated the water absorption properties of mixed and Cengal saw dust polyester composites. The minimum and maximum water absorption was noticed at 5% and 25% Cengal wood dust. The increase in water absorption was said to be due to increased number of micro-voids on the surface of the composites. This is a consequence of the bigger amount of poor adhesion area between hydrophilic saw dust and hydrophobic polymer matrix.

The hydrophilic property of cellulosic fibers, which is related to the existence of numerous H-bonds (OH– groups) between macromolecules of cell walls in fibers,

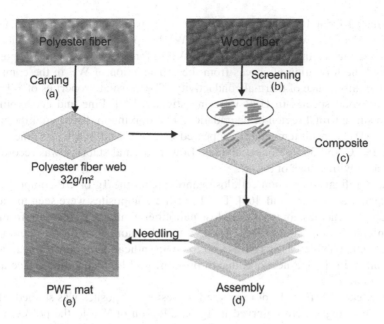

FIGURE 8.10 (a) Water absorption behavior of 10% (w/w) saw dust polyester composite; (b) water absorption behavior of 5% (w/w) saw dust polyester composite. KRC, GMC, P, and GRC denote kerosene wood particles, Gamari, pure polyester, and Garjan, respectively [64].

allows them to absorb moisture. By removing these groups, cellulosic fibers' water absorption can be reduced. This can be accomplished by using surface modification techniques and applying a polymer coating to the fibers' surfaces [68].

8.5.3 THERMAL PROPERTIES

The manufacturing temperature plays an important role in the production process of polymer composites, as the thermal stability of natural fibers composite is a relevant aspect to be considered because it plays an important role [69]. The maximum temperature at which any given fiber decomposes is regarded as its thermal stability [70]. The hemicellulose, cellulose, and lignin contents of natural fibers decompose at various temperatures, leading to the total decomposition of the fiber as a whole [71]. The thermal stability of fibers is considerably improved by eradicating as much hemicellulose, lignin, and other alkaline-soluble components from the fiber as possible, whether by physical, chemical, or biological means [52,72]. The thermal characteristics of natural fibers-filled composites are determined using different methods. This helps to understand the appropriateness of various fiber-reinforced composite for specific applications. Differential scanning calorimetry [73,74], dynamic mechanical analysis [75,76], and thermogravimetric analysis [76,77] are the methods reported to be used in literature for thermal analysis of composites.

Al-Anie, Hassan, and Hadithy [78] investigated the thermal conductivity of WF-incorporated polyester composite. The thermal conductivity of the composites was reduced from 10.56–8.06 w/mco when WF was increased from 5% to 20% wt.%. The decrease was due to the fact that WF has poor thermal conductivity and it was also suggested that the inability of the WF to form crosslinks with the polyester resin and the formation of spaces from the introduction of WF in the composites led to the impedance of thermal conductivity. The thermal properties of WFs from different wood species in UPR were investigated [79]. Pine and eucalyptus WF showed similar final region of degradation. The maximum differential thermogravimetry (DTG) peak temperature appeared at 285°C and 366°C for pine sample, and at 287°C and 364°C for eucalyptus. Lower thermal stability was recorded for Marmelero with a peak of 342°C.

The significance of wood cellulose nanofiber on the Tg of UP composites was investigated by Ansari et al. [63]. The Tg of the composites were seen to increase strongly with increase in wood cellulose nanofiber content. An increase from 66°C (unreinforced composite) to 78°C (for 45% wood nanofiber-reinforced composites) was noticed. The increased Tg was said to be an indication of constrained molecular mobility in the UP network, due to proximity of high specific surface area of wood nanofibrils.

The effect of WF on Tg of reinforced polyester composites was studied [18]. No noticeable changes were observed in Tg on addition of WF to the polyester resin. However, the storage modulus at the glassy stage increased to a maximum on addition of 2 wt% WF and to maximum at the rubbery state on addition of 5 wt% of WF. In a previous study [19], the glass transition temperature of the WF–polyester resin was greater than that of the pure resin (100°C). The glass transition temperature increased from 101°C to 105°C as the amount of WF in the composite increased.

8.5.4 PHYSICO-CHEMICAL PROPERTIES

The physico-chemical properties of wood-reinforced polyester composites have been documented in a number of studies.

The wettability and penetration of UPR into wood surface were studied through their contact angle and absorptivity [50]. This reflects the physical-chemical affinity between the fiber surface and the matrix resin, and it is crucial to examine the bond ability between wood and the matrix resin [80]. Dynamic contact angle is mostly used to study the wettability of wood surfaces and matrix. During a period of 10 minutes, Cheng et al. [50] found that the contact angle fell from 103° to 33° and that the absorptivity was 1.0×10^{-4} g/mm^2. Because of UPR's favorable wettability and penetrability, it was able to permeate wood and produce a good wood/UPR interface during fabrication.

As a strategy to improve environmental protection performance, the use of natural fibers in the production of sound absorbing materials has received some attention. Sound absorption performance of wood–polyester composite fiber needling mat were analyzed [51]. The study showed that increase in flow resistivity and density increased the sound absorption characteristics of the polyester composites. Also, the improvements of sound absorption characteristics of the composites were dependent on the wood fiber surface roughness and small cracks. The sound insulation capabilities of natural fiber Galam wood–polyester composites with respect to filler composition and shape were studied [51]. The studies investigated the sound insulation of WF and wood shavings polyester composites. The best insulation properties were observed in composites containing 7% saw dust filler at 800 Hz sound frequency [51,67].

Abdul Mun'aim et al. [67] studied the thickness swelling of Cengal and mixed saw dust polyester composite. The highest thickness swelling was recorded at 25% Cengal saw dust with 0.6242% thickness swelling. The observation was ascribed to be due to ample porosity or the presence of voids on the surface of the polyester composites leading to the dimensional change of cellulose composites.

The density of Kraft fiber-reinforced UPR composites was investigated [81]. The composite had comparable densities, ranging from 1371 to 1456 kg/m^3 with maximum difference of 6.2%. The contact angles were studied for three different wood species (Douglas fir, maple, and oak wood). The wettability of the wood with UP was studied before and after surface roughening [82]; the minimum average contact angle, 14°, was noticed after 5 seconds for Douglas fir sanded with 120 grit. When compared to un-sanded veneers, maple, sanding Douglas fir, and oak veneer sheets with 120 grit enhanced their wettability by decreasing contact angles. However, the Douglas fir veneer didn't show significant improvement. Dust and contaminants were cleaned from the surface of the veneer by 120 grits, creating a fresh, smooth surface, subsequently increasing veneer wettability [83,84].

8.5.5 Morphological Properties

Morphology is a phenomenon that focuses on studying the structure and relationships of polymer composites. It studies the arrangement of the wood fibers in the polyester matrix. These structures and arrangements affect the performance of the polyester composites. Some researchers have examined the morphological characteristics of wood-reinforced polyester composites.

The morphology of Bangkirai wood fibers in UPR was studied [44]. Scanning electron microscope (SEM) showed that there were voids in the 40% and 60% volume fraction wood-reinforced composites. This indicated trapped air bubbles (Figure 8.11). The studies showed that 60% filler fraction has lesser voids than 40% filler fraction. When hand lay-up method of composite production was used,

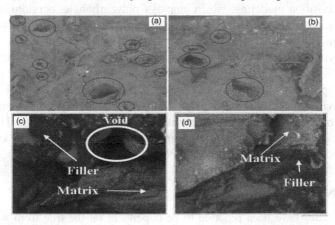

FIGURE 8.11   SEM observations showing void marks (a and b) and SEM observations after bending tests on 40% and 60% wood filler fraction (c and d) [44].

voids are usually unavoidable because it interacts directly with open air. Though the matrix transfers load to the reinforcer, however, the stress areas will move to the void areas and cause cracks when the composites receive load, thereby leading to early failure of the composite. Void content also depends on the degree of fiber orientation [85].

The morphology of Galam wood powder and wood shavings-reinforced polyester was investigated. The composite with wood powder appeared to be more evenly distributed, and the smaller particle size encouraged greater binding between the wood powders than the composite with wood shavings, which had bigger particle size. During application, the wood shavings reinforced composite cracks foremost due to encountered stress. Air bubbles were discovered in the samples, which may be remedied using the proper hand-lay approach [86]. The fracture morphology of wood particle-reinforced polyester composites was investigated using SEM [64]. The results indicated that fracture was initiated in the pure polyester sample due to dust particles picked up during molding or fabrication period. However, the fracture in wood particle composites was initiated by the air bubbles. Increase in wood fiber leads to wood agglomeration which Hossain et al. also thought to have contributed to the reduced strength of the wood particle-reinforced composites.

8.6 APPLICATIONS OF WOOD–POLYESTER COMPOSITES

WPCs are widely used in construction components such as flooring planks and screens, as well as in automotive components [87–89]. Aside from these, some more specific applications for WPCs, such as decking, cladding, paneling, fencing, and furniture, have been identified and tested in recent years. WPCs, for example, have been used in the production of pallets [90]. When compared to other synthetic structures such as fiber cement composites, WPCs can be considered value-added materials due to their wide range of applications.

Wood–polyester composites have found a special application in 3D printing. It has been determined that 3D printing is a critical component of the next-level industrial revolution [91–93]. The most common utilized process for 3D printing strips is fused deposition modeling, which employs polycarbonate, acrylonitrile–butadiene–styrene, polyamide, and thermoplastic material blends [94–96]. However, some of these materials are harmful to the environment, and harmful substances can be released during printing, causing significant harm to the environment and the human body [97]. Environmentally friendly aliphatic polyesters, on the other hand, have been developed [98]. Polyhydroxyalkanoate (PHA) is one of the most interesting of these environmentally friendly polyesters [99]. PHA has good biocompatibility, biodegradability, nontoxicity, processability, and mechanical properties [100]. When PHA/WF composite is made into 3D printing strips, for example, the composite's value and competitiveness are increased [99].

8.7 CONCLUSION

Wood properties have been reported to be dependent on the tree from which it has been obtained. Wood fibers or WF derived from wood waste are utilized in the right fiber to matrix ratio to produce wood polymer composites, having properties similar to or even better than wood. Wood fiber polymer composites are known for their eco-friendly nature in addition to their higher modulus to weight ratio when likened to synthetic fiber composites. However, the incompatibility of the wood fibers with polymer matrices and their high tendency to absorb moisture were reported to often reduce the tendency of bio-fibers to act as reinforcements and thus limit their utilization. Fiber chemical treatments such as isocyanate treatment, alkaline treatment, silane treatment, acrylation, benzoyl treatment, maleated coupling agents, and peroxide treatment have been reported to have introduced various levels of improvement in fiber strengths, water resistance, fiber–matrix adhesion, and thermal resistance in wood fiber composites, and are therefore used for final composite applications. Incorporation of wood fibers in the right proportion into polymer matrices for composite manufacture has been understood to modify various properties of the polyester composite materials.

REFERENCES

1. Ganesan, V. and B. Kaliyamoorthy, Utilization of Taguchi technique to enhance the inter-laminar shear strength of wood dust filled Woven Jute fiber reinforced polyester composites in cryogenic environment. *Journal of Natural Fibers* 2020. **19**(6): pp. 1–12.
2. Espert, A., et al., Comparison of water absorption in natural cellulosic fibers from wood and one- year crops in polypropylene composites and its influence on their mechanical properties. *Composites Part A: Applied Science and Manufacturing* 2004. **35**(11): pp. 1267–1276.
3. Farhan, A.J. and H.I.J.B.S.J. Jaffer, Effect of water on some mechanical properties for sawdust and chopped reeds/UPE composite. *Baghdad Science Journal.* 2011. **8**: p. 2.
4. Kumar, S. and K. K. Singh, Tribological behaviour of fiber-reinforced thermoset polymer composites: A review. *Part L: Journal of Materials: Design, and Applications* 2020. **234**(11): pp. 1439–1449.
5. Mallick, P.K., *Fiber-Reinforced Composites: Materials, Manufacturing, and Design.* 2007: CRC Press, Boca Raton, FL.
6. Chairman, C.A. and S.P. Kumaresh Babu, Mechanical and abrasive wear behavior of glass and basalt fabric-reinforced epoxy composites. *Journal of Applied Polymer Science* 2013. **130**(1): pp. 120–130.
7. Kumar, R., et al., Industrial applications of natural fiber-reinforced polymer composites– challenges and opportunities. *International Journal of Sustainable Engineering* 2019. **12**(3): pp. 212–220.
8. Sanjay, M., et al., Characterization and properties of natural fiber polymer composites: A comprehensive review. *Journal of Cleaner Production* 2018. **172**: pp. 566–581.
9. Sanjay, M., et al., A comprehensive review of techniques for natural fibers as reinforcement in composites: Preparation, processing and characterization. *Carbohydrate Polymers* 2019. **207**: pp. 108–121.
10. Sanjay, M. and B.J. Yogesha, Studies on hybridization effect of jute/kenaf/E-glass woven fabric epoxy composites for potential applications: Effect of laminate stacking sequences. *Journal of Industrial Textile* 2018. **47**(7): pp. 1830–1848.
11. Singh, K.K., N.K. Singh, and R.J. Jha, Analysis of symmetric and asymmetric glass fiber reinforced plastic laminates subjected to low-velocity impact. *Journal of Composite Materials* 2016. **50**(14): pp. 1853–1863.
12. Gaurav, A. and K.K. Singh, Effect of pristine MWCNTs on the fatigue life of GFRP laminates-an experimental and statistical evaluation. *Composites Part B: Engineering* 2019. **172**: pp. 83–96.
13. Ansari, M.T.A., et al., Fatigue damage analysis of fiber-reinforced polymer composites— A review. *Journal of Reinforced Plastics and Composites* 2018. **37**(9): pp. 636–654.

14. Rowell, R.M., Chemical modification of wood for improved adhesion in composites, in Alfred W. Christiansen (Editor), *Proceedings of Wood Adhesives*. United States Department of Agriculture, Wisconsin, 1995: pp. 55–60.

15. Nenkova, S., et al., Modification of wood flour with maleic anhydride for manufacture of wood- polymer composites. *Polymers and Polymer Composites* 2006. **14**(2): pp. 185–194.

16. Rahman, M.R., et al., Physical, mechanical, and thermal properties of wood flour reinforced maleic anhydride grafted unsaturated polyester (UP) biocomposites. *BioResources* 2015. **10**(3): pp. 4557–4568.

17. Vedrtnam, A., S. Kumar, and S. Chaturvedi, Experimental study on mechanical behavior, biodegradability, and resistance to natural weathering and ultraviolet radiation of wood-plastic composites. *Composites Part B: Engineering* 2019. **176**: pp. 107282.

18. Pączkowski, P., et al., Synthesis, characterization and testing of antimicrobial activity of composites of unsaturated polyester resins with wood flour and silver nanoparticles. *Materials* 2021. **14**(5): pp. 1122.

19. Pączkowski, P., A. Puszka, and B.J.P. Gawdzik, Green composites based on unsaturated polyester resin from recycled poly(ethylene terephthalate) with wood flour as filler—synthesis, characterization and aging effect. *Polymers* 2020. **12**(12): pp. 2966.

20. Baeza, J.J.W., Chemical characterization of wood and its components, in David N.S. Hon a, Nobuo Shiaishi (Editors), *Wood and Cellulosic Chemistry*. CRC Press, Boca Raton, FL 2001: pp. 275–384.

21. Wiedenhoeft, A.C., Wood handbook, Chapter 03: Structure and function of wood, in *Review Process: Informally Refereed (Peer-Reviewed)*. 2010.

22. Matuana, L. and N. Stark, The use of wood fibers as reinforcements in composites, in Omar Faruk and Mohini Sain (Editors) *Biofiber Reinforcements in Composite Materials*. Woodhead Publishing, Toronto 2015: pp. 648–688.

23. González, O.M., et al., Representative hardwood and softwood green tissue-microstructure transitions per age group and their inherent relationships with physical–mechanical properties and potential applications. *Forests*, 2020. **11**(5): pp. 569.

24. Rowell, R.M., *Handbook of Wood Chemistry and Wood Composites*. CRC Press, Boca Raton, FL, 2005.

25. Arzola-Villegas, X., et al. Wood moisture-induced swelling at the cellular scale—Ab intra. *Forests* 2019. **10**(11): p. 996.

26. Shah, D.U., Developing plant fiber composites for structural applications by optimising composite parameters: A critical review. *Journal of Materials Science* 2013. **48**(18): pp. 6083–6107.

27. Gwon, J.G., et al., Effect of chemical treatments of wood fibers on the physical strength of polypropylene based composites. *Korean Journal of Chemical Engineering* 2010. **27**(2): pp. 651–657.

28. Oh, S.Y., et al., Crystalline structure analysis of cellulose treated with sodium hydroxide and carbon dioxide by means of X-ray diffraction and FTIR spectroscopy. *Carbohydrate Research* 2005. **340**(15): pp. 2376–2391.

29. Suwanruji, P., et al., Modification of pineapple leaf fiber surfaces with silane and isocyanate for reinforcing thermoplastic. *Journal of Thermoplastic Composite Materials* 2017. **30**(10): pp. 1344–1360.

30. Xie, Y., et al., Silane coupling agents used for natural fiber/polymer composites: A review. *Composites Part A: Applied Science and Manufacturing* 2010. **41**(7): pp. 806–819.

31. Olakanmi, E.O., et al., Mechanism of fiber/matrix bond and properties of wood polymer composites produced from alkaline-treated *Daniella oliveri* wood flour. *Polymer Composites* 2016. **37**(9): pp. 2657–2672.

32. Li, X., et al., Chemical treatments of natural fiber for use in natural fiber-reinforced composites: A review. *Journal of Polymers and the Environment* 2007. **15**(1): pp. 25–33.

33. Gonçalves, F.A., et al., The potential of unsaturated polyesters in biomedicine and tissue engineering: Synthesis, structure-properties relationships and additive manufacturing. *Progress in Polymer Science*, 2017. **68**: pp. 1–34.

34. Higgins, C., et al., Polyester resins, in Swen Malte John, Jeanne Duus Johansen, Thomas Rustemeyer, Peter Elsner, and Howard I. Maibach (Editors), *Kanerva's Occupational Dermatology*. Springer Cham, Switzerland, 2020: pp. 809–819.

35. Ye, H., et al., Polyester elastomers for soft tissue engineering. *Chemical Society Reviews*, 2018. **47**(12): pp. 4545–4580.

36. Scheirs, J. and T.E. Long, *Modern Polyesters: Chemistry and Technology of Polyesters and Copolyesters*. John Wiley & Sons, West Sussex, 2005.

37. Mohd Nurazzi, N., et al., A review: Fibers, polymer matrices and composites. *Pertanika Journal of Science & Technology*, 2017. **25**(4): pp. 1085–1102.

38. Vilas, J., et al., Unsaturated polyester resins cure: Kinetic, rheologic, and mechanical-dynamical analysis. I. Cure kinetics by DSC and TSR. *Journal of Applied Polymer Science*, 2001. **79**(3): pp. 447–457.

39. Malik, M., V. Choudhary, and I. Varma, Current status of unsaturated polyester resins. *Journal of Macromolecular Science, Part C: Polymer Reviews*, 2000. **40**(2–3): pp. 139–165.

40. LS, J.K.Y., *Preparation, Properties and Applications of Unsaturated Polyesters. Modern Polyesters: Chemistry and Technology of Polyesters and Copolyesters*. New York: Wiley, 2003: pp. 699–713.

41. Dholakiya, B., Unsaturated polyester resin for specialty applications. *Polyester*, 2012. **7**: pp. 167–202.

42. Johannes, K., *Reactive Polymers: Fundamentals and Applications*. William Andrew Publishing, Amsterdam, 2017.

43. Lim, H.C., Thermoplastic polyesters, in Abbas Fadhil Aljuboori (Editor), *Brydson's Plastics Materials*. Elsevier, Bhopal, 2017: pp. 527–543.

44. Jani, S.P., et al., Mechanical behaviour of coir and wood dust particulate reinforced hybrid polymer composites. *International Journal of Innovative Technology and Exploring Engineering* 2019. **8**: pp. 3205–3209.

45. Kholil, A., et al. Characteristics of wood powder, coconut fiber and green mussel shell composite for motorcycle centrifugal clutch pads, in *IOP Conference Series: Materials Science and Engineering*. 2021. IOP Publishing.

46. Ahmad, I. and T.M. Mei, Mechanical and morphological studies of rubber wood sawdust-filled UPR composite based on recycled PET. *Polymer-Plastics Technology and Engineering* 2009. **48**(12): pp. 1262–1268.

47. Nuñez, A.J., M.I. Aranguren, and L.A. Berglund, Toughening of wood particle composites—Effects of sisal fibers. *Journal of Applied Polymer Science* 2006. **101**(3): pp. 1982–1987.

48. Nagavally, R.R., Composite materials-history, types, fabrication techniques, advantages, and applications. *International Journal of Mechanical and Production Engineering* 2017. **5**(9): pp. 82–87.

49. Cheng, F., et al., Interfacial properties of glass fiber/unsaturated polyester resin/poplar wood composites prepared with the prepreg/press process. *Fibers and Polymers* 2015. **16**(4): pp. 911–917.

50. Dong, W., et al., Analysis of polyester/wood composite fiber needling mat sound absorption performance. *Polymer Composites*, 2018. **39**(11): pp. 3823–3830.

51. Shesan, O.J., et al., Improving the mechanical properties of natural fiber composites for structural and biomedical applications, in *Renewable and Sustainable Composites*. Intech Open, London, 2019.

52. Edoziuno, F.O., et al., Experimental study on tribological (dry sliding wear) behaviour of polyester matrix hybrid composite reinforced with particulate wood charcoal and periwinkle shell. *Journal of King Saud University-Engineering Sciences* 2021. **33**(5): pp. 318–331.

53. Toyeebah, O. and O. Jamiu, Investigation of flexural and microstructural properties of coir fiber and wood dust reinforced polyester composite. *International Journal of Progressive Research in Science and Engineering* 2021. **2**(6): pp. 94–100.

54. Barbosa, A.D.P., et al. Charpy toughness behavior of eucalyptus fiber reinforced polyester matrix composites, in Gerard G. Dumancas (Editor), *Materials Science Forum*. Trans Tech Publications Ltd., Birmingham, 2016: pp. 227–232.

55. Olodo, E., et al., Long term behavior of composite material polyester-wood reinforced glass fiber. *Journal of Applied Sciences, Engineering and Technology* 2013. **6**(2): pp. 196–201.

56. Leong, Y., et al., Comparison of the mechanical properties and interfacial interactions between talc, kaolin, and calcium carbonate filled polypropylene composites. *Journal of Applied Polymer Science* 2004. **91**(5): pp. 3315–3326.

57. Marcovich, N.E., et al., Resin–sisal and wood flour composites made from unsaturated polyester thermosets. *Composite Interfaces*, 2009. **16**(7–9): pp. 639–657.

58. Marcovich, N.E., M.M. Reboredo, and M.I. Aranguren, Mechanical properties of wood flour unsaturated polyester composites. *Journal of Applied Polymer Science* 1998. **70**(11): pp. 2121–2131.

59. Alomayri, T., et al., Effect of water absorption on the mechanical properties of cotton fabric- reinforced geopolymer composites. *Journal of Asian Ceramic Societies* 2014. **2**(3): pp. 223–230.

60. Dhakal, H.N., et al., Effect of water absorption on the mechanical properties of hemp fiber reinforced unsaturated polyester composites. *Composites Science and Technology* 2007. **67**(7–8): pp. 1674–1683.

61. Meenalochani, K. and B.V. Reddy, A review on water absorption behavior and its effect on mechanical properties of natural fiber reinforced composites. *International Journal of Innovative Research in Advanced Engineering* 2017. **4**: pp. 143–147.

62. Ansari, F., et al., Nanostructured biocomposites based on unsaturated polyester resin and a cellulose nanofiber network. *Composites Science and Technology* 2015. **117**: pp. 298–306.

63. Hossain, M., F. Hossain, and M.A. Islam, Effects of wood properties on the behaviors of wood particle reinforced polymer matrix composites. *Journal of Scientific Research* 2014. **6**(3): pp. 431–443.

64. Panthapulakkal, S. and M. Sain, Agro-residue reinforced high-density polyethylene composites: Fiber characterization and analysis of composite properties. *Composites Part A: Applied Science and Manufacturing* 2007. **38**(6): pp. 1445–1454.

65. Bao, L.-R. and A.F. Yee, Effect of temperature on moisture absorption in a bismaleimide resin and its carbon fiber composites. *Polymer* 2002. **43**(14): pp. 3987–3997.

66. Idrus, M., M.A. Mun'aim, and M.F.S. Othman. Physical and mechanical properties of waste sawdust polymer composite for marine application, in Dong Zhang (Editor), *Advanced Materials Research*. Trans Tech Publications Ltd., Newbury Park, CA, 2015: pp. 292–295.

67. Sahu, P. and M.K. Gupta, Water absorption behavior of cellulosic fibers polymer composites: a review on its effects and remedies. *Journal of Industrial Textiles* 2020: p. 1528083720974424.

68. Neto, J.S., et al., A review on the thermal characterisation of natural and hybrid fiber composites. *Polymers* 2021. **13**(24): p. 4425.

69. Joseph, S., M.S. Sreekala, and S. Thomas, Effect of chemical modifications on the thermal stability and degradation of banana fiber and banana fiber-reinforced phenol formaldehyde composites. *Journal of Applied Polymer Science* 2008. **110**(4): pp. 2305–2314.

70. Azwa, Z., et al., A review on the degradability of polymeric composites based on natural fibers. *Materials & Design* 2013. **47**: pp. 424–442.
71. Kalia, S., et al., Surface modification of plant fibers using environment friendly methods for their application in polymer composites, textile industry and antimicrobial activities: A review. *Journal of Environmental Chemical Engineering* 2013. **1**(3): pp. 97–112.
72. Chung, T.-J., et al., The improvement of mechanical properties, thermal stability, and water absorption resistance of an eco-friendly PLA/kenaf biocomposite using acetylation. *Applied Sciences* 2018. **8**(3): p. 376.
73. Biswas, B., et al., Thermal stability, swelling and degradation behaviour of natural fiber based hybrid polymer composites. *Cellulose* 2019. **26**(7): pp. 4445–4461.
74. Gupta, M.K., Investigations on jute fiber-reinforced polyester composites: Effect of alkali treatment and poly (lactic acid) coating. *Journal of Industrial Textiles* 2020. **49**(7): pp. 923–942.
75. Chin, S.C., et al., Thermal and mechanical properties of bamboo fiber reinforced composites. *Materials Today Communications* 2020. **23**: p. 100876.
76. Wang, W., et al., Mechanical and thermal behavior analysis of wood–polypropylene composites. *Textile Research Journal* 2021. **91**(3–4): pp. 347–357.
77. Al-Anie, T.A., K.T. Hassan, and A.R. Al-Hadithy, Preparation and study hardness and thermal conductivity (Tc) to polyester resin composite with (titanium dioxide, zinc oxide, acrylonitril, wood flour coconut). *Baghdad Science Journal* 2010. **7**(4): pp. 1400–1409.
78. Aranguren, M.I., et al., Composites made from lignocellulosics and thermoset polymers. *Molecular Crystals and Liquid Crystals Science and Technology. Section A. Molecular Crystals and Liquid Crystals* 2000. **353**(1): pp. 95–108.
79. Vázquez, G., et al., Effect of veneer side wettability on bonding quality of Eucalyptus globulus plywoods prepared using a tannin–phenol–formaldehyde adhesive. *Bioresource Technology* 2003. **87**(3): pp. 349–353.
80. Gao, Z.H., et al. Effects of initiator level on performances of kraft fiber reinforced unsaturated polyester composites, in Lucian A. Lucia and Martin A. Hubbe (Editors), *Advanced Materials Research*. Trans Tech Publications Ltd., Charlotte, NC, 2011.
81. Haghdan, S., T. Tannert, and G.D. Smith, Wettability and impact performance of wood veneer/polyester composites. *BioResources* 2015. **10**(3): pp. 5633–5654.
82. Cappelletto, E., et al., Wood surface protection with different alkoxysilanes: A hydrophobic barrier. *Cellulose* 2013. **20**(6): pp. 3131–3141.
83. Wan, J., et al., Experimental investigation on FRP-to-timber bonded interfaces. *Journal of Composites for Construction* 2014. **18**(3): p. A4013006.
84. Hayashi, T. and J. Takahashi, Influence of void content on the flexural fracture behaviour of carbon fiber reinforced polypropylene. *Journal of Composite Materials* 2017. **51**(29): pp. 4067–4078.
85. Haliq, R., A.F. Rayudi, and A. Suprayitno. The effect of filler composition and shape to sound capability insulation and modulus elasticity natural fiber Galam wood (*Melaleuca leucadendra*)-polyester, in *IOP Conference Series: Materials Science and Engineering*. IOP Publishing, 2019.
86. Bolin, C.A. and S. Smith, Life cycle assessment of ACQ-treated lumber with comparison to wood plastic composite decking. *Journal of Cleaner Production*, 2011. **19**(6–7): pp. 620–629.
87. Ramesh, M., L. Rajeshkumar, and D. Balaji, Influence of process parameters on the properties of additively manufactured fiber-reinforced polymer composite materials: A review. *Journal of Materials Engineering and Performance*, 2021. **30**(7): pp. 4792–4807.
88. Sun, G., et al., Laboratory and exterior decay of wood–plastic composite boards: Voids analysis and computed tomography. *Wood Material Science & Engineering*, 2017. **12**(5): pp. 263–278.

89. Korol, J., D. Burchart-Korol, and M. Pichlak, Expansion of environmental impact assessment for eco-efficiency evaluation of biocomposites for industrial application. *Journal of Cleaner Production*, 2016. **113**: pp. 144–152.

90. Stansbury, J.W. and M.J. Idacavage, 3D printing with polymers: Challenges among expanding options and opportunities. *Dental Materials*, 2016. **32**(1): pp. 54–64.

91. Rayna, T. and L. Striukova, From rapid prototyping to home fabrication: How 3D printing is changing business model innovation. *Technological Forecasting and Social Change*, 2016. **102**: pp. 214–224.

92. Berman, B., 3-D printing: The new industrial revolution. *Business Horizons*, 2012. **55**(2): pp. 155–162.

93. Postiglione, G., et al., Conductive 3D microstructures by direct 3D printing of polymer/ carbon nanotube nanocomposites via liquid deposition modeling. *Composites Part A: Applied Science and Manufacturing*, 2015. **76**: pp. 110–114.

94. Casavola, C., et al., Orthotropic mechanical properties of fused deposition modelling parts described by classical laminate theory. *Materials & Design*, 2016. **90**: pp. 453–458.

95. Ning, F., et al., Additive manufacturing of carbon fiber reinforced thermoplastic composites using fused deposition modeling. *Composites Part B: Engineering*, 2015. **80**: pp. 369–378.

96. Nuñez, P., et al., Dimensional and surface texture characterization in fused deposition modelling (FDM) with ABS plus. *Procedia Engineering*, 2015. **132**: pp. 856–863.

97. Wu, C.-S. and H.-T. Liao, Fabrication, characterization, and application of polyester/ wood flour composites. *Journal of Polymer Engineering*, 2017. **37**(7): pp. 689–698.

98. Kumar, M., A. Gupta, and I.S. Thakur, Carbon dioxide sequestration by chemolithotrophic oleaginous bacteria for production and optimization of polyhydroxyalkanoate. *Bioresource Technology*, 2016. **213**: pp. 249–256.

99. Laycock, B., et al., The chemomechanical properties of microbial polyhydroxyalkanoates. *Progress in Polymer Science*, 2013. **38**(3–4): pp. 536–583.

100. Laksono, A. D., et al., Flexural and fractography behavior of unsaturated polyester composite filled with bangkirai wood fiber. *Teknika: Jurnal Sains dan Teknologi*, 2020. **16**(1): pp. 12–17.

9 Polyester-Based Composites Reinforced with Rice Husk Fillers

Vishnupriya Subramaniyan and
Periyar Selvam Sellamuthu
SRM Institute of Science and Technology

Emmanuel Rotimi Sadiku
Tshwane University of Technology

CONTENTS

9.1   INTRODUCTION

Polyesters are polymers fabricated from a diol and dicarboxylic acid. Polyesters are very resilient thermoplastics and they have a wide variety of applications, based on how they have been synthesized and the resultant position of polymer chains (Kobayashi and Uyama, 2019). In the last few decades, various types of synthetic polymers have been coupled with numerous cellulose-based fillers in order to enhance the

DOI: 10.1201/9781003270980-9

mechanical characteristics of the resultant composites (Gutierrez Cisneros, Bloemen and Mignon, 2021). Synthetic or petroleum-based composites have been largely used in various fields such as medical equipment, plastic bags, consumer goods, construction areas and the automotive domain (Aℓ-Maadeed, Shabana and Khanam, 2014; Noorunnisa Khanam and AlMaadeed, 2014). The worldwide production of plastics is increasing yearly, owing to the increasing usage of plastics. Generally, polyester-based composites comprise polymer resins as the matrix and one or several fillers are incorporated into the resin in order to enhance the requirement of a specific objective. For example, composites for sports and aerospace applications need high thermal and mechanical characteristics (Rajak et al., 2019). Conventionally, synthetic fibres, such as glass, silicone and carbon, have been utilized as composite reinforcements. However, synthetic fillers have excellent characteristics but their processing cost is little high and they are toxic to environment and human health (Rajak, Wagh and Linul, 2021). Due to this fact, environmentally friendly materials, as fillers in polymer composite, are attained with great recognition in the industries. Natural fillers have many beneficial characteristics, such as low cost, lightweight, nontoxic effect and decrease of the machinery abrasion. The usage of natural fillers for the strengthening of composites has gained considerable attention from both the industry and academia. There are many categories of natural fillers, such as oats, wood, barely, wood, hemp, rice, wheat, barley and flax (Sethulekshmi, Saritha and Joseph, 2022).

Rice (*Oryza sativa* L.) is one of the main crops; it surrounds ~1% of the earth surface and it is the primary food source for billions of the world's population (FAOSTAT, n.d.). According to the United States Department of Agriculture, 500 million tonnes of husked or milled rice were produced globally in the year 2020–2021 (USDA ERS - Agricultural Trade, n.d.). Statistical studies showed that Asia produces ~90% of rice that is required for the global populations as illustrated in Figure 9.1. During the rice processing operation, rice husk is the by-product and it is segregated from rice grains during the milling process (Muthayya et al., 2014). Rice milling is a major industry in countries such as China, Pakistan, Indonesia, Malaysia, Bangladesh and India. Often, rice husks are eliminated by the incineration process and the incineration of rice husk produces toxic gases and ash fumes, which cause the severe air pollution (Thomas, 2018). The incorporation of rice husk as a filler in polyester aids in overcoming these disadvantages and acts as natural filler in the composite produced. In this chapter, we detailed the use of rice husk filler in polyester composites.

9.2 NATURAL FILLER AS POLYESTER COMPOSITES

Polymers or polyesters are generally filled with blended solid materials, e.g., glass or mineral. The filled polymers are not commonly accepted as composite materials, despite the fact that they comprise more than one component (Kroczek et al., 2022). Polymeric matrix composite materials are generally reinforced with long or short fibres. Natural fibre-reinforced polyester composites show a means to partially improve the negative environmental effects by incorporating biodegradable filler into the polyester-based composites (Nurazzi et al., 2021). The most common bio-wastes are hemp, rice and husk, banana fibres, wood flour, kenaf, flax, hemp, coir,

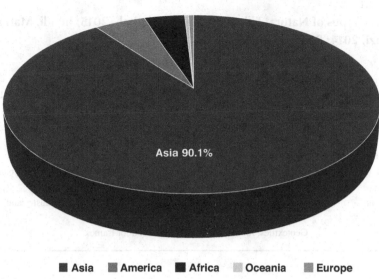

America 5.1%   Africa 4.2%   Oceania 0.1%   Europe 0.5%

Asia 90.1%

■ Asia   ■ America   ■ Africa   Oceania   ■ Europe

FIGURE 9.1   Worldwide rice production chart represents that compared to other continents, Asia is the largest producer of rice (copyrights reserved, 5312650211007).

jute wood flour, etc., often used as replacements for synthetic filler, namely steel, carbon, calcium hydroxyapatite and other petroleum-based filler (Andrew and Dhakal, 2022). Certainly, the utilization of natural filler, which is a type of organic filler, is growing, and it has many advantages over synthetic fillers (Aℓ-Oqla and Salit, 2017). Cotton fibres were used as reinforced material in radar aircraft. Thereafter, several cellulosic products wastes have been utilized as natural filler in polyester-based products (Soutis, 2005). Because of their good mechanical characteristics, low density, significant processing benefits and low cost, the growth of natural fillers-reinforced polyester composites has steadily been recorded (Rinawa et al., 2021). Natural fibres as fillers in composites are used mainly in the automotive, electromagnetic, aerospace and biomedical applications.

These fillers could confer some advantages to the materials in respect of density control, fire-proofing, better optical, electrical, mechanical, magnetic and thermal properties (Aℓ-Oqla et al., 2015). Scientists around the world have researched on the merits of natural fibres reinforced with polyester composites, and some of the broadly used natural fibres as fillers in polyester composites are listed in Table 9.1 (Esa and Abdul Rahim, 2013; Korotkova et al., 2016).

RH is one of the most valuable agricultural by-products or wastes and it has the potential to reinforce polyester matrices for the growth of added-value products. The major components of RH are tabulated in Table 9.2 and the pictorial representation of RH is given in Figure 9.2. RH is the inedible guarding layer of the grain and it is always eliminated or separated from the grain (Chan, 2016). In addition to this, the environmental benefits of integrating RH with polyester or polymer materials are

TABLE 9.1

Different Types of Natural Fillers (Mohammed et al., 2015; Imadi, Mahmood and Kazi, 2014; Reddy et al., 2020)

Fibre	Applications
Coconut coir	Household and automobile applications
Hemp fibre	Construction, packaging, electrical, paper, pipe manufacturing and textiles industries
Kenaf fibre	Mobile cases, bags, insulation material, packing material, mobile cases
Bamboo fibre	Attires, sanitary napkin, masks and bandages
Bagasse fibre	Door and window panels, fencing, decking and railing systems
Pineapple leaf fibre	Packaging trays, seat backs, automotive industries, dashboards, door panels
Flax fibre	Snowboard, deckling, fork, laptop cases, fencing
Ramie fibre	Sewing thread, fishing nets, packaging materials, clothing industry and household furniture
Jute fibre	Geotextiles, roofing sheets, clip boards, door frames

TABLE 9.2
Chemical Composition of RH

Composition of RH	Percentage
Lignin	26–31
Cellulose	25–35
Moisture content	5–10
Hemicellulose	18–21
Silica	15–17
Water	7.86
Mineral ash	13.87
Extractives	2.33

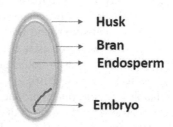

FIGURE 9.2 Structure of rice grain: rice husk is the outermost layer of rice, embryo and endosperm constitute the innermost layer and bran is the middle layer (copyrights reserved, 5312650211007).

increasing due to the unique and unmatched composition with other types of natural fillers, high availability and low cost (Aℓ-Oqla et al., 2015). Even though RH has numerous advantages, the major demerit in synthesizing RH composites is their low compatibility to the hydrophobic matrices due to their hydrophilic nature (Majeed et al., 2017). The low compatibility of RH hinders its interfacial interaction between the polymer matrices and RH filler (Aℓ-Oqla et al., 2015). A study reported that RH-reinforced polypropylene (PP) composite showed low tensile strength due to the low compatibility between the matrix material and the RH filler. The study suggested that the use of a coupling agent would improve the compatibility within the matrix material and the filler (Arjmandi, Hassan and Zakaria, 2017).

The reinforcing potential of RH filler mostly depends on its interfacial interaction with the matrix material. Improved interfacial interactions produce greater mechanical properties (Malkapuram, Kumar and Negi, 2009). However, the intrinsic variations between hydrophilic, the RH and hydrophobic, the polymer matrices hinder the interfacial interaction and compatibility between the RH filler and the polymer matrices (Yap et al., 2020). In order to improve the compatibility, physical and chemical pre-treatment methods were employed by researchers to reduce RH surface, with the motivation to activate new reactive site or region for interfacial interactions.

9.3 PRE-TREATMENT OF RICE HUSK

9.3.1 PHYSICAL PRE-TREATMENT

Physical pre-treatment methods are often used to improve compatibility between the hydrophobic matrices and the hydrophilic RH filler (Ciannamea, Stefani and Ruseckaite, 2010). Steam treatment is a physical treatment in which high-pressure steam is applied in a steam explosion unit. The unit is quickly depressurized, which leads to steam explosion comprising the application of high-pressure steam in a steam explosion unit (Ndazi et al., 2007). The unit is rapidly depressurized resulting in steam-exploded fibre. This process separates the cellulose (structural element) from lignin and hemicellulose (non-structural elements), as well as the reactive site of the natural fibre becomes apparent. An improved availability of filler's reactive region will enhance the interfacial interaction, which leads to better mechanical characteristics (Hakeem et al., 2015).

9.3.2 CHEMICAL TREATMENT

Like physical treatment, chemical methods also help to enhance the interfacial interaction between the components, making up the composites. The additions of coupling agent or compatibilizer, alkaline, acetylation and silane treatment are the most practised chemical methods. In comparison to all methods, the usage of compatibilizer has increased the compatibility of RH and the matrix. However, the alkaline method of RH is the most preferable than other treatment methods, due to its low cost and high effectiveness.

Soaking RH in a NaOH solution helps to remove the non-structural parts of the RH. The removal of the non-structural parts improves the interfacial bonding and increases the mechanical properties of the resulting composites (Ndazi, Nyahumwa and Tesha, 2008).

Acetic anhydride treatment of RH improves the hydrophobicity by reducing the hydroxyl (OH) groups. During this treatment, acetic anhydride cross-links with the RH hydroxyl groups. This results in the OH group being substituted by the acetyl group. This is similar to acetylation process, which lessens the hydrophilic nature of RH in order to elevate its affinity with polyethylene (PE) (Fávaro et al., 2010; Hill, Khalil and Hale, 1998).

Silane treatment is another effective method to improve the RH compatibility with polymer matrices. Silane has two bifunctional structures and during this treatment the silane's one reactive group combines with the OH group of the natural fibre; meanwhile, the second reactive group is polymerized with the matrix component (Xie et al., 2010). This interaction causes the formation of covalent bond between the natural fibre and polymer matrices. Thus, this interaction improves the characteristics of the filler in polymer matrices. Another research was reported on rice husk ash (RHA)-reinforced high-density polyethylene (HDPE). The RHA-HDPE composites showed ~18% higher tensile and higher thermostability than virgin the HDPE did. The study proved that RHA is one of the best reinforcing materials (Abdelmouleh et al., 2007). Zhao et al. (2009) analyzed the flammability and thermal stability of HDPE-RH in e-composites. The study was evident that the flammability of the composites increased with increase in RH content and that the incorporation of RH had retarded thermal-oxidation process.

9.4   RH-REINFORCED POLYMER COMPOSITES

A polymer is a material that comprises large molecules, consisting of numerous repeating subunits, and it can be produced through the polymerization of monomer. Both the natural and synthetic polymers play vital roles in the everyday human life. Herein, we discussed some of the most important polymers that are often reinforced with rice husk.

PE is a polymer also known as polythene; the IUPAC name is poly(methylene or polyethene). It is a commonly used plastic for many and everyday purposes. PE is classified into HDPE and low-density polyethylene (LDPE) and they vary in their molecular weights, thermal properties and crystallization nature. When compared to other polymers, polyethene is the most commonly used polymer (Ronca, 2017).

HDPE is a type of PE that is synthesized from the ethylene monomer. HDPE is a thermoplastic, known for its high strength-to-density ratio. HPDE has very broad range of applications, ranging from storage container to pipes. Relatively, the melting point of HPDE is high when compared to other plastics. The chemical formula of HPDE is $(C_2H_4)_n$ (Kim et al., 1996). The density of HDPE varies between 930 and 970 kg/m^3 (Rodrigue, Kavianiboroujeni and Cloutier, 2017). Even though HDPE is non-biodegradable, it has been used in recycled products for automobile parts, etc., (Subramanian, 2000). For the production of HDPE plastics, many synthetic fillers are being utilized. RH filler is used in many studies to replace the synthetic fillers (Medina and Dzalto, 2018). An investigation reported that RH-incorporated HDPE composites resulted in adequate rigidity and strength by integrating various coupling agents, such as ethylene-(acrylic ester)-(glycidyl methacrylate) terpolymers and ethylene-(acrylic ester)-(maleic anhydride) terpolymers. The results showed that the incorporation of coupling agent enhanced the strength of the composites (Panthapulakkal, Sain and Law, 2005). Rahman et al. studied the RH-filled HPDE composites for injection moulding purposes. Ultra-Plast TP01 and UltraPlast TP10 were used as lubricant and coupling agent, respectively. The melting flow rate of the composites was reduced with increased composition of RH (Rahman et al., 2010). Similarly, low filler size decreased the impact strength of the composite and high filler content and also reduced the impact strength. Gejo et al., 2010 reported that 30% of RH was found to have increased the rheological characteristics and the impact strength of the bio-composites produced. RH was used as filler in recycled high-density polyethylene (rHPDE) and in addition to the RH, wood dust was also used as secondary filler, antioxidant and fire retardants as mineral filler. The overall performance of the RH- and sawdust-reinforced composites was increased. Besides, the antioxidant and fire retardants enhanced the ageing and durability properties, respectively. Similar to the above-mentioned study, RH filler-treated with maleic acid PE-reinforced rHDPE enhanced the interfacial bonding (Tong et al., 2014). In another study on rice husk, an 8% alkaline (NaOH) was employed to treat rice husk flour and the result was an improvement in the interaction between RH and matrix. The treatment also exhibited positive impact on the mechanical characteristics (Bisht and Gope, 2018). Further studies also investigated RH-reinforced rHDPE in which the RH surface was modified through alkali, UV/O_3 (ultraviolet) and acid treatment. The result of the study indicated that alkali treatment resulted in higher tensile strength and other mechanical properties than the other treatments employed (Hamid, Ab Ghani and Ahmad, 2012).

Like HDPE, LDPE is a thermoplastic, which is synthesized from the monomer ethylene (Goswami and Mangaraj, 2011). LDPE is the first grade of PE and it is produced by the high-pressure process (Butler, 2010). It is broadly utilized for the manufacturing of numerous items, such as bottle, plastic bags, containers, laboratory equipment and computer parts. LDPE has higher ductility but low strength, stiffness and hardness when compared to HDPE with lower hardness (Achilias et al., 2007). When compared to HDPE, only few investigations on LDPE-RH composite have been reported. A report stated that 30% of RH-loaded LDPE composites showed an increase in tensile strength and other mechanical properties of the composites are marginally enhanced (U. et al., 2013). Santiagoo et al., 2015 developed LDPE nano-composite based modified atmospheric packaging films which consist of rice husk (RH), and montmorillonite (MMT).The outcome of the study found that an increased concentration of RH in the hybrid filler enhanced the tensile modulus, crystallinity and the melting temperature, whereas decreases in the tear and tensile strength were recorded. Likewise, various concentrations of RH filler and recycled LDPE and virgin PE amalgamation was carried out. The tensile modulus, Brinell hardness and flexural strength increased with RH filler loadings, and in contrast to this observation, increased filler loading reduced the impact strength. Along with this, the tensile strength improved with a 10% filler loading (Bertin and Robin, 2002). MMT nanoclay and RH-filled LDPE were studied. A study conducted by Hwang, Kwon and Lee, 2018 stated that maleic anhydride-modified PE was exploited as a compatibilizer. The study resulted in an improved interfacial adhesion between the matrix and the filler, which produced better barrier and tensile characteristics of the resulting composites.

Medium-density polyethylene (MDPE) is also another type of PE, whose density is lower than the LDPE (density range: 0.926–0.940 g/cm^3). MDPE also has better drop-and-shock resistance characteristics (Crawford and Throne, 2002). Additionally, MDPE has better stress-cracking resistance and with low notch sensitivity when compared to HDPE (Ayyer, Hiltner and Baer, 2007). MDPE is commonly used in packaging films, sacks, screw closure, gas pipes, etc. (AℓMaadeed, Ouederni and Noorunnisa Khanam, 2013). A research aimed at analyzing the effect of nanoclay-loaded in MDPE/RH-based composites was carried out. RH was treated with NaOH in order to improve the characteristics of RH; in addition to this treatment, maleic anhydride-grafted PE was added to the composite in order to improve the interfacial bonding between RH and MDPE matrix. The results exhibited the fact that the composite recorded better tensile strength, flexural strength and other mechanical properties when compared to the untreated MDPE (Thungphotrakul, Somboonwanna and Prapainainar, 2020).

Linear low-density polyethylene (LLDPE) is a linear PE, with substantial range of short branches. It is generally produced via the copolymerization of ethylene monomers with longer chain olefins. LLDPE is structurally different from LDPE (Ranjan, n.d.). Like other polyethene, LLDPE has a wide variety of applications, such as plastic bags, pouches, lids, toys and pipes, just as the usage of LLDPE is increasing (Satapathy et al., 2018). The rice husk powder filler-loaded LLDPE/recycled acrylonitrile butadiene rubber composites were analyzed. Santiagoo et al., 2015 reported that epoxidized natural rubber-50 (ENR-50) was used as the compatibilizer. The structural outcome shows that ENR-50 improved the interactions between LLDPE and RHP (rice husk particle) filler. In addition, the incorporation of RHP filler improved the mechanical characteristics of the resulting composites. In another study, rice husk-loaded LLDPE/styrene butadiene rubber (SBR) composite blends with maleic anhydride as a compatibilizer was prepared. Herewith, virgin SBR-LLDPE was also compared with RH-loaded SBR-LLDPE composite. Similar to other studies, the RH-filled LLDPE/SBR exhibited increased mechanical properties than the virgin SBR-LLDPE composites (Zurina, Ismail and Bakar, 2004; Wang et al., 2014). The NaOH treated RH reinforced with LLDPE had a positive influence on the composite constituents' compatibility, tear strength, composite and the Young's modulus (Hwang, Kwon and Lee, 2018). Further examination also indicated the fact that the surface-modified RH had better interaction with the LLDPE matrix and displayed higher rigidity and compressive strength. In the same way as the previous study, the surface-modified RH filler in the LLDPE recorded elevated stiffness and strength characteristics (Abdul Wahab et al., 2016).

PP, also known as polypropene, is synthesized from the monomer, propylene, and PP has been used in a wide variety of applications. A research was carried out on maleic anhydride-altered PP reinforced with RH. The result showed that the composite recorded superior flexural and tensile properties (Raghu et al., 2018). Many investigations have shown that RH-reinforced PP has improved overall characteristics.

Polystyrene (PS) is a synthetic aromatic monomer, also known as styrene. The Young's modulus of RH/waste PS composite was found to have increased with increasing RH content, up to ~40 wt.% when compared to pure PS. Like PE, it is one of the most commonly used plastics. PS is naturally transparent in nature and it can be coloured by using colourant. Generally, it is used in food packaging, trays, bottles, etc. RH contents of 30% and 40%-reinforced PS have greater moisture diffusivity and better Young's modulus, respectively, than the unfilled PS (Farhan Zafar and Siddiqui, 2018).

Polyurethane (PU or PUR) is composed of organic groups that are combined by carbamate. PU is synthesized from a broad range of monomers (Kemona and Piotrowska, 2020). PUs are used in varnishes, coating, fibres, rigid and flexible foams, etc. (Somarathna et al., 2018). An investigation showed that PU foams, containing modified tung oil and RH filler have higher thermal property than the unfilled PU (Ribeiro da Silva et al., 2013).

Poly vinyl chloride (PVC) is the third most widely used and largest polymer in the world. PVC is used in doors, panels, construction pipes, plastic bottles, non-food packaging, etc. Maleic anhydride- and aminosilane-treated rice husk was used to reinforce PVC. The RH positively influenced the tribological and mechanical properties of the PVC composites and especially, the tensile properties of the resulting composites (Chand, Sharma and Fahim, 2010; Ramle, Romli and Abidin, 2013).

9.5 RH-REINFORCED POLYESTER MATRICES

9.5.1 POLYESTER

Polyester is a group of polymers that consists of ester functional groups in each of the repeat unit of its main chain. The naturally obtained polyesters are biodegradable, but the synthetic polyesters are not biodegradable. Synthetic polyesters are widely used in the clothing industry (Campbell, 2001). Synthetic polyester has a high environmental concern and it is more water-resistant than the natural polyesters. Polyester synthesis is commonly obtained *via* the polycondensation reaction. Azeotrope esterification, alcoholic transesterification, melt esterification and acylation are the most commonly used esterification methods for the production of polyesters (Otera, 1993). The industrial applications of polyester are illustrated in Figure 9.3. India, Indonesia, Korea, United States, Japan and Taiwan are the major polyester-producing countries. When compared to the Western countries, Asian countries are the most dominant in polyester production, as represented in Figure 9.4. Different types of polyester are used widely in various applications. Many investigations have proved that the incorporation of RH improved the characteristics of polyester matrices. The RH-reinforced polyester matrices is discussed in detail below.

Delamination is one of the major defects of machining composites and it is a main form of breakdown in drilled composite, which causes less bearing strength, surface finish and structural integrity of machines (Babu et al., 2016). In order to reduce the delamination process, nature fibre, e.g., RH fibre, was incorporated into polyester matrices of machines. Drilling process was carried out at three different speeds, geometries and feed rates. The results obtained from the work displayed that RH-loaded polyester matrices have less delamination value than the glass fibre-loaded fibre matrices. The mechanical and physical properties of RHA-filled polyester resin composites were analyzed. From the experiment, it was found that the bulk density and flexural strength were reduced with increase in the RHA content; however, the

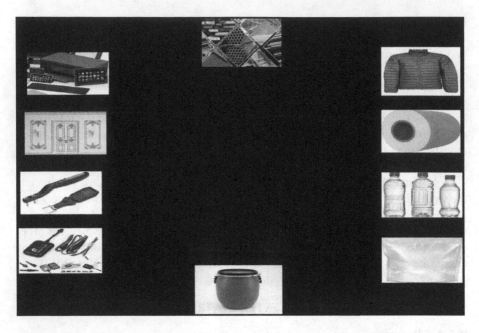

FIGURE 9.3 General industrial applications of polyester, including construction material, electronic devices, clothing material, fabrics, PET bottles, packaging material, storage containers, automobile parts, biomedical devices and furniture.

FIGURE 9.4 Polyester-dominant countries across the globe.

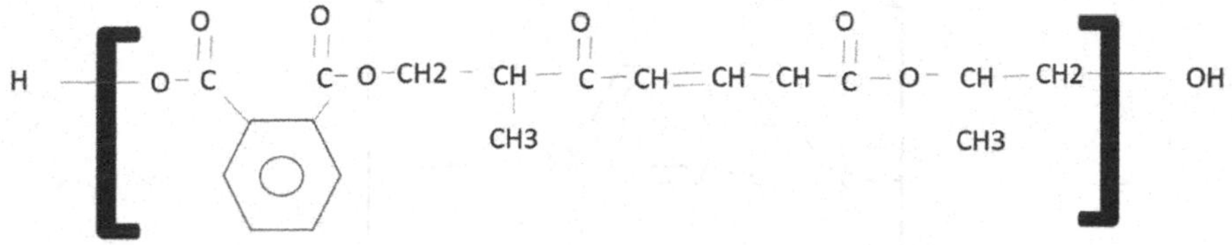

FIGURE 9.5 The chemical structure of unsaturated polyester resins.

water absorption rate was increased with increase in the RH addition. However, e-modulus of the composites was initially increased, but it suddenly decreased after the addition of 10% and 15% of RHA. The hardness of the composites also decreased with increase in RHA content. In contrast to this, a 5% of RHA increased the overall mechanical properties of composites (Barreno-Avila, Moya-Moya and Pérez-Salinas, 2022; Azuan, Juraidi and Muhamad, 2012; Ahmad et al., 2005).

9.5.2 UNSATURATED POLYESTERS RESINS

Unsaturated polyesters resins (UPRs) are also called polyester resins or synthetic resin produced by polyhydric alcohol and dibasic organic acids. The chemical structure of UPR is given in Figure 9.5. UPRs are thermosetting resins and used in fibreglass, wall panels, printer tonners, bulk and sheet moulding compounds. An UPR, based on vinyl silane reinforced with RHA, was prepared and characterized. The RHA-reinforced composites have better mechanical and thermal properties than their unreinforced counterparts (Kanimozhi et al., 2014; Islam, 2016). RHA (5% to 25%)-loaded unsaturated polyesters were evaluated; in this study methyl ethyl ketone peroxide was utilized as the hardener. It was observed that 10% of RH inclusion in a UPR matrix improved flexural and hardness properties. Besides, a 10% RH decreased the tensile strength; in contrast to this, the impact strength improved with a 5% RH. RH was reinforced into the unsaturated polyester matrices. The study found out that the flexural strength, flexural modulus, tensile strength and tensile modulus were increased with increase in RH fibres (Islam et al., 2015; Surata et al., 2014).

9.5.3 POLYETHYLENE TEREPHTHALATE

Polyethylene terephthalate, abbreviated as PET, PETP and PETE, is the most customary resin of the polyester group, and this is used in thermoforming, packaging, storage container, etc. PET ($C_{10}H_8O_4$) comprises of units of the monomer, ethylene terephthalate (De Vos et al., 2021). PET can be synthesized from ethylene glycol or monoethylene glycol and dimethyl terephthalate (Fink, 2013). Based on the processing technique, PET could be available in semi-crystalline and amorphous forms (Arnoult, Dargent and Mano, 2007). The chemical structure of PET is given in Figure 9.6.

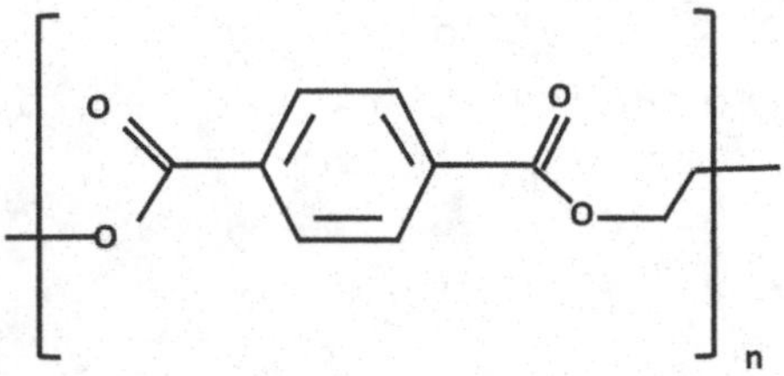

FIGURE 9.6 The chemical structure of polyethylene terephthalate (Copyrights reserved, 5312650765510).

High amount of RH was loaded in biocomposites of recycled polyethylene terephthalate (rPET) matrices. Ethylene-glycidyl methacrylate was used as a copolymer in order to enhance the compatibility of the immiscible rHDPE/rPET blend. Maleic anhydride polyethylene was utilized as a coupling agent in order to improve the interaction of matrix and fibre. The effect of natural fibre loadings on rHDPE/rPET blends was examined. The biocomposites loaded with RH filler have greater tensile strength (58%–172%), storage modulus and flexural modulus (80%–305%), when compared with the matrices without RH filler. These investigations concluded that RH filler can perform efficiently with rPET and rHDPE for the production of biocomposite products (Chen et al., 2015). The RH was combined with polyester resin, produced from a glycolyzed product of PET. This polyester reinforced with RH increased the hardness and tensile modulus (Ahmad et al., 2005). According to the study conducted by Chen, Ahmad and Gan, 2016, recycled thermoplastic blend (rPET/rHDPE) was reinforced with RH via the moulding and extrusion processes. The composites recorded better improvements in storage, loss modulus, storage and thermal stability. The composites submerged in tap water, distilled water and seawater revealed lowest absorption of water. The thickness dimension of the composite specimens exhibited the highest swelling values, followed by the width and then, the length dimensions. The elastic modulus and tensile strength showed higher values than without RH-loaded composites.

This study focused on the thermal, morphological and rheological characteristics of PET and polyamide-6 composites. de Matos Costa et al., 2021, stated that RHA was used as filler. This study concluded that 20% RHA filler improved the rate of degradation of polyester matrices. In addition to this, during processing, the thermal stability of the composites and particle agglomeration size were increased in the RHA filler-incorporated matrices. The 70% RH-reinforced rPET and rHDPE composites recorded increased tensile strength, elongation and flexural characteristics (Chen, Ahmad and Gan, 2018). Recycled PET loaded with RH filler was investigated. The study reported reduced tensile strength, impact strength and elongation-at-break, but with increased hardness and tensile modulus (Ahmad et al., 2005).

9.5.4 POLY-3-HYDROXYBUTYRATE

Polyhydroxybutyrate (PHB) is also known as polyhydroxyalkanoate (PHA) and belongs to the polyesters family. PHB is a biodegradable plastic and it is mainly

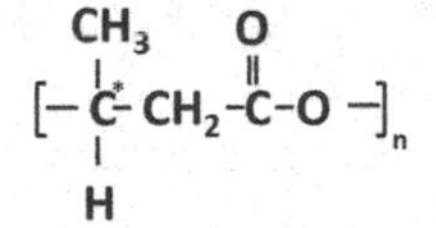

$$[-\overset{\overset{\textstyle CH_3}{|}}{\underset{\underset{\textstyle H}{|}}{\overset{*}{C}}}-CH_2-\overset{\overset{\textstyle O}{\|}}{C}-O-]_n$$

FIGURE 9.7 The chemical structure of polyhydroxybutyrate (Copyrights reserved, 5312661236994).

synthesized by *Methylobacterium, Bacillus megaterium* and *Cupriavidus necator* bacteria (McAdam et al., 2020). The structure of PHB is given in Figure 9.7. Due to its biodegradable nature, PHB is used in many industrial applications.

Poly-3-hydroxybutyrate (PHB), loaded with 10% RH flour with dicumyl peroxide (DCP), used as compatibilizer, was studied. Simultaneously, poly(3-hydroxybutyrate-*co*-3-hydroxyvalerate) (PHBV) synthesized by mixture of bacterial cultures obtained from fruit pulp waste was reinforced into a green composite in the range between 5% and 50%. The results revealed that the incorporation of 20% of PHBV produced a green composite that has a low crystallinity, medium barrier function to flavour, high thermal stability, transparency and mechanical ductility. The PHB/PHBV/RHF green composite films could be a good alternative in food packaging industry. The main aim of this investigation was to assess the effect of RH undergoing hot water treatment in PHB matrix. The mechanical, chemical, morphological and physical properties were characterized. The RH-reinforced composites showed a better performance than the non-reinforced matrix (Melendez-Rodriguez et al., 2019; Moura et al., 2018). Poly(3-hydroxybutyrate-*co*-4-hydroxybutyrate) (P34HB) is a fourth-generation PHA family compound. Wu et al., 2018 stated that RHA-reinforced P34HB composites were examined, and melamine phosphate-modified lignin was used as additive in the RHA-P34HB composites. RHA-P34HB displayed improved tensile tress, storage modulus and elongation-at-break. Besides, RH is often used as one of the sources to enhance the production of PHB in many studies.

9.5.5 POLYLACTIC ACID

Polylactic acid (PLA) is a biodegradable polyester and is also called polylactide (PLA). PLA is produced from the condensation of lactic acid. PLA is one of the most widely used bioplastics throughout the world. It has a wide range of applications and majorly used in 3D printing of plastic filament material (Nofar et al., 2019). The structure of PLA is illustrated in Figure 9.8.

Three-dimensional printing filaments were made from biodegradable PLA composited filled with renewable RH. PLA was grafted with acrylic acid (PLA-*g*-AA) and a coupling agent-treated RH (TRH) was incorporated into the PLA-RH biocomposites in order to improve the characteristics. TRH was uniformly spread in the PLA-*g*-AA matrix. The PLA-*g*-AA-TRH improved the tensile and other characteristics compared to the PLA-RH composites. Therefore, PLA-*g*-AA-TRH biocomposites could be utilized as biodegradable filaments in 3D printing (Wu and Tsou, 2019). The other study aimed to develop biocomposites, based on the polybutylene succinate (PBS)/PLA matrix with the incorporation of RH, was synthesized. RH

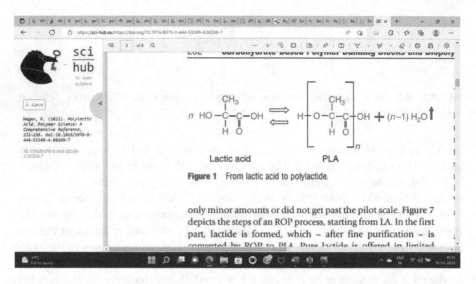

FIGURE 9.8   The chemical structure of polylactic acid (Copyrights reserved, 5312660197616).

underwent chemical modifications in order to improve the compatibility with the PLA/PBS matrix. RH increased impact strength and elongation-at-break. The study revealed the fact that RH is one of the effective fillers needed to improve properties of PLA/PBA composites (Sirichalarmkul and Kaewpirom, 2021). PLA grafted with maleic anhydride-grafted (PLA-g-MAH) was blended with RH. Methylenediphenyl diisocyanate was utilized as the compatibilizing agent in order to increase the compatibility between the RH filler and the PLA-g-MAH matrix. The results showed that PLA-g-MAH/RH composites positively influenced the thermal characteristics and tensile strength (Tsou et al., 2014). The RH and cellulose fibres in PLA matrix enhanced the composite stiffness, flexural modulus and crystallization kinetics (Běhálek et al., 2020).

9.5.6   POLY(BUTYLENE ADIPATE-CO-TEREPHTHALATE)
AND POLYBUTYLENE SUCCINATE

Poly(butylene adipate-co-terephthalate) (PBAT) is otherwise known as polybutyrate or polybutyrate-adipate-terephthalate; it is a biodegradable, semi aromatic and thermoplastic copolyester (Azevedo et al., 2022). PBAT has many appealing characteristics, similar to PE. PBAT, PBS or polytetramethylene succinate is an aliphatic polyester and biodegradable in nature. PBS has similar characteristics to PP. The scope of PBS industrial application is still growing, and it is envisaged that in future, it will be a suitable alternative candidate to non-degradable plastics (Siracusa et al., 2015). The structures of PBAT and PBS are presented in Figure 9.9.

This work presents the fabrication of RH in polyester matrices of PBAT and PBS in order to obtain better mechanical characteristics. The main challenge in this work is the incorporation of the RH, considering its hydrophilic nature, into the PBS/

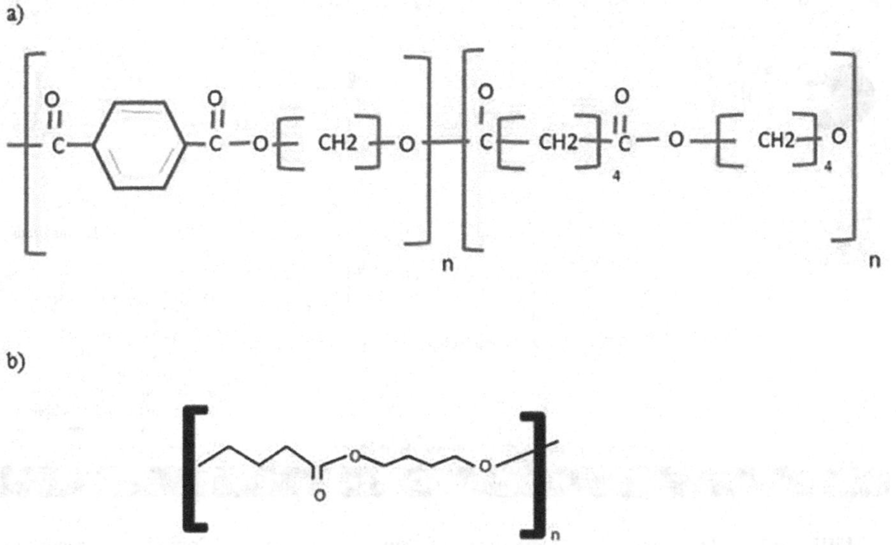

a)

b)

FIGURE 9.9 The chemical structure of (a) poly(butylene adipate-*co*-terephthalate) and (b) polybutylene succinate.

PBAT matrices. This issue was addressed by adding calcium stearate and glycerol as plasticizers. The maleic anhydride and DCP components were utilized to elevate the miscibility between hydrophilic the RH filler and the hydrophobic PBS/PBAT matrices. The PBAT/PBS/TPRH (Thermo Plastic Rice Husk) composites possessed desirable elongation-at-break, tensile strength and modulus of 12.99%, 14.27 MPa and 200.43 MPa, respectively. This composite was found suitable for the production of straw, lunchbox and tray. In addition, this composite exemplified good biodegradability (Yap et al., 2020).

9.5.7 POLYCAPROLACTONE

Polycaprolactone (PCL) is a polyester with biodegradable nature, and it has been approved by the Food and Drug Administration for application in drug delivery and biomedical devices. PCL is also used for filament fabrications of 3D printer (Peña et al., 2016). The structure of PCL is illustrated in Figure 9.10.

PCL/RHP composites were prepared, specifically, for increase in the performance of electronic components. The study showed that RHP (%) helps to improve the hardness of composites. In contrast, elongation-at-break, tensile, hardness and flexural strength were affected. However, increase in the RHP particle size decreased the impact strength and increased the tensile strength. In addition, the efficiency of the shielding effectiveness increases with increase in the RHP filler and the dielectric characteristics of the composites increased with decreasing RH particle size. The study concluded that the PCL/RHP composites are more appropriate for a small shielding application in telecommunication and military devices (Yakubu, Abbas

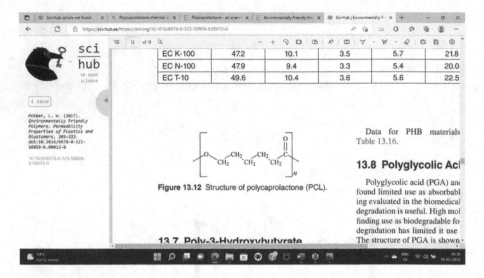

EC K-100	47.2	10.1	3.5	5.7	21.8
EC N-100	47.9	9.4	3.3	5.4	20.0
EC T-10	49.6	10.4	3.6	5.6	22.5

Figure 13.12 Structure of polycaprolactone (PCL).

Data for PHB materials
Table 13.16.

13.8 Polyglycolic Aci

Polyglycolic acid (PGA) and
found limited use as absorbabl
ing evaluated in the biomedical
degradation is useful. High mol
finding use as biodegradable fo
degradation has limited it use
The structure of PGA is shown

13.7 Poly-3-Hydroxybutyrate

FIGURE 9.10 The chemical structure of polycaprolactone (Copyrights reserved, 5312660624885).

and Abdullahi, 2020). Novel e-composites, synthesized based on biodegradable PCL and RH, have been prepared. The findings of the study concluded that RH enhanced the biodegradability of e-composites (Zhao et al., 2008).

9.6   CONCLUSION

Rice is one of the major food sources throughout the world. As a result, millions of tons of RH are produced annually, during the milling process of rice. In recent years, RH-reinforced polyester composites have gained attention due to its numerous advantageous effects. RH-reinforced composites have been used in many industrial applications, such as the construction, automobile, biomedical, aircraft and textile industries. There is a wide range of investigations that have explored the efficiency of RH in polymer matrices. However, only few studies focused on the RH-filled polyester matrices. Besides, these RH-filled polyester composites are destined to discover many applications in the near future. The extent of RH–polyester interaction determines the properties of final composite; therefore, most of the studies have emphasized the use of compatibilizer, coupling agent and grafter polyester. The chemical and physical treatments are the primary choices, required to enhance the interfacial interaction between RH and the polyester matrix. Along with this, polyester blend could be used to improve the characteristics of RH composites.

In addition to this, RH mixed with other natural fibres, such as hemp and coconut coir as hybrid fillers, silica and cellulose, obtained from rice husk, have also been investigated by many researchers in order to improve the properties of the polyester matrices. This chapter has briefly discussed the RH-reinforced polyester matrices. Hence, it is envisaged that a better understanding on the use of RH-reinforced for different polyester matrices would provide novel insights into many industrial applications.

ACKNOWLEDGEMENT

The authors Emmanuel Rotimi Sadiku and Periyar selvam Sellamuthu acknowledge the financial support from the NIHSS for Digital APP and BRICS Mobility Teaching and Research Projects, the Joint International Collaborations Research Projects (Project Ref: JNI21/1010).

REFERENCES

Abdelmouleh, M. et al. (2007) 'Short natural-fibre reinforced polyethylene and natural rubber composites: Effect of silane coupling agents and fibres loading', *Composites Science and Technology*, 67(7), pp. 1627–1639. doi:10.1016/j.compscitech.2006.07.003.

Abdul Wahab, N.S. et al. (2016) 'Effect of surface modification on rice husk (RH)/linear low density polyethylene (LLDPE) composites under various loading rates', *Materials Science Forum*, 840, pp. 3–7. doi:10.4028/www.scientific.net/MSF.840.3.

Achilias, D.S. et al. (2007) 'Chemical recycling of plastic wastes made from polyethylene (LDPE and HDPE) and polypropylene (PP)', *Journal of Hazardous Materials*, 149(3), pp. 536–542. doi:10.1016/j.jhazmat.2007.06.076.

Ahmad, I. et al. (2005) 'Recycled pet for rice husk/polyester composites', *ASEAN Journal on Science and Technology for Development*, 22(4), pp. 345–353. doi:10.29037/ajstd.170.

Aℓ-Maadeed, M.A., Ouederni, M. and Noorunnisa Khanam, P. (2013) 'Effect of chain structure on the properties of Glass fibre/polyethylene composites', *Materials & Design*, 47, pp. 725–730. doi:10.1016/j.matdes.2012.11.063.

Aℓ-Maadeed, M.A., Shabana, Y.M. and Khanam, P.N. (2014) 'Processing, characterization and modeling of recycled polypropylene/glass fibre/wood flour composites', *Materials & Design*, 58, pp. 374–380. doi:10.1016/j.matdes.2014.02.044.

Aℓ-Oqla, F.M. et al. (2015) 'Natural fiber reinforced conductive polymer composites as functional materials: A review', *Synthetic Metals*, 206, pp. 42–54. doi:10.1016/j. synthmet.2015.04.014.

Aℓ-Oqla, F.M. and Salit, M.S. (2017) '2- Natural fiber composites', in AL-Oqla, F.M. and Salit, M.S. (eds.) *Materials Selection for Natural Fiber Composites*. Woodhead Publishing, pp. 23–48. doi:10.1016/B978-0-08-100958-1.00002-5.

Andrew, J.J. and Dhakal, H.N. (2022) 'Sustainable biobased composites for advanced applications: Recent trends and future opportunities – A critical review', *Composites Part C: Open Access*, 7, p. 100220. doi:10.1016/j.jcomc.2021.100220.

Arjmandi, R., Hassan, A. and Zakaria, Z. (2017) '5-Rice husk and kenaf fiber reinforced polypropylene biocomposites', in Jawaid, M., Md Tahir, P., and Saba, N. (eds.) *Lignocellulosic Fibre and Biomass-Based Composite Materials*. (Woodhead Publishing Series in Composites Science and Engineering). Woodhead Publishing, pp. 77–94. doi:10.1016/B978-0-08-100959-8.00005-6.

Arnoult, M., Dargent, E. and Mano, J.F. (2007) 'Mobile amorphous phase fragility in semi-crystalline polymers: Comparison of PET and PLLA', *Polymer*, 48(4), pp. 1012–1019. doi:10.1016/j.polymer.2006.12.053.

Atuanya, C.U. et al. (2013) 'Effect of rice husk filler on mechanical properties of polyethylene matrix composite', *International Journal of Current Research and Review*, 5, pp. 111–118.

Ayyer, R., Hiltner, A. and Baer, E. (2007) 'A fatigue-to-creep correlation in air for application to environmental stress cracking of polyethylene', *Journal of Materials Science*, 42(16), pp. 7004–7015. doi:10.1007/s10853-006-1108-2.

Azevedo, J.V.C. et al. (2022) 'Process-induced morphology of poly(butylene adipate terephthalate)/poly(lactic acid) blown extrusion films modified with chain-extending crosslinkers', *Polymers*, 14(10), p. 1939. doi:10.3390/polym14101939.

Azuan, S.A.S., Juraidi, J.M. and Muhamad, W.M.W. (2012) 'Evaluation of delamination in drilling rice husk reinforced polyester composites', *Applied Mechanics and Materials,* 232, pp. 106–110. doi:10.4028/www.scientific.net/AMM.232.106.

Babu, J. et al. (2016) 'Assessment of delamination in composite materials: A review', *Proceedings of the Institution of Mechanical Engineers, Part B: Journal of Engineering Manufacture,* 230(11), pp. 1990–2003. doi:10.1177/0954405415619343.

Barreno-Avila, E., Moya-Moya, E. and Pérez-Salinas, C. (2022) 'Rice-husk fiber reinforced composite (RFRC) drilling parameters optimization using RSM based desirability function approach', *Materials Today: Proceedings,* 49, pp. 167–174. doi:10.1016/j.matpr.2021.07.498.

Běhálek, L. et al. (2020) 'Thermal properties and non-isothermal crystallization kinetics of biocomposites based on poly(lactic acid), rice husks and cellulose fibres', *Journal of Thermal Analysis and Calorimetry,* 142(2), pp. 629–649. doi:10.1007/s10973-020-09894-3.

Bertin, S. and Robin, J.-J. (2002) 'Study and characterization of virgin and recycled LDPE/PP blends', *European Polymer Journal,* 38(11), pp. 2255–2264. doi:10.1016/S0014-3057(02)00111-8.

Bisht, N. and Gope, P.C. (2018) 'Effect of alkali treatment on mechanical properties of rice husk flour reinforced epoxy bio-composite', *Materials Today: Proceedings,* 5(11, Part 3), pp. 24330–24338. doi:10.1016/j.matpr.2018.10.228.

Butler, T.I. (2010) 'Chapter 2- PE processes', in Wagner, J.R. (ed.) *Multilayer Flexible Packaging.* Boston, MA: William Andrew Publishing, pp. 15–30. doi:10.1016/B978-0-8155-2021-4.10002-4.

Campbell, R.W. (2001) 'Polyesters', in Buschow, K.H.J. et al. (eds.) *Encyclopedia of Materials: Science and Technology.* Oxford: Elsevier, pp. 7171–7172. doi:10.1016/B0-08-043152-6/01271-7.

Chan, A. (2016) 'Rice husk – a useful by-product for rice growing countries', *Myanmar Insider,* 21 February. Available at: https://www.myanmarinsider.com/rice-husk-a-useful-by-product-for-rice-growing-countries/ (Accessed: 18 May 2022).

Chand, N., Sharma, P. and Fahim, M. (2010) 'Tribology of maleic anhydride modified rice-husk filled polyvinylchloride', *Wear,* 269(11), pp. 847–853. doi:10.1016/j.wear.2010.08.014.

Chen, R.S. et al. (2015) 'Rice husk flour biocomposites based on recycled high-density polyethylene/polyethylene terephthalate blend: Effect of high filler loading on physical, mechanical and thermal properties', *Journal of Composite Materials,* 49(10), pp. 1241–1253. doi:10.1177/0021998314533361.

Chen, R.S., Ahmad, S. and Gan, S. (2016) 'Characterization of rice husk-incorporated recycled thermoplastic blend composites', *BioResources,* 11(4), pp. 8470–8482.

Chen, R.S., Ahmad, S. and Gan, S. (2018) 'Rice husk bio-filler reinforced polymer blends of recycled HDPE/PET: Three-dimensional stability under water immersion and mechanical performance', *Polymer Composites,* 39(8), pp. 2695–2704. doi:10.1002/pc.24260.

Ciannamea, E.M., Stefani, P.M. and Ruseckaite, R.A. (2010) 'Medium-density particleboards from modified rice husks and soybean protein concentrate-based adhesives', *Bioresource Technology,* 101(2), pp. 818–825. doi:10.1016/j.biortech.2009.08.084.

Crawford, R.J. and Throne, J.L. (2002) '2- Rotational molding polymers', in Crawford, R.J. and Throne, J.L. (eds.) *Rotational Molding Technology.* Norwich, NY: William Andrew Publishing, pp. 19–68. doi:10.1016/B978-188420785-3.50004-6.

da Silva, V.R. et al. (2013) 'Polyurethane foams based on modified tung oil and reinforced with rice husk ash II: Mechanical characterization', *Polymer Testing,* 32(4), pp. 665–672. doi:10.1016/j.polymertesting.2013.03.010.

de Matos Costa, A.R. et al. (2021) 'Rheological, thermal and morphological properties of polyethylene terephthalate/polyamide 6/rice husk ash composites', *Journal of Applied Polymer Science,* 138(36), p. 50916. doi:10.1002/app.50916.

De Vos, L. et al. (2021) 'Poly(alkylene terephthalate)s: From current developments in synthetic strategies towards applications', *European Polymer Journal,* 161, p. 110840. doi:10.1016/j.eurpolymj.2021.110840.

Esa, Y. and Abdul Rahim, K.A. (2013) 'Genetic structure and preliminary findings of cryptic diversity of the Malaysian mahseer (tor tambroides valenciennes: cyprinidae) inferred from mitochondrial DNA and microsatellite analyses', *BioMed Research International*, 2013, p. e170980. doi:10.1155/2013/170980.

FAOSTAT (n.d.). Available at: https://www.fao.org/faostat/en/#home (Accessed: 18 May 2022).

Farhan Zafar, M. and Siddiqui, M.A. (2018) 'Raw natural fiber reinforced polystyrene composites: Effect of fiber size and loading', *Materials Today: Proceedings*, 5(2, Part 1), pp. 5908–5917. doi:10.1016/j.matpr.2017.12.190.

Fávaro, S.L. et al. (2010) 'Chemical, morphological, and mechanical analysis of rice husk/ post-consumer polyethylene composites', *Composites Part A: Applied Science and Manufacturing*, 41(1), pp. 154–160. doi:10.1016/j.compositesa.2009.09.021.

Fink, J.K. (2013) 'Chapter 16- Compatibilization', in Fink, J.K. (ed.) *Reactive Polymers Fundamentals and Applications (Second Edition)*. Oxford: William Andrew Publishing, pp. 373–409. doi:10.1016/B978-1-4557-3149-7.00016-4.

Gejo, G. et al. (2010) 'Recent advances in green composites', *Key Engineering Materials,* 425, pp. 107–166. doi:10.4028/www.scientific.net/KEM.425.107.

Goswami, T.K. and Mangaraj, S. (2011) '8- Advances in polymeric materials for modified atmosphere packaging (MAP)', in Lagarón, J.-M. (ed.) *Multifunctional and Nanoreinforced Polymers for Food Packaging*. Woodhead Publishing, pp. 163–242. doi:10.1533/97808 57092786.1.163.

Gutierrez Cisneros, C., Bloemen, V. and Mignon, A. (2021) 'Synthetic, natural, and semi-synthetic polymer carriers for controlled nitric oxide release in dermal applications: A review', *Polymers*, 13(5), p. 760. doi:10.3390/polym13050760.

Hakeem, K.R. et al. (2015) 'Bamboo biomass: Various studies and potential applications for value-added products', in Hakeem, K.R., Jawaid, M., and Y. Alothman, O. (eds.) *Agricultural Biomass Based Potential Materials*. Cham: Springer International Publishing, pp. 231–243. doi:10.1007/978-3-319-13847-3_11.

Hamid, M.R.Y., Ab Ghani, M.H. and Ahmad, S. (2012) 'Effect of antioxidants and fire retardants as mineral fillers on the physical and mechanical properties of high loading hybrid biocomposites reinforced with rice husks and sawdust', *Industrial Crops and Products*, 40, pp. 96–102. doi:10.1016/j.indcrop.2012.02.019.

Hill, C.A.S., Khalil, H.P.S.A. and Hale, M.D. (1998) 'A study of the potential of acetylation to improve the properties of plant fibres', *Industrial Crops and Products*, 8(1), pp. 53–63. doi:10.1016/S0926-6690(97)10012-7.

Hwang, K.S., Kwon, H.J. and Lee, J.-Y. (2018) 'Physicochemical analysis of linear low-density polyethylene composite films containing chemically treated rice husk', *Korean Journal of Chemical Engineering*, 35(2), pp. 594–601. doi:10.1007/s11814-017-0304-x.

Imadi, S.R., Mahmood, I. and Kazi, A.G. (2014) 'Bamboo fiber processing, properties, and applications', in Hakeem, K.R., Jawaid, M., and Rashid, U. (eds.) *Biomass and Bioenergy: Processing and Properties*. Cham: Springer International Publishing, pp. 27–46. doi:10.1007/978-3-319-07641-6_2.

Islam, M.M. et al. (2015) 'Study on physio-mechanical properties of rice husk ash polyester resin composite', *International Letters of Chemistry, Physics and Astronomy*, 53, pp. 95–105. doi:10.18052/www.scipress.com/ILCPA.53.95.

Kabir, H. et al. (2016) 'Study and characterization of rice husk ash with polyester resin composite', Daffodil *International University Journal of Science and Technology*, 11(1), pp.67–73.

Kanimozhi, K. et al. (2014) 'Vinyl silane-functionalized rice husk ash-reinforced unsaturated polyester nanocomposites', *RSC Advances*, 4(35), pp. 18157–18163. doi:10.1039/ C4RA01125B.

Kemona, A. and Piotrowska, M. (2020) 'Polyurethane recycling and disposal: Methods and prospects', *Polymers*, 12(8), p. 1752. doi:10.3390/polym12081752.

Kim, Y.S. et al. (1996) 'Melt rheological and thermodynamic properties of polyethylene homopolymers and poly(ethylene/α-olefin) copolymers with respect to molecular composition and structure', *Journal of Applied Polymer Science*, 59(1), pp. 125–137. doi:10.1002/(SICI)1097-4628(19960103)59:1<125::AID-APP18>3.0.CO;2-Z.

Kobayashi, S. and Uyama, H. (2019) 'Synthesis of polyesters I: Hydrolase as catalyst for polycondensation (condensation polymerization)', in Kobayashi, S., Uyama, H., and Kadokawa, J. (eds.) *Enzymatic Polymerization towards Green Polymer Chemistry* (Green Chemistry and Sustainable Technology). Singapore: Springer, pp. 105–163. doi:10.1007/978-981-13-3813-7_5.

Korotkova, T.G. et al. (2016) 'Physical properties and chemical composition of the rice husk and dust', *Oriental Journal of Chemistry*, 32(6), pp. 3213–3219.

Kroczek, K. et al. (2022) 'Characterisation of selected materials in medical applications', *Polymers*, 14(8), p. 1526. doi:10.3390/polym14081526.

Majeed, K. et al. (2017) '22- Structural properties of rice husk and its polymer matrix composites: An overview', in Jawaid, M., Md Tahir, P., and Saba, N. (eds.) *Lignocellulosic Fibre and Biomass-Based Composite Materials* (Woodhead Publishing Series in Composites Science and Engineering). Woodhead Publishing, pp. 473–490. doi:10.1016/B978-0-08-100959-8.00022-6.

Malkapuram, R., Kumar, V. and Negi, Y.S. (2009) 'Recent development in natural fiber reinforced polypropylene composites', *Journal of Reinforced Plastics and Composites*, 28(10), pp. 1169–1189. doi:10.1177/0731684407087759.

McAdam, B. et al. (2020) 'Production of polyhydroxybutyrate (PHB) and factors impacting its chemical and mechanical characteristics', *Polymers*, 12(12), p. 2908. doi:10.3390/polym12122908.

Medina, L.A. and Dzalto, J. (2018) '1.11 Natural fibers', in Beaumont, P.W.R. and Zweben, C.H. (eds.) *Comprehensive Composite Materials II*. Oxford: Elsevier, pp. 269–294. doi:10.1016/B978-0-12-803581-8.09877-5.

Melendez-Rodriguez, B. et al. (2019) 'Reactive melt mixing of poly(3-hydroxybutyrate)/rice husk flour composites with purified biosustainably produced poly(3-hydroxybutyrate-co-3-hydroxyvalerate)', *Materials (Basel, Switzerland)*, 12(13), p. E2152. doi:10.3390/ma12132152.

Mohammed, L. et al. (2015) 'A review on natural fiber reinforced polymer composite and its Applications', *International Journal of Polymer Science*, 2015, p. e243947. doi:10.1155/2015/243947.

Moura, A. et al. (2018) 'Effect of rice husk treatment with hot water on mechanical performance in poly(hydroxybutyrate)/rice husk biocomposite', *Journal of Polymers and the Environment*, 26(6), pp. 2632–2639. doi:10.1007/s10924-017-1156-5.

Muthayya, S. et al. (2014) 'An overview of global rice production, supply, trade, and consumption', *Annals of the New York Academy of Sciences*, 1324(1), pp. 7–14. doi:10.1111/nyas.12540.

Ndazi, B.S. et al. (2007) 'Chemical and physical modifications of rice husks for use as composite panels', *Composites Part A: Applied Science and Manufacturing*, 38(3), pp. 925–935. doi:10.1016/j.compositesa.2006.07.004.

Ndazi, B.S., Nyahumwa, C.W. and Tesha, J. (2008) 'Chemical and thermal stability of rice husks against alkali treatment', *BioResources*, 3(4), pp. 1267–1277.

Nofar, M. et al. (2019) 'Poly (lactic acid) blends: Processing, properties and applications', *International Journal of Biological Macromolecules*, 125, pp. 307–360. doi:10.1016/j.ijbiomac.2018.12.002.

Noorunnisa Khanam, P. and AlMaadeed, M.A. (2014) 'Improvement of ternary recycled polymer blend reinforced with date palm fibre', *Materials & Design*, 60, pp. 532–539. doi:10.1016/j.matdes.2014.04.033.

Nurazzi, N.M. et al. (2021) 'A review on natural fiber reinforced polymer composite for bullet proof and ballistic applications', *Polymers*, 13(4), p. 646. doi:10.3390/polym13040646.

Otera, J. (1993) 'Transesterification', *Chemical Reviews*, 93(4), pp. 1449–1470. doi:10.1021/cr00020a004.

Panthapulakkal, S., Sain, M. and Law, S. (2005) 'Effect of coupling agents on rice-husk-filled HDPE extruded profiles', *Polymer International*, 54(1), pp. 137–142. doi:10.1002/pi.1657.

Peña, J.A. et al. (2016) 'Policaprolactone/polyvinylpyrrolidone/siloxane hybrid materials: Synthesis and in vitro delivery of diclofenac and biocompatibility with periodontal ligament fibroblasts', *Materials Science and Engineering: C*, 58, pp. 60–69. doi:10.1016/j.msec.2015.08.007.

Raghu, N. et al. (2018) 'Rice husk reinforced polypropylene composites: Mechanical, morphological and thermal properties', *Journal of the Indian Academy of Wood Science*, 15(1), pp. 96–104. doi:10.1007/s13196-018-0212-7.

Rahman, W.A.W.A. et al. (2010) 'Rice husk/high density polyethylene bio-composite: Effect of rice husk filler size and composition on injection molding processability with respect to impact property', *Advanced Materials Research*, 83–86, pp. 367–374. doi:10.4028/www.scientific.net/AMR.83-86.367.

Rajak, D.K. et al. (2019) 'Fiber-reinforced polymer composites: Manufacturing, properties, and applications', *Polymers*, 11(10), p. 1667. doi:10.3390/polym11101667.

Rajak, D.K., Wagh, P.H. and Linul, E. (2021) 'Manufacturing technologies of carbon/glass fiber-reinforced polymer composites and their properties: A review', *Polymers*, 13(21), p. 3721. doi:10.3390/polym13213721.

Ramle, M.S., Romli, A.Z. and Abidin, M.H. (2013) 'Tensile properties of aminosilane treated rice husk/recycled PVC composite', *Advanced Materials Research*, 812, pp. 151–156. doi:10.4028/www.scientific.net/AMR.812.151.

Ranjan, S. (n.d.) 'Linear Low Density Polyethylene'. Available at: https://polymeracademy.com/linear-low-density-polyethylene-lldpe/ (Accessed: 18 May 2022).

Reddy, B.S. et al. (2020) 'Pineapple leaf fibres for automotive applications', in Jawaid, M. et al. (eds.) *Pineapple Leaf Fibers: Processing, Properties and Applications*. Singapore: Springer, pp. 279–296. doi:10.1007/978-981-15-1416-6_14.

Rinawa, M., Chauhan, P., Suresh, D., Kumar, S. and Kumar, R.S. (2021) 'A review on mechanical properties of natural fiber reinforced polymer (NFRP) composites', *Materials Today: Proceedings*. doi:10.1016/j.matpr.2021.07.275.

Rodrigue, D., Kavianiboroujeni, A. and Cloutier, A. (2017) 'Determination of the optimum coupling agent content for composites based on hemp and high density polyethylene', *AIP Conference Proceedings*, 1914(1), p. 030003. doi:10.1063/1.5016690.

Ronca, S. (2017) 'Chapter 10- Polyethylene', in Gilbert, M. (ed.) *Brydson's Plastics Materials (Eighth Edition)*. Butterworth-Heinemann, pp. 247–278. doi:10.1016/B978-0-323-35824-8.00010-4.

Santiagoo, R. et al. (2015) 'Tensile properties of linear low density polyethylene (LLDPE)/recycled acrylonitrile butadiene rubber (NBRr)/ rice husk powder (RHP) composites', *Applied Mechanics and Materials*, 754–755, pp. 210–214. doi:10.4028/www.scientific.net/AMM.754-755.210.

Satapathy, M. et al. (2018) 'Fabrication of durable porous and non-porous superhydrophobic LLDPE/SiO$_2$ nanoparticles coatings with excellent self-cleaning property', *Surface and Coatings Technology*, 341, pp. 31–39. doi:10.1016/j.surfcoat.2017.07.025.

Sethulekshmi, A.S., Saritha, A. and Joseph, K. (2022) 'A comprehensive review on the recent advancements in natural rubber nanocomposites', *International Journal of Biological Macromolecules*, 194, pp. 819–842. doi:10.1016/j.ijbiomac.2021.11.134.

Siracusa, V. et al. (2015) 'Poly(butylene succinate) and poly(butylene succinate-co-adipate) for food packaging applications: Gas barrier properties after stressed treatments', *Polymer Degradation and Stability*, 119, pp. 35–45. doi:10.1016/j.polymdegradstab.2015.04.026.

Sirichalarmkul, A. and Kaewpirom, S. (2021) 'Enhanced biodegradation and processability of bio-degradable package from poly(lactic acid)/poly(butylene succinate)/rice-husk green composites', *Journal of Applied Polymer Science*, 138(27), p. 50652. doi:10.1002/app.50652.

Somarathna, H.M.C.C. et al. (2018) 'The use of polyurethane for structural and infrastructural engineering applications: A state-of-the-art review', *Construction and Building Materials*, 190, pp. 995–1014. doi:10.1016/j.conbuildmat.2018.09.166.

Soutis, C. (2005) 'Fibre reinforced composites in aircraft construction', *Progress in Aerospace Sciences*, 41(2), pp. 143–151. doi:10.1016/j.paerosci.2005.02.004.

Subramanian, P.M. (2000) 'Plastics recycling and waste management in the US', *Resources, Conservation and Recycling*, 28(3), pp. 253–263. doi:10.1016/S0921-3449(99)00049-X.

Surata, I.W., Suriadi, I.G.A.K. and Arnis, K. (2014) 'Mechanical properties of rice husks fiber reinforced polyester composites', *International Journal of Materials, Mechanics and Manufacturing*, 2(2), pp. 165–168. doi:10.7763/IJMMM.2014.V2.121.

Thomas, B.S. (2018) 'Green concrete partially comprised of rice husk ash as a supplementary cementitious material – A comprehensive review', *Renewable and Sustainable Energy Reviews*, 82, pp. 3913–3923. doi:10.1016/j.rser.2017.10.081.

Thungphotrakul, N., Somboonwanna, J. and Prapainainar, P. (2020) 'Effect of nanoclay on properties of medium density polyethylene (MDPE)/rice husk flour (RHF) composites', *AIP Conference Proceedings*, 2279(1), p. 070001. doi:10.1063/5.0023271.

Tong, J.Y. et al. (2014) 'Study of the mechanical and morphology properties of recycled HDPE composite using rice husk filler', *Advances in Materials Science and Engineering*, 2014, p. e938961. doi:10.1155/2014/938961.

Tsou, C.-H. et al. (2014) 'New composition of maleic-anhydride-grafted poly(lactic acid)/rice husk with methylenediphenyl diisocyanate', *Materials Science*, 20(4), pp. 446–451. doi:10.5755/j01.ms.20.4.6034.

USDA ERS - Agricultural Trade (n.d.). Available at: https://www.ers.usda.gov/data-products/ag-and-food-statistics-charting-the-essentials/agricultural-trade/ (Accessed: 18 May 2022).

Wang, W. et al. (2014) 'Properties of rice husk-HDPE composites after exposure to thermo-treatment', *Polymer Composites*, 35(11), pp. 2180–2186. doi:10.1002/pc.22882.

Wu, C.-S. and Tsou, C.-H. (2019) 'Fabrication, characterization, and application of biocomposites from poly(lactic acid) with renewable rice husk as reinforcement', *Journal of Polymer Research*, 26(2), p. 44. doi:10.1007/s10965-019-1710-z.

Wu, W. et al. (2018) 'Synergetic enhancement on flame retardancy by melamine phosphate modified lignin in rice husk ash filled P34HB biocomposites', *Composites Science and Technology*, 168, pp. 246–254. doi:10.1016/j.compscitech.2018.09.024.

Xie, Y. et al. (2010) 'Silane coupling agents used for natural fiber/polymer composites: A review', *Composites Part A: Applied Science and Manufacturing*, 41(7), pp. 806–819. doi:10.1016/j.compositesa.2010.03.005.

Yakubu, A., Abbas, Z. and Abdullahi, S. (2020) 'Mechanical, dielectric and shielding performance of rice husk/polycaprolactone composites enhanced via rice husk particles inclusion', *Open Access Library Journal*, 7(7), pp. 1–12. doi:10.4236/oalib.1106514.

Yap, S.Y. et al. (2020) 'Characterization and biodegradability of rice husk-filled polymer composites', *Polymers*, 13(1), p. 104. doi:10.3390/polym13010104.

Zhao, Q. et al. (2008) 'Biodegradation behavior of polycaprolactone/rice husk ecocomposites in simulated soil medium', *Polymer Degradation and Stability*, 93(8), pp. 1571–1576. doi:10.1016/j.polymdegradstab.2008.05.002.

Zhao, Q. et al. (2009) 'Flame retardancy of rice husk-filled high-density polyethylene ecocomposites', *Composites Science and Technology*, 69(15), pp. 2675–2681. doi:10.1016/j.compscitech.2009.08.009.

Zurina, M., Ismail, H. and Bakar, A.A. (2004) 'Rice husk powder–filled polystyrene/styrene butadiene rubber blends', *Journal of Applied Polymer Science*, 92(5), pp. 3320–3332. doi:10.1002/app.20321.

10 Polyester-Based Bio-Nanocomposites

Siby Isac, Ishwariya A., Dharini V.,
and Periyar Selvam S.
SRM Institute of Science and Technology

Emmanuel Rotimi Sadiku
Tshwane University of Technology

CONTENTS

10.1 INTRODUCTION

Plastics are manufactured by using non-biodegradable polymers, such as polyethene, poly(vinyl chloride), polypropylene, poly(ethylene terephthalate), polycarbonate resin, phenolic resin and polystyrene. Being non-biodegradable, these polymers persist in the surrounding nature for years, far beyond their end dates, leaving a significant plastic garbage collection in the environment and see beds. Plastics are released into the environment *via* dumping, landfilling, consumption by animals and incineration. The consumption of plastics by the marine species is considered a big problem since the ocean is the largest sink or dumping site for polymeric wastes. Bio-magnification causes the mortality of fishes, birds, reptiles and other endangered species. The incineration of plastics releases CO_2 that significantly contributes to

DOI: 10.1201/9781003270980-10

global warming and endangers human health. Due to their versatility, plastics usage is difficult to scale down, although biodegradable polymers may help (Alothman et al., 2020; Chandrasekar et al., 2020).

Plastics that degrade *via* metabolism in the presence of naturally existing organisms are known as biodegradable plastics. Biodegradable plastics are a class of polymeric materials that may be broken down, enzymatically, into smaller pieces over time. Biodegradation results in the fragmentation or disintegration of plastics under the right circumstances of moisture, temperature and the availability of oxygen (Chandrasekar, Senthilkumar, Jawaid, Alamery et al., 2022a,b; Chandrasekar, Senthilkumar, Jawaid, Mahmoud, et al., 2022a,b). This process leaves no toxic or ecologically detrimental residue. Biodegradable polymers or biopolymers are the most promising alternatives to replace plastics that are derived from petroleum. According to the Association of European Bioplastics, biopolymers are considered to be derived from renewable resources and are completely degradable. However, the term "biodegradation" essentially refers to the breakdown of substances by enzymatic activity. Biodegradable natural polymers, extracted from renewable natural resources, such as chitosan, polylactic acid, starch, pectin, cellulose derivatives or polycaprolactone (PCL), are gradually replacing synthetic plastics, which are derived from petroleum, for the production of bioplastics and used in packaging applications. This substitution occurs most frequently in the production of bioplastics (Nasimudeen et al., 2021; Senthilkumar et al., 2015, 2021, 2022). Numerous research organizations have devoted their attention to the potential of replacing synthetic polymers, generated from petroleum, with biodegradable materials that are natural, plentiful and relatively inexpensive and that are acquired from renewable sources. The sources of these materials determine their (biodegradable polymers) classifications:

1. Biomass-derived polymers (i.e., all types of polysaccharides, various proteins, polypeptides and polynucleotides);
2. Chemically synthesized polymeric materials that employ bio-based monomers, petroleum-derived (i.e., polylactic acid or bio-polyester);
3. Microorganism- or GMO (Genetically Modified Organism)-developed polymers (polyhydroxybutyrate, bacterial cellulose [BC], xanthan gum, pullan).

There are three challenges that are related to biodegradable polymers: performance, processing and cost. Despite the fact that these factors are related, in part to one another, efficiency and manufacturing concerns are same for all biodegradable polymers, independent of their origin. Nanotechnology may improve these polymers' characteristics and cost–performance ratio. Dispersed nanoparticles provide nanocomposites with better mechanical, thermal, barrier and physicochemical characteristics compared to the neat polymers and the traditional (microscale) composites.

Over the last several years, the phrase "bio-nano composites" has emerged as a popular one. A new class of nanostructured hybrid materials, known as bio-nanocomposites, is now coming into being. Composite is said to be a "nanocomposite", if at least one phase has dimensions smaller than 1 nm (10^{-9} m). There are a number of advantages to using nanocomposite materials in place of microcomposites and

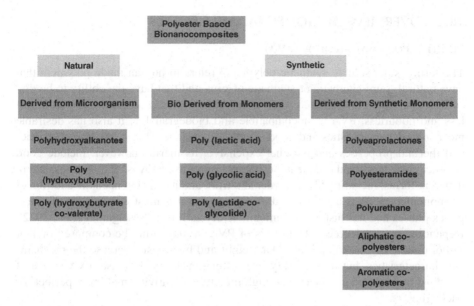

FIGURE 10.1 Different polyester-based bio-nanocomposites and their classifications (Sivakanthan et al., 2020).

monolithics; however, the nanocluster phase presents unique preparation issues, pertaining to the management of the chemical composition of the composite as well as its stoichiometry. From the standpoint of holding the design originality and the different properties that are not present in the ordinary composites, they are believed to be the materials of the twenty-first century. The "critical size" has been proposed as a threshold below which changes in the particle characteristics may be conspicuously observed. Additionally, when dimensions reduce to the nanoscale range, the interactions at the phase interfaces increase dramatically, which is critical for improving the characteristics of the materials in question. The understanding of the link between nanocomposites' structures and properties hinges on the surface area-to-volume ratio of reinforcing materials used in its development.

Nanoparticles have at least one dimension that is on the nanoscale, and their inclusion in bio-based polymer matrices produces natural polymers and inorganic solids that are combined. In a manner similar to that of traditional nanocomposites, which are composed of synthetic polymers, these biodegradable hybrid composites also have enhanced morphological and functional features that are of considerable significance for several applications. Because of the amazing capabilities that they carry, bio-nanocomposites are becoming more common in the modern world. Researchers have learned to identify proper matrices (such as synthetic fibre, polypeptides and proteins (such as gluten, casein, lecithin), polysaccharides (such as starch, chitosan, pectin, cellulose) and polynucleic acids and fillers (such as cellulose nanocrystals, chitosan nanocrystals, pectin nanocrystals, nanotubes, nanofibre varieties, clay nanoparticles, hydroxyapatite and metal nanoparticles, e.g., silver nanoparticles) and reconfigure their chemical properties and framework to fit the target field.

10.2 OVERVIEW OF BIODEGRADABLE POLYESTERS

10.2.1 POLYVINYL ALCOHOL (PVA)

The term "water-soluble synthetic polymers" refers to human-made polymers that have natural, semi-synthetic or synthetic origins and that have the ability to be dissolved, distributed and inflated when exposed to water. In addition to being harmless and odourless, PVA is biocompatible and biodegradable. It also has desirable mechanical characteristics and good chemical resistance, but it has limited barrier and thermal properties and is rather expensive. Its merits, however, include good chemical resistance and excellent mechanical qualities. PVA is readily accessible in different forms, including fibre, powder and film, and it does not need the addition of co-monomers. PVA is a semicrystalline polymer that is mostly composed of amorphous phases and has just a trace amount of crystallinity (Rescignano et al., 2022). Depending on the degree of hydrolysis of PVA, it may either be composed of 1 or 3-diol units (vinyl acetate). Molecular weight and hydrolysis increase the mechanical characteristics, but its flexibility and water sensitivity decrease. PVA is one of the most hydrophilic polymers with high moisture sensitivity, making it perfect for packaging.

In 1924, the saponification technique of poly(vinyl acetate), abbreviated as PVAc, was used to synthesis PVA for the very first time. PVA cannot be produced by using the direct polymerization of PVAc monomer. Therefore, Herrmann and Haehnel synthesized PVA in 1924, by initially combining, at ambient temperature, ethanol solutions containing potassium hydroxyl and polyvinyl acetate; these solutions were then heated in the presence of hydraulic acid for between 1 and 2 hours in order to form PVA (Alghunaim et al., 2016). The reaction routes for the synthesis of PVA are shown in the chemical equations below:

$$-CH-CH_2- + CH_3OH \rightarrow -CH-CH_2- + CH_3OAC \ldots\ldots \text{(Reaction A)}$$

$$\quad\quad\quad |\quad\quad\quad\quad\quad\quad |$$

$$\quad\quad\quad OAc\quad\quad\quad\quad\quad OH$$

$$-CH-CH_2- + NaOH \rightarrow -CH-CH_2- + NaOAc \ldots\ldots\ldots \text{(Reaction B)}$$

$$\quad\quad\quad |\quad\quad\quad\quad\quad\quad |$$

$$\quad\quad\quad OAc\quad\quad\quad\quad\quad OH$$

$$CH_3OAc + NaOH \rightarrow CH_3-OH + NaOAc \ldots\ldots\ldots\ldots \text{(Reaction C)}$$

Alcoholysis of polyvinyl acetate is described in Reaction A. The production of methyl acetate is accomplished in Reaction C, whereas the saponification of polyvinyl acetate is represented in Reaction B.

Industries use acetylene or ethylene as basic materials in the presence of acetic acid and/or oxygen, to produce vinyl acetate. This process requires the major presence of both of these elements. After that, the purified vinyl acetate is brought into contact with a solution containing methanol and heat is applied in order to commence the polymerization process. During the polymerization process, more than 70% of the monomers are typically transformed into polyvinyl acetate. This is done in order to make PVA go through the saponification process. In the presence of sodium hydroxide or anhydrous sodium methylate, the ester groups of polyvinyl acetate are partly or totally replaced by hydroxyl groups during the hydrolysis process. The degree of hydrolysis is usually adjusted by the concentration of the catalyst and the hydrolysis temperature. When compared to other biodegradable polymers, PVA may be synthesized easily, from the suspension and water and from the melt by extrusion or injection moulding at a reduced cost. Casting from a water solution, injection moulding and blast extrusion are all viable methods for producing PVA films for packaging applications, despite the heat sensitivity of the material. The use of plasticizers as well as lubricants is a potential strategy for overcoming this major problem. Plasticizers may lower the melting temperature and viscosity of PVA in order to prevent thermal deterioration during the extrusion process.

10.2.2 POLYLACTIDE (PLA)

PLA, a polymer derived from lactic acid (2-hydroxy propionic acid), has been intensively investigated for its therapeutic uses, owing to its bioresorbable and biocompatible properties in the human body. Due to its high cost, PLA was first used primarily for high-value films, stiff thermoforms, food and beverage containers and coating papers (Yang et al., 2016). As contemporary and upcoming manufacturing methods may reduce the cost of producing PLA, they may be used for packaging a wider variety of items. Furthermore, the production of PLA has numerous advantages, including the fact that it can be made from corn, which is in abundance, and it is a renewable agricultural resource, carbon-neutral, and recyclable and compostable. Moreover, its physical and mechanical properties can be exploited through appropriate polymer design. There are several uses for the polymer PLA, including containers, ice cream and sundae cups, as well as lamination and overwrap films and blister packs for short-lived items. Furthermore, thermoformed PLA containers for fresh fruits and vegetables are already being utilized in retail shops.

Lactic acid, the integral element of PLA, is usually synthesized by the fermentation of carbohydrates or chemical synthesis. The vast majority of lactic acid is produced by the bacterial fermentation of carbohydrates. Lactic acid production processes may be categorized depending on the type of bacteria utilized in the fermentation. It is estimated that the hetero-fermentative process yields less than one mole of lactic acid for every eighteen moles of hexose consumed, with significant quantities of other acids and alcohols, carbon dioxide and other waste products produced during the process. An average of 1.8 moles of lactic acid per mole of hexose is produced by the homo-fermentative method, which also generates trace quantities of other metabolites. Homo-fermentative processes are preferred by the industrial sector because they produce more lactic acid and produce fewer by-products. Nowadays, *Lactococcus* is used in the vast majority of fermentation activities because of its ability to create a considerable quantity of lactic acid. Homo-fermentative bacteria need a pH of between 5.4 and 6.4, a temperature of between 38 and 42.8°C and a low oxygen level for digestion. In most circumstances, simple sugars are employed as a source of energy. Sucrose comes from sugar beets or sugarcane. Other examples include corn- and potato-derived glucose and maltose and cheese-derived lactose. Besides carbohydrates, corn steep liquor is a rich source of B vitamins, amino acids and nucleotides.

PLA with high molecular weight may be produced by a variety of methods. The chemical synthesis method is one that can be employed to manufacture huge amounts of racemic lactic acid; however, this method is not commercially feasible. As a direct result of this, the use of fermentation as a method for the generation of *L*-lactic acid is gaining popularity. In order to manufacture PLA with a high molecular mass, one may often choose between three different production methods: (1) polymerization by means of direct condensation, (2) polymerization by means of azeotropic dehydrative condensation and (3) polymerization by use of lactide formation.

PLA is primarily produced *via* polymerization through lactide production. Polymerization by direct condensation is the cheapest method. However, it is very difficult to create a solvent-free poly(lactic acid) with high molecular weight. This procedure becomes more expensive and more difficult since chain coupling agents and adjuvants are used. Lactic acid may be converted into high-molecular-weight poly(lactic acid) *via* an azeotropic dehydrative condensation without the need for adjuvants or chain extenders. In general, the process entails the lowering of the distillation pressure of lactic acid for 2–3 hours at a temperature of 130°C. After that, the great majority of the water that had condensed is drained. In addition to the diphenyl ester, a catalyst is also added. After the solvent has been passed through the molecular sieves and back into the reaction vessel for a further 30–40 hours at 130°C, the tube containing the molecular sieves is then linked to the reaction vessel. At this point, the polymer may either be left in its natural state or dissolved and precipitated for further purification.

10.2.3 POLYCAPROLACTONE (PCL)

Polycaprolactone is a biodegradable polyester used in wound dressings, contraceptives, fastening machinery and drug passage systems. PCL is used to deliver drugs, amino acids, peptides, vaccinations and other bioactive compounds. A monomer and an initiator are the two major components for the polymerization of PCL. The monomer and the initiator are combined at high temperatures under a purging nitrogen gas in order to create the polymer. After cooling, dissolving in an organic solvent and washing are done and the unreacted molecules are removed from the polymer. The freeze-dried polymer is used to make nanoparticles, microparticles and scaffolds. The PCL's T_g is 213 K and T_m is between 332K and 337K, depending on its crystallinity. Its low melting point facilitates its mixing with other polymers. Chemically, PCL is formed of hexanoate repeating units with the formula $(C_6H_{10}O_2)_n$. It is also called 2-oxepanone homopolymer or 6-caprolactone polymer (Espinoza et al., 2019).

It is possible to copolymerize PCL in order to alter its physiological, biochemical and mechanical behaviours. PCL copolymerization influences its crystallinity, dissolution, breakdown pattern and permeability, resulting in customized polymers for drug delivery. PCL may be produced by using either the polycondensation of 6-hydroxyhexanoic acid or the ring-opening polymerization (ROP) of ε-caprolactone. Polycondensation is a method for the production of polymers that involves the sequential development of condensation processes. Two molecules that possess functional groups that are complementary to one another, react, which enables the formation of the polymeric chain, while simultaneously releasing the by-products that have low molecular weight. An esterification reaction takes place during the PCL synthesis *via* polycondensation reaction. This reaction takes place either between the acidic function of 6-hydroxyhexanoic acid or the growing polymeric chain and the hydroxyl functional group of another molecule of 6-hydroxyhexanoic acid. This results in the formation of PCL. Throughout the course of the process, one water molecule is produced as a by-product. During the synthesis process, the elimination of water shifts the equilibrium of the reaction in a direction that favours the creation of the polymer. The production of PCL by polycondensation has been established either through the use of *Candida antartica*-derived lipases or that derived from *Pseudomonas* sp. or through the complete elimination of the need for a catalyst. In spite of the fact that polycondensation is used in the production of a number of aliphatic polyesters, the production of PCL *via* the use of this technique is quite uncommon. On the other hand, the ROP synthesis process has been proposed as a superior alternative to the PCL synthesis process since it makes it possible to develop polymers with higher molecular weights and lower polydispersity values.

The ROP process may be carried out in either bulk or solution and involves the cyclic monomeric addition to the structure of a polymeric chain that already has a propagating centre. This depends on the steric concerns rather than the conversion of multiple bonds into a single bond. When it comes to ROP, the microstructure of the products is significantly impacted by both the starting monomer content and the temperature.

10.2.4 Polyhydroxybutyrate (PHB)

Poly-β-hydroxy alkanoates, often known as PHA, are thermoplastics that are biocompatible and biodegradable, and they have recently attracted considerable interest as a result of their prospective application areas (Policastro et al., 2021). Generally, PHBs (poly-hydroxybutyrate) are considered as one of the well-known biodegradable polymers under the common PHA (Poly-hydroxy alkanoates) group. PHB is a naturally occurring thermoplastic polyester that has many mechanical attributes similar to the synthetically manufactured degradable polyesters, e.g., poly-*L*-lactides. PHB is a highly crystalline thermoplastic with low water vapour permeability, similar to LDPE (Low Density Poly Ethylene). PHB homopolymer's undesirable ageing process is its biggest commercial downside. As an intracellular reserve material, PHB has been synthesized *via* bacterial fermentation in the presence of a broad range of microorganisms. The intracellular granules of PHB have been found in, at least, 75 distinct bacterium species. Despite the fact that this polymer can only be formed under highly constrained culture conditions, studies on its synthesis have focused on bacteria from *Alcaligenes*, *Azotobacter*, *Bacillus* and *Pseudomonas* genera. Some bacteria may develop up to 80% of their body weight in PHB under conditions of restricted nitrogen and ample carbon. In order to produce PHB, *Alcaligenes eutrophus* is the most often employed organism because it is easy to culture, can accumulate significant quantities of PHB (up to 80% of dry cell weight on a simple medium) and its physiology and biochemistry are well characterized. Various microorganisms (bacteria, fungus and algae) break down PHB films into a variety of habitats. The enzymes secreted by microbes when they come into contact with the polymer break it down into smaller pieces, thereby lowering its average molecular weight. PHB has been utilized as packaging materials and tiny disposable items. However, very little is known regarding the use of PHB in food packaging. The food packaging industry has difficulty in its attempts to manufacture biodegradable primary packaging by balancing the product shelf life with the packaging durability. Stability in the mechanical and/or barrier integrities, as well as the functionality, is essential for biological packing materials. Throughout the storage period, the circumstances that promote biodegradation must be avoided, whereas optimal conditions for biodegradation must exist after discarding the food product.

10.3   POLYESTER-BASED HYBRID BIOCOMPOSITES

10.3.1   POLYESTER AND NANOCELLULOSE-BASED HYBRIDS

Nanocelluloses have significant promise since cellulose is the most prevalent biological raw material and can self-assemble into well-defined designs from micro- to nanoscale. Cellulose is renewable and versatile, and it is predicted to replace numerous non-renewable materials. Over the past several decades, the creation of cellulose nanofibres (CNFs) has garnered a lot of attention because of its unique properties such as high surface area-to-volume proportion, high specific surface area, high elastic modulus, high tensile strength and low coefficient of thermal expansion. CNFs are isolated from algae, tunicates and BC, although plant cell walls are their principal sources. Numerous researchers have investigated the extraction of nanofibres from natural plant cellulose fibres due to their regenerative, abundance and low-cost characteristics. Ultra-pure CNFs have been isolated from several plant sources, including wood derivatives (Abe et al. 2007). Different parts of bamboo (Abe and Yano 2010), unused soy hulls (Alemdar and Sain 2008) and the remains of pineapple leaf fibres are good sources for CNC (Cellulose Nano Crystals) generation (Cherian et al. 2010. The extraction methods include cryo-crushing (Chakraborty et al. 2005), grinding (Abe et al. 2007), high-pressure homogenization (Herrick et al., 1983) and acid hydrolysis (Araki et al., 2000). Depending on the cellulose sources, its preliminary treatment, and the hydrolysis process, various approaches can be employed to produce distinct nano-fibrillar materials. CNF is used as a nanofiller for various biodegradable polyesters such as PVA and PLA. This is because polyester alone has many limitations of its own. PLA, being highly biocompatible and biodegradable, has few drawbacks such as high production cost, brittleness, low mechanical strength and poor thermal stability, when used in packaging applications. Fillers, such as cellulose or cellulose derivatives, which are renewable and biodegradable, might increase the mechanical and thermal properties and improve the cost–performance ratio. It is becoming more popular to employ natural fibres in polymer composites, instead of synthetic fibres, e.g., glass. This is because of their availability, renewability, high stiffness and non-abrasiveness to processing equipment, as well as the capability to be incinerated, lightweight and reduced cost (Shahroze et al., 2021; Thomas et al., 2021). In this regard, CNFs have been used to reinforce various polymeric matrices, including starch (Wang et al., 2009), PVA (Lee et al., 2009) and polylactic acid (Dobreva et al., 2010).

Cellulose nanocrystals, due to their distinctive structural qualities and impressive physicochemical properties, such as compatibility, biodegradability, compostability, lightweight, adaptive surface composition, optical transmittance and improved mechanical properties and tensile strength, have attracted considerable interest from both the industries and the academia (Frone et al., 2013). This nanomaterial is a prospective contender for use in biomedical, pharmaceutical, electrical, barrier film, nanocomposites and membrane and supercapacitor applications, among others. In order to meet the rising demand for the production of new kinds of cellulose nanocrystals-based products on an industrial scale, new sources, isolation methods and various other treatments are constantly being developed. CNCs are nanometric or rod-like particles that have a needle-like configuration and have at least one dimension that is less than 100 nm. They have a very high crystalline structure. They may be made from a wide variety of plant-based materials, including wood pulp, BC, algal cellulose, bast fibres, cotton linters, microcrystalline cellulose and tunicin. These nanocrystals have attractive combinations of bio-physicochemical properties, such as compatibility, biodegradability, compact size, non-toxic nature, stuffiness, recyclability and sustainability. They also have absorptivity, heat resistance, gas barrier properties, adaptable surface composition and improved mechanical properties. These nanocrystals may also replace certain goods that are based on petrochemicals, and they are more cost-effective than some comparative high-performance nanomaterials. Strong acids, e.g., sulfuric and hydrochloric acid, are used in the process of acid hydrolysis, which is one of the most common methods for synthesizing CNCs from a wide variety of starting materials that are based on cellulose. In more recent times, the production of CNCs has also made use of a variety of different mineral and organic acids. Several alternative methods of preparation, such as enzymatic hydrolysis, mechanical refining, ionic liquid treatment, subcritical water hydrolysis, the oxidation technique and combination procedures, have also been developed. During the synthesis process, cellulose chains are joined, largely into fascicular microfibrils. Due to the kinetic considerations and the lower steric hindrance, amorphous domains disperse as chain dislocations on segments of the elementary fibril that are more accessible to acid and more prone to hydrolytic action, while the crystalline domains have a stronger resistance to acid assault. Consequently, CNCs may be generated by removing the amorphous areas from microfibrils. When compared to nanofillers made of minerals or metals that are commercially accessible, cellulose nanocrystals are manufactured from numerous feedstocks; they have low densities, their prices are comparatively inexpensive and they do not compromise biodegradability.

As a result of these benefits, CNCs are now undergoing intensive research in order to determine whether or not they are ideal fillers for the mechanical reinforcement of a polymer matrix. Because of their nanoscale size and high inherent rigidity, they are very desirable for the purpose of improving the stiffness of polymer composites. A percolating CN network will cause gas molecules to follow a convoluted route in order for them to pass through the polymer film, which will result in the layer acting as a barrier. The enhancement in the permeability property of a PVA membrane was established by Paralikar et al. (2008), who observed a decrease in the transmission of water vapour and organic vapour (i.e., trichloroethylene) with a 10% loading of cotton CNC in comparison to the plain (neat) polymer. Additionally, they recorded a reduction in the transmission of trichloroethylene. In research conducted by Suryanegara, Nakagaito and Yano (2009), differential scanning calorimetry studies and dynamic mechanical analysis were conducted to shed light on the use of micro-fibrillated cellulose (MFC) as fillers that might expand the application range of PLA to high temperature-exposed products. In this study, the effect of MFC on polylactic acid's (PLA) thermal and mechanical properties was investigated (Suryanegara, Nakagaito & Yano, 2009). It was also shown that PLA's tensile characteristics were improved when 20 wt.% MFC was added, compared to the PLA that had no MFC inclusion (Yano et al., 2009).

10.3.2 CARBON-BASED POLYESTER BIO-NANOCOMPOSITES

10.3.2.1 Graphene

Graphene, a two-dimensional carbon sheet, has a thickness of one atom and is one of the lightest substances in the universe, and it has sparked a great deal of curiosity in physics, materials engineering, chemistry and biology (Chang et al., 2013). The superior mechanical characteristics of graphene, including its high Young's modulus, tensile strength and its heat conduction capability, are also advantageous for a variety of graphene applications. Despite the fact that graphene sheets were first studied, primarily in basic physics, graphene nanocomposites and their hybrids are garnering an increasing number of real-world applications. By incorporating controlled functional building blocks into graphene, the current emphasis on graphene as a generic substrate for nanocomposites has spawned several energy and environmental applications. In order to have increases in the characteristics of polymer nanocomposites, it is necessary to have uniform dispersion of nanofiller throughout the polymer matrix and to ensure that there are no agglomerates present. The characteristic analysis of polymer nanocomposites is significantly connected to their microstructures. During the reduction process, a greater repulsion was observed for graphene oxide (GO) sheets towards the water, and it has a propensity to quickly agglomerate into flakes of single-layered sheets, which then stack up to produce graphite and eventually precipitate. The in situ reduction of GO that is disseminated in a polymer matrix, is one of the most successful routes that can be employed to create stable graphene dispersions. The in situ intercalative polymerizations, solution intercalation and melt intercalation methods are the primary approaches that are used during the creation of polymer/graphene nanocomposites. The solution casting approach is the simplest method and in comparison, to the melt mixing method, it is more effective to obtain excellent dispersion of the graphene throughout the polymer matrix.

The incorporation of graphene and its derivatives into PVA has the potential to bring about significant improvements in the material's characteristics and to expand the range of applications for which it may be deplored. PVA nanocomposites have been prepared by using a wide variety of graphene derivatives and reports have shown that these preparations result in an improvement in the bulk physical characteristics.

10.3.2.2   Carbon Nanotubes (CNTs)

The structural, mechanical and electrical characteristics of CNTs (multiwalled carbon nanotubes, MWNTs, and single-wall carbon nanotubes, SWNTs), as well as the prospective technological uses of these nanotubes, have sparked a great deal of research. As a result, the production of composite materials by incorporating CNTs has piqued a lot of scientific interest, whether for the understanding of their physical and chemical characteristics or for the development of new usages. Improved electrical, mechanical and thermal characteristics have been achieved by the incorporation of CNTs into diverse matrices (Gan et al., 2020). The complexes resulting from the combination of polymeric nanomaterials with appropriate nanofillers, such as CNTs, are beneficial for the design of ultra-high-energy devices, such as battery cells, fuel cells and display systems, due to their ease of fabrication into thin films of desirable dimensions. This is one reason why polymeric nanomaterials are of particular interest. It is possible for CNTs to have a significant impact on the characteristics of polymers if they are incorporated into the matrix of the polymeric material. The characteristics of the polymer matrix are significantly enhanced when just a trace amount of CNTs is included in the formulation. Therefore, CNT/polymer composites make use of the remarkable qualities of the CNTs along with the characteristic property of the polymeric matrix. According to the findings of certain investigations, the characteristics of polymer and CNT nanocomposites are significantly improved. It was discovered by scientist Dalton that fibres synthesized with PVA and the addition of a high proportion of CNTs resulted in the development of a film, which had a tensile strength and toughness far greater when compared to that of Kevlar or even spider silk. In addition, in comparison to pure PVA films, PVA composite films that include poly(vinyl pyrrolidone) and completely dispersed CNTs demonstrate enhanced tensile strength and modulus. Experiments carried out recently on PVA combined with CNTs and subjected to non-isothermal crystallization procedure have shown that nanotubes are capable of increasing the crystallinity of the mixture. It is critical to choose the right CNTs when producing nanocomposites containing CNTs in order to obtain the most reinforcement. Due to their high modulus and compact size, high-quality SWNTs are typically considered the best choice. Arc-discharged CNT, on the other hand, is often regarded as superior to chemical vapour deposition CNT. The mechanical characteristics of pure PVA composites containing various types of nanotubes have been studied by Cadek et al. (2004) and Coleman et al. (2004), who revealed that the reinforcement increased with the inverse of the nanotube diameter for MWNTs, whereas SWNTs showed very weak reinforcement, which was due to uneven bundle formation.

10.3.3 Polyester-Based Active Biocomposite Films

Active packaging film for food is a growing alternative to the common procedures, such as the addition of antioxidants or essential oils or nanoparticles directly in food samples in combination with a vacuum or modified atmosphere for the protection of sensitive foods from oxidation (Chen et al., 2021). In most cases, the foundation for these systems is made up of materials in which certain additives, exhibiting antioxidant and antibacterial capabilities, have been directly incorporated into the polymer matrix. These additives have the potential to serve two purposes: firstly, they may prevent oxidation of fat and colour components in foods by controlling how quickly they are released, and secondly, they can shield polymers from damage caused during processing. Nanotechnology-enhanced materials for packaging have the potential to function through two distinct modes of action. The first advantage is an improvement in packaging. A high interfacial area between the polymer matrix and the nanofillers is achieved due to the uniform distribution of nanofillers. Structural, thermodynamic and barrier characteristics of the matrix are enhanced due to the presence of this interfacial region. The second technique is called the active packaging and it offers improved food safety by virtue of the nanofillers' direct interaction with the food that is packaged and with the environment in which it is stored. It is possible to increase the functional qualities of polyester-based bio-nanocomposite films by using a wide range of naturally occurring bioactive substances. In particular, there has been an increased emphasis on components that may prevent the deterioration of foods, owing to the antioxidant and antibacterial qualities that they possess. There has already been a lot of research done on the topic of inclusion of natural bioactive compounds, including polyphenols, essential oils, curcumin and carotenoids into edible films for food packaging purposes. The antioxidant, antibacterial, colouring, flavouring and/or nutritional properties of these components are often put to use in a variety of recipes. It is important to note that the insertion of these active components has the potential to have an effect, not only on the structural organization of the films, but also on their functional qualities. For example, when added to bio-nanocomposite films, essential oils not only have an antioxidant and antimicrobial impacts, but also change the films' mechanical and barrier qualities. Before the film is formed, the active components are often simply combined with the materials that will be used to produce the film. In certain circumstances, the active substances have to be enclosed inside colloidal particles first, before they can be combined with the components that are used to produce the film. The technique that is applied is determined by whether or not the active components are soluble in the solution that forms the film. In order to attain the necessary functional qualities, such as an antioxidant and an antimicrobial, the films may be supplemented with a single kind of active ingredient or with multiple types of active compounds.

10.4 CONCLUSION

Nanocomposites, made from bionanomaterials, are a new category of hybrid materials that have a wide range of uses in a variety of sectors. In the modern age of innovative materials, a large variety of biodegradable polyester-based composites

TABLE 10.1

The Applications and Functions of Polymer-Based Bio-Nanocomposite Functions

Biopolymer	Functional Properties	Applications	Reference
Polyvinyl alcohol (PVA)	a. High dissolution b. Good biodegradability c. Non-toxic d. Biocompatible	a. Food packaging b. Textile industry c. Paper industry	Rescignano et al. (2022) Gaaz et al. (2015)
Polylactide (PLA)	a. Poor dissolution b. Highly biodegradable c. Non-toxic d. Biocompatible and bioresorbable	a. Food packaging b. Tissue engineering c. Drug carriers d. Medical implants e. Textile	Yang et al. (2016) De Stefano et al. (2020)
Polycaprolactone (PCL)	a. Poor dissolution b. Good biodegradability c. Non-toxic at molecular level d. Highly biocompatible	a. Drug delivery b. Medical implants	Espinoza et al. (2019) Manivasagam et al. (2019)
Poly-b-hydroxy alkanoates (PHB)	a. Poor dissolution b. Highly biodegradable c. Non-toxic d. Highly biocompatible	a. Speciality packaging b. Drug delivery c. Surgical implants	Policastro et al. (2021) Briso et al. (2018)
Nanocellulose-based bio-nanocomposites	a. Poor dissolution b. Biodegradable and renewable c. Non-toxic d. Biocompatible e. High mechanical strength f. Good barrier properties	a. Reinforcing agent b. Packaging c. Coating	Abe et al. (2007) Li et al. (2009)
Graphene-based bio-nanocomposites	a. Very poor dissolution b. Non-biodegradable (natural means) c. Concentration-dependent toxicity d. Good biocompatibility e. Superior mechanical characteristics f. Heat-conductive	a. Packaging b. Medical instruments c. Drug delivery d. Medical implants	Chang et al. (2013)
CNT-based bio-nanocomposites	a. Poor dissolution b. Non-biodegradable (natural means) c. Toxic d. High mechanical and structural characteristics e. Heat-conductive	a. Drug delivery b. Textile and fabric industry c. Electrical types of equipment	Coleman et al. (2004) Cadek et al. (2004) Gan et al. (2020)
Active bio-nanocomposites	a. Dissolution depends on the active agent used b. Biodegradable c. High antioxidant property d. High antimicrobial and antifungal activity	a. Food packaging b. Delays food spoilage c. Delays fruit ripening d. Prevents microbial spoilage of food e. Provides mechanical strength	Chen et al. (2021)

are accessible. Active agents/carbon and natural fibres, e.g., cellulose, are often used to reinforce composites that are developed using polyesters. In the recent decades, the use of natural fibres derived from natural resources has been growing in usage, as reinforcement agents, because of their easy availability and rapid degradation due to their biodegradable nature. Polyester composites reinforced with natural fibres typically have adequate mechanical qualities, including tensile properties, impact resistance and elasticity, making them to be appropriate for a variety of applications. It has been shown that aliphatic polyesters are more biodegradable than other types of polymers. Biodegradable composite materials are often made from PLAs, PVAs, and PHBs. In addition to being easy to handle, polyester-based biodegradable composites have excellent mechanical and functional qualities. Biodegradable polyester-based composite materials first appeared on the market in the medical industry, specifically in the tissue engineering and packaging industries. Its performance relies heavily on its biodegradability, mechanical characteristics and piezoelectric qualities. Numerous breakthroughs and innovations have occurred in the last several decades. The environmental advantages of exploiting renewable resources are a major driving force behind the biopolymer industry's bright future. The biodegradable nanocomposite of the next era will most probably concentrate on producing materials with superior material characteristics and other functional properties, as well as a significant level of biodegradability. Certainly, the biopolymer industry has a very prominent future, which is mainly due to the environmental advantages of using renewable resources. Future research studies and experimental development of biodegradable nanocomposite materials are envisaged to concentrate completely on developing materials with optimal strength and barrier qualities, as well as their capability to inhibit pathogenic organisms by providing antimicrobial or antioxidant ability. This is shown by the widespread usage of biodegradable composite materials in several applications, including the food packaging, pharmaceutical, automobile, industrial, construction and cosmetics industries. The relevance and value of biodegradable composite materials cannot be emphasized from an ecological standpoint. From this perspective, polyester-based biodegradable composites will play a major role in the development of emerging products.

ACKNOWLEDGEMENT

The authors Emmanuel Rotimi Sadiku and Periyar selvam Sellamuthu acknowledge the financial support from the NIHSS for the Digital APP and BRICS Mobility Teaching and Research Projects, the Joint International Collaborations Research Projects, Project Ref: JNI21/1010.

REFERENCES

Abdullah, Z., Dong, Y., Davies, I., & Barbhuiya, S. (2017). PVA, PVA blends and their nanocomposites for biodegradable packaging application. *Polymer-Plastics Technology and Engineering*, *56*(12), 1307–1344. https://doi.org/10.1080/03602559.2016.1275684

Alghunaim, N. (2016). Optimization and spectroscopic studies on carbon nanotubes/PVA nanocomposites. *Results in Physics*, *6*, 456–460. https://doi.org/10.1016/j.rinp.2016.08.002

Alothman, O. Y., Jawaid, M., Senthilkumar, K., Chandrasekar, M., Alshammari, B. A., Fouad, H., Hashem, M., & Siengchin, S. (2020). Thermal characterization of date palm/epoxy composites with fillers from different parts of the tree. *Journal of Materials Research and Technology*, 9(6), 15537–15546. https://doi.org/10.1016/j.jmrt.2020.11.020

Amin, K., Partila, A., Abd El-Rehim, H., & Deghiedy, N. (2020). Antimicrobial ZnO nanoparticle–doped polyvinyl alcohol/pluronic blends as active food packaging films. *Particle & Particle Systems Characterization*, 37(4), 2000006. https://doi.org/10.1002/ppsc.202000006

Anžlovar, A., Krajnc, A., & Žagar, E. (2020). Silane modified cellulose nanocrystals and nanocomposites with LLDPE prepared by melt processing. *Cellulose*, 27(10), 5785–5800. https://doi.org/10.1007/s10570-020-03181-y

Auras, R., Harte, B., & Selke, S. (2004). An overview of polylactides as packaging materials. *Macromolecular Bioscience*, 4(9), 835–864. https://doi.org/10.1002/mabi.200400043

Aziz, T., Fan, H., Zhang, X., Haq, F., Ullah, A., & Ullah, R. et al. (2020). Advance study of cellulose nanocrystals properties and applications. *Journal of Polymers and the Environment*, 28(4), 1117–1128. https://doi.org/10.1007/s10924-020-01674-2

Azizi Samir, M., Alloin, F., Sanchez, J., & Dufresne, A. (2004). Cellulose nanocrystals reinforced poly(oxyethylene). *Polymer*, 45(12), 4149–4157. https://doi.org/10.1016/j.polymer.2004.03.094

Bucci, D., Tavares, L., & Sell, I. (2005). PHB packaging for the storage of food products. *Polymer Testing*, 24(5), 564–571. https://doi.org/10.1016/j.polymertesting.2005.02.008

Chandrasekar, M., Senthilkumar, K., Jawaid, M., Alamery, S., Fouad, H., & Midani, M. (2022a). Tensile, thermal and physical properties of washightonia trunk fibres/pineapple fibre biophenolic hybrid composites. *Journal of Polymers and the Environment*, 30(10), 4427–4434. https://doi.org/10.1007/s10924-022-02524-z.

Chandrasekar, M., Senthilkumar, K., Jawaid, M., Mahmoud, M. H., Fouad, H., & Sain, M. (2022b). Mechanical, morphological and dynamic mechanical analysis of pineapple leaf/washingtonia trunk fibres based biophenolic hybrid composites. *Journal of Polymers and the Environment*, 30(10), 4157–4165. https://doi.org/10.1007/s10924-022-02482-6.

Chandrasekar, M., Siva, I., Kumar, T. S. M., Senthilkumar, K., Siengchin, S., & Rajini, N. (2020). Influence of fibre inter-ply orientation on the mechanical and free vibration properties of banana fibre reinforced polyester composite laminates. *Journal of Polymers and the Environment*, 28(11), 2789–2800. https://doi.org/10.1007/s10924-020-01814-8.

Chang, H., & Wu, H. (2013). Graphene-based nanocomposites: Preparation, functionalization, and energy and environmental applications. *Energy &Amp; Environmental Science*, 6(12), 3483. https://doi.org/10.1039/c3ee42518e

Chen, D., Lawton, D., Thompson, M., & Liu, Q. (2012). Biocomposites reinforced with cellulose nanocrystals derived from potato peel waste. *Carbohydrate Polymers*, 90(1), 709–716. https://doi.org/10.1016/j.carbpol.2012.06.002

Chen, W., Ma, S., Wang, Q., McClements, D., Liu, X., Ngai, T., & Liu, F. (2021). Fortification of edible films with bioactive agents: A review of their formation, properties, and application in food preservation. *Critical Reviews in Food Science and Nutrition*, 1–27. https://doi.org/10.1080/10408398.2021.1881435

Chen, W., Yu, H., Liu, Y., Hai, Y., Zhang, M., & Chen, P. (2011). Isolation and characterization of cellulose nanofibers from four plant cellulose fibers using a chemical-ultrasonic process. *Cellulose*, 18(2), 433–442. https://doi.org/10.1007/s10570-011-9497-z

Ciambelli, P., Sarno, M., Gorrasi, G., Sannino, D., Tortora, M., & Vittoria, V. (2005). Preparation and physical properties of carbon nanotubes–PVA nanocomposites. *Journal of Macromolecular Science, Part B*, 44(5), 779–795.

Cobos, M., Fernández, M., & Fernández, M. (2018). Graphene-based poly(vinyl alcohol) nanocomposites prepared by *in-situ* green reduction of graphene oxide by ascorbic acid: Influence of graphene content and glycerol plasticizer on properties. *Nanomaterials*, 8(12), 1013. https://doi.org/10.3390/nano8121013

Conzatti, L., Giunco, F., Stagnaro, P., Capobianco, M., Castellano, M., & Marsano, E. (2012). Polyester-based biocomposites containing wool fibres. *Composites Part A: Applied Science and Manufacturing*, *43*(7), 1113–1119. https://doi.org/10.1016/j.compositesa.2012.02.019

DeStefano, V., Khan, S., & Tabada, A. (2020). Applications of PLA in modern medicine. *Engineered Regeneration*, *1*, 76–87. https://doi.org/10.1016/j.engreg.2020.08.002

Espinoza, S., Patil, H., San Martin Martinez, E., Casañas Pimentel, R., & Ige, P. (2019). Poly-ε-caprolactone (PCL), a promising polymer for pharmaceutical and biomedical applications: Focus on nanomedicine in cancer. *International Journal of Polymeric Materials and Polymeric Biomaterials*, *69*(2), 85–126. https://doi.org/10.1080/00914037.2018.15 39990.

Fraschini, C., Chauve, G., Le Berre, J., Ellis, S., Méthot, M., O'Connor, B., & Bouchard, J. (2014). Critical discussion of light scattering and microscopy techniques for CNC particle sizing. *Nordic Pulp & Amp; Paper Research Journal*, *29*(1), 31–40. https://doi.org/10.3183/npprj-2014-29-01-p031-040

Frone, A., Berlioz, S., Chailan, J., & Panaitescu, D. (2013). Morphology and thermal properties of PLA–cellulose nanofibers composites. *Carbohydrate Polymers*, *91*(1), 377–384. https://doi.org/10.1016/j.carbpol.2012.08.054

Gaaz, T., Sulong, A., Akhtar, M., Kadhum, A., Mohamad, A., & Al-Amiery, A. (2015). Properties and applications of polyvinyl alcohol, halloysite nanotubes and their nanocomposites. *Molecules*, *20*(12), 22833–22847. https://doi.org/10.3390/molecules201219884

Gan, D., Dou, J., Huang, Q., Huang, H., Chen, J., & Liu, M. et al. (2020). Carbon nanotubes-based polymer nanocomposites: Bio-mimic preparation and methylene blue adsorption. *Journal of Environmental Chemical Engineering*, *8*(2), 103525. https://doi.org/10.1016/j.jece.2019.103525

García-García, D., Balart, R., Lopez-Martinez, J., Ek, M., & Moriana, R. (2018). Optimizing the yield and physico-chemical properties of pine cone cellulose nanocrystals by different hydrolysis time. *Cellulose*, *25*(5), 2925–2938. https://doi.org/10.1007/s10570-018-1760-0

Gupta, V., Ramakanth, D., Verma, C., Maji, P., & Gaikwad, K. (2021). Isolation and characterization of cellulose nanocrystals from amla (*Phyllanthus emblica*) pomace. *Biomass Conversion and Biorefinery*. https://doi.org/10.1007/s13399-021-01852-9

Inácio, E., Souza, D., & Dias, M. (2020). Thermal and crystallization behavior of PLA/PLLA-grafting cellulose nanocrystal. *Materials Sciences and Applications*, *11*(1), 44–57. https://doi.org/10.4236/msa.2020.111004

Jonoobi, M., Harun, J., Mathew, A., Hussein, M., & Oksman, K. (2009). Preparation of cellulose nanofibers with hydrophobic surface characteristics. *Cellulose*, *17*(2), 299–307. https://doi.org/10.1007/s10570-009-9387-9

Liu, H., Li, C., Wang, B., Sui, X., Wang, L., & Yan, X. et al. (2017). Self-healing and injectable polysaccharide hydrogels with tunable mechanical properties. *Cellulose*, *25*(1), 559–571. https://doi.org/10.1007/s10570-017-1546-9

Manivasagam, G., Reddy, A., Sen, D., Nayak, S., Mathew, M., & Rajamanikam, A. (2019). Dentistry: Restorative and regenerative approaches. In *Encyclopedia of Biomedical Engineering*, pp. 332–347. https://doi.org/10.1016/b978-0-12-801238-3.11017-7

Nasimudeen, N. A., Karounamourthy, S., Selvarathinam, J., Kumar Thiagamani, S. M., Pulikkalparambil, H., Krishnasamy, S., & Muthukumar, C. (2021). Mechanical, absorption and swelling properties of vinyl ester based natural fibre hybrid composites. *Applied Science and Engineering Progress*. https://doi.org/10.14416/j.asep.2021.08.006

Policastro, G., Panico, A., & Fabbricino, M. (2021). Improving biological production of poly(3-hydroxybutyrate-co-3-hydroxyvalerate) (PHBV) co-polymer: A critical review. *Reviews in Environmental Science and Bio/Technology*, *20*(2), 479–513. https://doi.org/10.1007/s11157-021-09575

Rescignano, N., Fortunati, E., Montesano, S., Emiliani, C., Kenny, J., Martino, S., & Armentano, I. (2022). PVA bio-nanocomposites: A new take-off using cellulose nanocrystals and PLGA nanoparticles. *Carbohydrate Polymers*, *99*, 47–58. https://doi.org/10.1016/j.carbpol.2013.08.061

Rivera-Briso, A., & Serrano-Aroca, Á. (2018). Poly(3-hydroxybutyrate-co-3-hydroxyvalerate): Enhancement strategies for advanced applications. *Polymers*, *10*(7), 732. https://doi.org/10.3390/polym10070732

Saravanan, N., Rajasekar, R., Mahalakshmi, S., Sathishkumar, T., Sasikumar, K., & Sahoo, S. (2014). Graphene and modified graphene-based polymer nanocomposites–A review. *Journal of Reinforced Plastics and Composites, 33*(12), 1158–1170. https://doi.org/10.1177/0731684414524847

Sathishkumar, T., Naveen, J., Navaneethakrishnan, P., Satheeshkumar, S., & Rajini, N. (2016). Characterization of sisal/cotton fibre woven mat reinforced polymer hybrid composites. *Journal of Industrial Textiles, 47*(4), 429–452. https://doi.org/10.1177/1528083716648764

Senthilkumar, K., Saba, N., Chandrasekar, M., Jawaid, M., Rajini, N., Siengchin, S., Ayrilmis, N., Mohammad, F., & Al-Lohedan, H. A. (2021). Compressive, dynamic and thermo-mechanical properties of cellulosic pineapple leaf fibre/polyester composites: Influence of alkali treatment on adhesion. *International Journal of Adhesion and Adhesives, 106*, 102823. https://doi.org/10.1016/j.ijadhadh.2021.102823

Senthilkumar, K., Siva, I., Rajini, N., & Jeyaraj, P. (2015). Effect of fibre length and weight percentage on mechanical properties of short sisal/polyester composite. *International Journal of Computer Aided Engineering and Technology, 7*(1), 60. https://doi.org/10.1504/IJCAET.2015.066168

Senthilkumar, K., Subramaniam, S., Ungtrakul, T., Kumar, T. S. M., Chandrasekar, M., Rajini, N., Siengchin, S., & Parameswaranpillai, J. (2022). Dual cantilever creep and recovery behavior of sisal/hemp fibre reinforced hybrid biocomposites: Effects of layering sequence, accelerated weathering and temperature. *Journal of Industrial Textiles, 51*(2_suppl), 2372S-2390S. https://doi.org/10.1177/1528083720961416

Shahroze, R. M., Chandrasekar, M., Senthilkumar, K., Senthil Muthu Kumar, T., Ishak, M. R., Rajini, N., Siengchin, S., & Ismail, S. O. (2021). Mechanical, interfacial and thermal properties of silica aerogel-infused flax/epoxy composites. *International Polymer Processing, 36*(1), 53–59. https://doi.org/10.1515/ipp-2020-3964

Sivakanthan, S., Rajendran, S., Gamage, A., Madhujith, T., & Mani, S. (2020). Antioxidant and antimicrobial applications of biopolymers: A review. *Food Research International, 136*, 109327. https://doi.org/10.1016/j.foodres.2020.109327

Sorrentino, A., Gorrasi, G., & Vittoria, V. (2007). Potential perspectives of bio-nanocomposites for food packaging applications. *Trends in Food Science & Technology, 18*(2), 84–95. https://doi.org/10.1016/j.tifs.2006.09.004

Thomas, S. K., Parameswaranpillai, J., Krishnasamy, S., Begum, P. M. S., Nandi, D., Siengchin, S., George, J. J., Hameed, N., Salim, Nisa. V., & Sienkiewicz, N. (2021). A comprehensive review on cellulose, chitin, and starch as fillers in natural rubber biocomposites. *Carbohydrate Polymer Technologies and Applications, 2*, 100095. https://doi.org/10.1016/j.carpta.2021.100095

Usmani, M., Khan, I., Gazal, U., Mohamad Haafiz, M., & Bhat, A. (2018). Interplay of polymer bionanocomposites and significance of ionic liquids for heavy metal removal. *Polymer-Based Nanocomposites for Energy and Environmental Applications*, 441–463. https://doi.org/10.1016/b978-0-08-102262-7.00016-7

Yang, W., Fortunati, E., Dominici, F., Giovanale, G., Mazzaglia, A., & Balestra, G. et al. (2016). Effect of cellulose and lignin on disintegration, antimicrobial and antioxidant properties of PLA active films. *International Journal of Biological Macromolecules, 89*, 360–368. https://doi.org/10.1016/j.ijbiomac.2016.04.068

11 Hybrid Polyester and Bio-Polyester Composites

Siti Noor Hidayah Mustapha and Rohani Mustapha
Universiti Malaysia Pahang

CONTENTS

11.1   INTRODUCTION

Polyester is a polymer with an ester functional group in every repeating unit of its main chain. It is synthetically produced by condensation reactions between dicarboxylic acids and diols. The polyester group includes polyhydroxyalkanoates, which are produced in nature by numerous microorganisms and also through bacteria fermentation of sugars or lipids, and poly(alkylene dicarboxylates) or also known as bio-polyester [1]. Depending on its structure and behavior toward heat, polyester can be classified as thermoset and thermoplastic. Examples of thermoplastic polyester are polyethylene terephthalate, polyglycolic acid, and polylactic acid, which can easily be recycled, while thermoset polyester is known as unsaturated polyester resin (UPR), which is commonly applied in composite industries and cannot be recycled. UPR is extremely versatile in meeting the end-user requirements, which have significantly been the driving force in their widespread growth. UPRs can be compounded with various types of reinforcement fillers and cured under heat or radiation. The wide-ranging chemical and physical properties of UPRs depend upon the selection of diacids and diols, cross-linking agents, initiators, and other additives [2]. The majority of UPRs are used in the marine industry, such as in the production of hulls, decks, hatch, and engine conveyers; in the construction industry, such as in the production of bathroom compartments and fixtures, pipes, tanks, panels, and window frames;

and in the transportation industry, such as in the production of automotive parts and structural components [3].

Synthetic and natural fibers-reinforced polyester is the major approach applied in composite industry. Besides their biodegradability, natural fibers are much cheaper, have a lower density, and have properties that are acceptable compared to synthetic fibers [4, 5]. Polyester gives high mechanical properties to the fabricated composites; while the associated fibers contribute excellent rigidity, low density, and biodegradability over polyester, in which this combination provides excellent properties of composites [6].

11.2   CONCEPT OF HYBRID POLYMER COMPOSITES

Hybrid polymer composites are not clearly defined in a material category. It is the result of the design method that associates the polymer integration in a polymer or composite system. It is understood that high-functionality materials are difficult to obtain using a single polymer matrix. Thus, hybrid polymer composite is understood as an effective strategy to achieving desired properties by combining more than single continuous phase (matrix) and discontinuous phase (reinforcement). In some composites, there may be two interpenetrating continuous phases, while in others, there needs to be a non-continuous phase or two non-continuous phases in a single continuous phase. It should be noted that there are no limits to hybridizing polymers and composites.

The common example of hybrid composites is laminated composite, in which two or more types of fiber are incorporated in a laminated form. The laminated fibers provide enhanced properties in comparison with the corresponding single fiber-reinforced composite [7]. Currently, increasing sustainability and environmental awareness have increased the interest to substitute synthetic fiber with natural fiber in engineering applications. However, natural fibers have variability in their characteristic properties; thus, hybridizing two fibers, including natural and natural fibers or natural and synthetic fibers, could improve specific performances and reduce the weakest aspect of both natural and synthetic fibers [7–9].

A hybrid system could also be designed on the matrix or a continuous system in a composite. Bio-based matrices have widely been studied, but their performance still could not compete with the synthetic polymer matrix. For example, vegetable oil (VO)-based thermoset resins gives various performances depending on the reactivity sites such as unsaturation and ester bonds. Lack of aromatic or cycloaliphatic compound in the VO structure resulted in lower stiffness and strength properties of the resulting resin [10]. Thus, blending VO partly with synthetic resin could maintain the stiffness and strength, and improve the thermal and toughness properties of the resin [11, 12]. However, providing higher toughness at higher VO loadings in the blends lowers the stiffness performance of the composite [13]. Thus, reinforcement filler is always added to the blends system, which provides higher stiffness and therefore results in a proper toughness–stiffness balance of the resulting resin [14, 15]. In addition, filler also provides improvement in specific properties of composite such as hydrophobic properties, thermal resistance and fire retardancy. In other words, composites with two or more fillers of the same matrix are also called hybrid composites [16].

11.3   HYBRID NATURAL FIBERS-REINFORCED POLYESTER COMPOSITES

Natural fiber-reinforced polyester composites have widely been applied specifically for low technology applications, including structural applications such as building, automotive, and sports equipment industry. Natural fibers have attracted more interest than synthetic fibers as it is abundantly available, renewable, and environmentally friendly. In addition, natural fiber possesses properties of lightweight, low cost, high flexibility, biodegradability, high specific strength, high toughness, and easy processing. Natural fiber is a fiber that is synthesized from natural sources, and can be classified according to its origin: vegetable fiber from cellulose, wood fiber from a tree, animal fiber such as spider silk and silkworm silk, mineral fiber including asbestos group, and biological fiber such as fibrous protein and collagen. Different types of fiber orientations such as random orientation, unidirectional for continuous or long fiber, and bidirectional such as woven fiber reinforcement, have a range of properties. Thus, natural fibers of specific type of orientation are selected as the reinforcement agent in polyester for a specific composite application [17]. Besides the advantages, natural fiber also shows low strength and high-water absorption properties. To meet the demand for advanced composite materials with higher mechanical properties, recently, there has been a growing interest in hybridizing different types of natural fibers as reinforcement in a polyester composite. Besides the properties of the fiber and matrix, the design of the hybrid system and interfacial adhesion between the fiber and matrix significantly affect the properties of the final composite [18]. Specific volume and weight percent of fiber are also important in determining the optimum behavior of the hybrid natural fiber-reinforced polyester composite.

Dixit and Verma [19] studied the mechanical performance of hybridized different natural fibers such as coir, sisal, and jute reinforced in polyester composite in sandwich construction using the hand lay-up technique. This study revealed that composite with the outer layer of jute and core and sisal in the middle has the highest modulus, tensile strength, and flexural strength properties as compared to other arrangements. The same fiber construction also provided the lowest water absorption after immersion in water for 24 hours. These findings demonstrate that the improvement of mechanical properties of polyester composite is highly dependent on the hybridization arrangement. Another study conducted by Shanmugam and Thiruchitrambalam [20] proved that the hybridization of fibers provided higher thermo-mechanical properties of the polyester composite. It was revealed by the increase of static and dynamic mechanical properties of the polyester composite when unidirectional continuous palmyra palm leaf stalk hybridized with jute fiber was reinforced in the composite. The tensile strength, tensile modulus, storage, and loss modulus, and tan d of the continuous unidirectional bi-layer hybrid composite improved with increasing jute content with an alkali treatment of fibers.

Athijayamani et al. [21], who studied randomly oriented roselle/sisal (1:1) hybrid polyester composite, found that fiber content and lengths significantly affect the water absorption and mechanical properties of the composite in dry and wet conditions. Fiber loadings and fiber length have different effects on moisture absorption, i.e., higher fiber loadings increase the moisture absorption, while higher fiber length gives the opposite effect. In dry condition, increasing hybrid fiber loading and length increases the tensile and flexural strength of the hybrid composite. Exposure to moisture (wet condition) degraded the fiber–matrix interface and caused a significant drop in the mechanical properties of the composite. In a previous study, Alavudeen et al. [22] explored the mechanical properties of banana/kenaf fiber-reinforced hybrid polyester composites, and in particular the effect of woven fabric of plain and twill type and random orientation. Based on the results, they found that the plain type of woven fiber showed higher tensile and flexural strength than the twill-type woven fiber and random-orientation fiber. The plain-woven hybrid banana/kenaf composite showed 54% and 15% higher tensile strength than the pure banana/polyester and kenaf/polyester composite, respectively. Moreover, the highest tensile modulus and flexural strength properties were also observed. The performance of random-orientation fiber was lower compared to woven fiber due to non-uniform distribution of stress transfer in both the longitudinal and transverse directions when tensile load is applied. Moreover, the plain type of banana/kenaf woven fiber shows highest mechanical performance due to the highest stress transfer when the fibers of greater strength (kenaf fiber) are oriented longitudinally which the tensile load could be bear along the direction of fiber.

When two types of fibers have the same length but different diameters, they will provide a different effective area for the fiber–matrix adhesion, which provides different stress transfer to the composite material [23]. An analysis on thermo-physical properties of banana/sisal hybrid fiber-reinforced polyester composite conducted by Idicula et al. [24] found that, at a relative 1:1 volume fraction of banana and sisal fiber, 40% fiber volume fraction (banana/sisal) showed the highest increment in storage modulus, tensile modulus, flexural strength, and glass transition temperature of the polyester composite. This condition indicates better fiber–matrix interaction at 40% fiber volume fractions. They also confirmed that banana/polyester and hybrid fibers/polyester showed almost the same storage modulus properties, which are higher than sisal/polyester composite. The high storage modulus indicates good interface bonding for the hybrid fiber with the polyester matrix. Based on the analysis, even though the cellulose content of banana and sisal fiber is almost similar, the lumen size, microfibrillar angle, and the diameter of banana fiber are smaller, which contributed to higher wetting and interactions with the matrix. This phenomenon affected better performances of hybridizing banana fiber with other fiber types with higher diameters. The same results were found by Venkateshwaran et al. [25] on epoxy banana/sisal-reinforced epoxy hybrid composite and Boopalan et al. [26] on hybridizing banana fiber in jute/epoxy composites. They proved that hybridizing up to 50 wt.% banana fibers in sisal/epoxy and jute/epoxy composite, respectively, result in higher mechanical and thermal properties and decreased moisture absorption properties.

11.4 HYBRID NATURAL-SYNTHETIC FIBERS-REINFORCED POLYESTER COMPOSITES

Although natural fibers are cheaper and lighter, they have low mechanical properties than synthetic fibers such as glass, aramid, and carbon fiber. The hybridization of different types of natural fibers-reinforced polyester composite still could not yield superior mechanical properties. This phenomenon limits the use of natural fibers and their hybridization of polyester composites in high-performance technology applications. However, the cost of synthetic fibers which have higher mechanical properties is high, and the higher production cost urges for an alternative approach. Due to this issue, hybridization of synthetic with natural fibers, which can reduce the production cost while not compromising its performances, is preferable [27]. For instance, hybridization of natural fiber with synthetic glass and carbon fiber can also improve the stiffness, strength, and moisture-resistant behavior of the natural fiber composite [28]. Due to the lower price, glass fibers are typically being selected to be hybridized with natural fibers in composite applications. Like natural–natural fibers hybridization process, the synthetic–natural fibers hybridization can also be done at different orientations and stagnation, loadings, and fiber sizes.

Ahmed and Vijayarangan [29] studied the stacking sequence of untreated woven jute and glass fabric-reinforced polyester hybrid composites using the hand lay-up technique. At a constant of 10 plies layers, there is a huge improvement in tensile and flexural properties of jute composite by the incorporation of glass fiber as extreme plies (outer surfaces). Based on the results, the tensile and flexural modulus and strength have improved by 30% and 53%, and 62% and 31%, respectively. In another study, Ramesh, Palanikumar, and Reddy [27] have proven that orientation and fiber selection are important to determine the best mechanical performance of the composite. Based on their study, jute/glass has the highest tensile strength as compared to sisal/glass and sisal/jute/glass-reinforced polyester composite at 5 plies of fiber with glass fiber as the extreme plies.

Mishra et al. [30] studied the effect of glass/pineapple leaf/sisal fibers in polyester composite. Based on their results, the ultimate tensile strength of pineapple leaf fibers (PALF)-polyester composite at 25 wt.% fiber volume fraction, V_f, is significantly improved by 66% when the PALF is hybridized with 8.5 wt.% glass fiber. While for sisal-polyester composite at 30 wt.% V_f, the ultimate tensile strength is increased by more than 50% when hybrid sisal/glass fiber is used. These results revealed that the properties of the composite are mainly dependent on the properties of individual reinforcing fiber and specific composition of synthetic-natural fiber hybridization is important to provide the ultimate mechanical properties of the composite. Besides the mechanical properties, hybridization of PALF/glass fiber also allows significant better heat transportability of composite. It can be observed by the 50% increment of thermal conductivity, and higher thermal diffusivity in the hybrid PALF/glass fiber as compared to PALF reinforced in polyester composite [31]. In a different study by Sapuan et al. [32], they found that addition of basalt in woven glass fiber polyester composite has improved the density, tensile and flexural strength of polyester composite. Hybridization of 75% basalt fiber with 25% of woven glass fiber-reinforced polyester composite showed the highest improvement in tensile strength which is more than 50% of woven glass fiber reinforcement composite.

11.5 BIO-BASED POLYESTER RESIN

The polyester resins synthesized directly from natural resources such as lipids, protein, polysaccharides, sugar, VOs, and other monomers usually exhibit low mechanical and thermal properties, and limit their use in the market. In the literature, the most reported polyester blending is blending with VO. VO is one of the potential materials to replace petroleum-based polymer due to its worldwide availability and is a major agricultural commodity. Therefore, VOs are a valuable source of bio-based polymer resins. VO is composed of various chemical structures and compositions, and it can be activated for condensation polymerization by adding a curing agent or a latent catalyst [33].

The use of VO in polyester resins has caught the attention of researchers due to its apparent toughening effect [34]. This indirectly overcomes the main limitations of polyester resin: brittleness and low impact resistance. VO is mainly composed of triglycerides and contains various fatty acids. The physical and chemical properties of VOs are different from each other, depending on the fatty acid composition and whether there are unsaturated or double bonds in the triglyceride chain [35]. VO in native form is not reactive. Therefore, several techniques have been used to functionalize VOs to increase their reactivity with polymer resins, such as epoxidation, acrylation, maleinization, amidation, and glycerolysis reactions [36].

There are several researchers conducted studies of blending polyester resin with VOs such as soybean oil [14, 37, 38], palm oil [9, 39], grapeseed oil [40], and linseed oil [41]. In conclusion, the experimental results suggest that the inclusion of VO in polyester resin enhanced the toughness properties and elasticity of brittle polyester resins [12, 14, 15]. The higher the amount of VO, the higher the ability of the bio-based resin to absorb the impact energy. This may be due to the long, flexible, and elastic network structure of the VO, which improves the mobility of the resulting resin and its resistance to cracking and propagation [42].

Liu et al. [43] studied the potential use of tung oil (TO) as a toughening agent in UPRs. They observed that the impact strength of the modified system was significantly improved with the optimum value of 20% TO content. The impact strength of the modified system was significantly improved with a maximum increase of 370% compared with that of the unmodified system. Das et al. [44] produced bio-based polymer resin by blending UPR and TO. They also concluded that the TO improved resin toughness at a very low content of TO.

Liu et al. [45] investigated the mechanical and thermal properties of UPR with the addition of castor oil. In their work, they found that the mechanical and thermal properties of the unsaturated polyester (UPE) blend were better than those of neat UPE. However, the tensile strength and flexural strength values decreased as castor oil increased. Therefore, to balance the strength or stiffness and toughness properties of bio-based thermoset resin, several researchers reported the addition of nanofiller is required.

11.6 NATURAL FIBER-REINFORCED POLYESTER/ VEGETABLE OIL HYBRID COMPOSITE

The development of bio-based polymer composites reinforced with natural fibers also known as Fiber Reinforced Polymer (FRP). Indeed, natural fibers are less expensive, very abundant, and have good mechanical properties. As a result, its bio-polyester resin reinforcement has contributed to the creation of green composite products, offering a promising compromise between respect for the environment and performance. The commonly used natural fibers are kenaf, flax, hemp, and sisal fibers. The employment of treated natural fibers was reported to give significant improvements in tensile strength, flexural strength, Izod impact, and storage modulus of bio-based composites.

The development of natural fiber-reinforced bio-based polyester resins was reported by a few researchers. One of the studies was conducted by Pfister, D. P. and Larock [46], who prepared composites from flaxseed oil-based conjugate resins using wheat straw as a reinforcement. They found that increasing the wheat straw loading has increased the mechanical properties and increased the thermal properties of the composite. This composite material is suitable for use in non-structural applications such as automotive panels, wall or floor panels, furniture, and residential building materials.

Hosur et al. [47], on the other hand, showed that the use of the commercial UPR/VO-based system exhibit similar properties to the commercial polyester resin. The study also showed that introducing a higher concentration of alkali solution up to 2.5% will increase the degree of crystallinity of the fibers. Liu and co-workers [48] have developed green composites from soybean thermoset resin using two different processing methods: injection molding and compression molding. The kenaf fibers have been used as reinforcement. The effects of different processing methods and fiber length on the physical properties of the composites were investigated. It was observed that the compression-molded composites showed a similar modulus to injection-molded composites. Haq et al. [14] employed natural fibers (industrial hemp) as reinforcement in UPE/epoxidized soybean oil (epoxidized methyl soyate (EMS)). Their works revealed that the UPE/EMS/hemp fiber composites experience improvement in impact strength with the increase in EMS loading.

11.7 HYBRID NATURAL FIBER-REINFORCED POLYESTER/VEGETABLE OIL COMPOSITE

More recently, the focus has also been on the development of hybrid composites that contain more than one type of reinforcing fiber in a single matrix. Hybrid composites have properties that cannot be achieved by single-fiber composites. They offer more advanced composites with reduced cost, acceptable corrosion resistance, good mechanical properties, i.e., flexural modulus, good strength properties, and good thermal stability. However, studies on hybrid natural fiber-reinforced bio-based polyester composites are rarely seen in the literature. There is one recent study conducted by Mustapha and co-workers [9] on the hybrid kenaf and empty fruit bunch (EFB) fibers-reinforced polyester/VO composites. They investigated the effect of different weight percentages of hybrid kenaf and EFB compositions and fiber fractions (wt.%) on the mechanical and thermal properties of UPE/epoxidized palm oil (EPO) composites. In their findings, they revealed that hybridization of kenaf and EFB fibers improved the mechanical properties such as strength, stiffness, and toughness of the composites and improved thermal stability. Dramatic improvements were observed at lower fibers fractions, i.e., 9 and 12% by weight.

11.8 NANOFILLER-FILLED NATURAL FIBER-REINFORCED POLYESTER HYBRID COMPOSITE

As reported earlier, most polymers, including polyester, require fillers to improve the properties of the resin. Nanomaterials such as nanoclays, graphene, carbon nanotubes, silica, and alumina have been utilized as reinforcing fillers in polymer nanocomposites. Incorporating small amounts of nanofillers into polymer matrices has been reported to significantly improve the tensile strength, modulus, and thermal stability of composites.

In the development of natural fiber-reinforced polyester nanocomposites, the most investigated nanofillers were nanoclays [12, 49], eggshell powder (ESP) [50, 51], and calcium carbonate [52]. The main challenges in the development of hybrid natural fiber-based nanocomposites are to achieve good dispersion of nanoparticles within the polyester resin and also to obtain good interfacial adhesion between polyester, fibers, and the nanofillers. Various researchers have used nanoclays as nanofillers in polymer nanocomposite systems. However, most of the work using nanoclays as reinforcing nanofillers has been hindered by obstacles related to achieving optimal dispersion of nanoclays in the polymer matrix and inadequate interfacial interaction between the nanoclays and the polymer matrix. The problem of inadequate dispersion of nanoclays arose from the aggregation of nanoclays by Van der Waals forces between the nanofillers. Therefore, various mechanical and chemical techniques have been used to improve the dispersion of fillers in the matrix. The most widely used technique in polyester/nanoclays is using the ultrasonication process. This method is reported to produce good dispersion of nanoclays in the polymer matrix which leads to improved mechanical and thermal properties [53].

On the other hand, the hydrophilicity of nanoclay hinders the ability of nanoclay to disperse well in many polymer matrices. To overcome this problem, various modifications of nanoclays such as amino acids, tetraorganic phosphonium or organic ammonium salts are used [53]. The addition of clay to a polymer resin system can be exfoliated, intercalated, or tactoid, depending on the method of preparation or synthesis, type of clay, amount of organic modifier, the polarity of the organic modifier, and the type of polymer matrix used [54].

Some researchers have carried out a study on utilizing nanofillers in hybrid natural fiber-reinforced composites. Ganesan et al. [55] studied the synergistic effect of jute fiber and coir fiber as hybrid reinforcements in polyester composites-filled hybrid eggshell powder (ESP) and montmorillonite nanoclay filler. The results show that the addition of the hybrid filler to the hybrid fiber composite improved the mechanical properties compared to the composite without the hybrid filler. Moreover, hybridization of fillers also resulted in reducing water uptake of the composites. The hybrid polyester composite reinforced with 1.5% by weight ESP/MMT hybrid filler shows the lowest water absorption of 3.8% and 4.0% in river and seawater environments, respectively.

The uses of nanofiller in VO-based composites are more significant as the addition of VO in polyester (UPE) resin improved the toughness properties of the VO/UPE resin; however, the stiffness of the VO/UPE resin decreased. The opposing stiffness and toughness performance of the VO/UPE resin needs a proper balance in order to get an efficient thermoset resin. One way to attain this balance is by adding nano-filler. In the literature, there are several researchers reported on the enhancement of the stiffness and mechanical properties of UPE/VO with the addition of nanofillers. The synergy between nanofiller and UPE/VO matrix may produce composites with properties equivalent to or improved from pure polyester resin, and it is expected to be used in a wide range of applications.

Haq et al. [12] studied the effects of nanoclays on epoxidized methyl linseedate/UPR. In this study nanoclays loadings varied between 2.5 and 5.0 wt.%. The addition of nanoclays reduced the strength and toughness properties of the resins. This effect is explained by the increased stiffness of the nanoclay, which in turn forms more brittle nanocomposites. Liu et al. [48] investigated the effects of clay content on mechanical and thermal properties of soybean oil/clay nanocomposites. It was observed that the stiffness properties of the resulted resins were improved by the addition of clay up to 8 wt.%. The further increases in clay loading up to 10 wt.% reduce the storage modulus of the resins.

11.9 FUTURE OUTLOOK

Overall, natural fiber-reinforced polyester composites have proven to be a viable alternative to fiberglass-reinforced polyester composites, especially in automotive and construction products. However, natural fiber-reinforced polyester composites are not environmentally friendly as the polymer matrix material is still petroleum-based and not biodegradable. Therefore, it is important to develop natural fiber-reinforced resins based on renewable resources in order to improve the sustainability of composite materials and solve environmental problems such as reducing dependence on non-renewable resources. The development of bio-based resins from renewable sources is a clear future outlook. For the past decade, natural oils have received a great deal of attention as a raw material for thermosetting resins due to their availability and low cost. However, it has been found that the use of VO-based resins in composite materials is limited due to their lower mechanical and thermo-physical properties compared to synthetic resins. Therefore, most of the research done so far has focused on blends of petroleum-based resins and VOs. Partial replacement of petroleum-based resins with functionalized VOs has been reported to improve the toughness, barrier, and thermal properties of the resulting bio-based resin system. However, the improvement in toughness performance generally goes against the stiffness performance of the material. Therefore, some fillers need to be added to achieve the proper balance of toughness and stiffness and to improve certain properties of the bio-based resin. The development of composites from entirely natural materials and reinforced with natural fibers is one of the goals in the world of collaborative research. For example, the development of polyester resins that are synthesized directly from VOs, reinforced with natural fibers, and filled with natural fillers, can be further investigated in the future. However, some research suggestions are needed to get a composite that gives satisfactory results in a real application.

11.10 CONCLUSION

In summary, this chapter has explained extensively the method, approach, detailed important parameters, and performances of the hybrid polyester composite. There are multiple hybridization approaches discussed in this chapter, including the hybridization of the reinforcing agent such as fibers and fillers. Bio-polyester successfully synthesized by blending polyester and VO reinforced with hybrid natural fibers has a good future in green composites. This chapter also highlights the importance of analyzing the interactions between the continuous and non-continuous, and also between all the hybrid entities contained in the composite to allow optimum performance of polyester composite. In summary, hybridization is the best method that could enhance the properties of polyester and bio-polyester composite to meet the high technology demand in future.

ACKNOWLEDGMENT

This work is supported by FRGS grant funded by Ministry of Higher Education (FRGS/1/2019/STG07/UMP/02/2; Universiti reference RDU1901105), Universiti Malaysia Pahang Research Grant (RDU1903136), and Universiti Malaysia Terengganu (Talent and Publication Enhancement-Research Grant (TAPE-RG); universiti reference Vot 55234), which the authors gratefully acknowledge.

REFERENCES

1. Network Group for Composites in Construction: *Technical Sheet 08. SSustainable FRPs - Naturally Derived Resins and Fibres*. NGCC, Chesterfield, UK (2008).
2. Dholakiya, B.: Unsaturated polyester resin for speciality applications. *Polyester*. 7, 167–202 (2012).
3. Long, T.E. and Scheirs, J. eds.: *Modern Polyesters: Chemistry and Technology of Polyesters and Copolyesters*. Wiley Series in Polymer Science Series Editor. John Wiley & Sons, West Sussex, England (2003).
4. Noor Azammi, A.M., Ilyas, R.A., Sapuan, S.M., Ibrahim, R., Atikah, M.S.N., Asrofi, M., Atiqah, A.: Characterization studies of biopolymeric matrix and cellulose fibres based composites related to functionalized fibre-matrix interface. In *Interfaces in Particle and Fibre Reinforced Composites* (pp. 29–93). Woodhead Publishing, Kidlington, United Kingdom (2019).
5. Nurazzi, N.M., Khalina, A., Sapuan, S.M., Ilyas, R.A., Rafiqah, S.A., Hanafee, Z.M.: Thermal properties of treated sugar palm yarn/glass fiber reinforced unsaturated polyester hybrid composites. *J. Mater. Res. Technol.* 9, 1606–1618 (2020). https://doi.org/10.1016/j.jmrt.2019.11.086
6. Azlin, M.N.., Sapuan, S.., Zainudin, E.., Zuhri, M.Y..: Natural fibre reinforced polyester composites: A review. In *7th Postgraduate Seminar on Natural Fibre Reinforced Polymer Composites* (2020).
7. Safri, S.N.A., Sultan, M.T.H., Jawaid, M., Jayakrishna, K.: Impact behaviour of hybrid composites for structural applications: A review. *Composi. B Eng.* 133, 112–121 (2018).
8. Ahmad, F., Choi, H.S., Park, M.K.: A review: Natural fiber composites selection in view of mechanical, light weight, and economic properties. *Macromol. Mater. Eng.* 10–24 (2015). https://onlinelibrary.wiley.com/doi/full/10.1002/mame.201400089

9. Mustapha, S.N.H., C W Norizan, C.W.N.F., Roslan, R., Mustapha, R.: Effect of kenaf/ empty fruit bunch (EFB) hybridization and weight fractions in palm oil blend polyester composite. *J. Nat. Fibers*. 1885–1898 (2020). https://doi.org/10.1080/15440478.2020.17 88686

10. Mustapha, R., Rahmat, A.R., Abdul Majid, R., Mustapha, S.N.H.: Vegetable oil-based epoxy resins and their composites with bio-based hardener: A short review. *Polym. Technol. Mater*. 58, 1311–1326 (2019). https://doi.org/10.1080/25740881.2018.1563119

11. Rahmat, A.R., Arsad, A., Alsagayar, Z.S., Shoot Kian, Y.: Flexural properties of MMT reinforced unsaturated polyester/epoxidized palm oil biobased resin. *Adv. Mater. Res*. 1112, 377–380 (2015). https://doi.org/10.4028/www.scientific.net/AMR.1112.377

12. Haq, M., Burgueño, R., Mohanty, A.K., Misra, M.: Bio-based polymer nanocomposites from UPE/EML blends and nanoclay: Development, experimental characterization and limits to synergistic performance. *Compos. Part A Appl. Sci. Manuf*. 42, 41–49 (2011).

13. Burgueño, R., Haq, M., Mohanty, A.K., Misra, M.: Hybrid bio-based composites from nano-reinforced bio-petro polymer blends and natural fibers. In *7th Annual Automotive Composites Conference and Exhibition*. ACCE, pp. 32–40 (2007).

14. Haq, M., Burgueño, R., Mohanty, A.K., Misra, M.: Hybrid bio-based composites from blends of unsaturated polyester and soybean oil reinforced with nanoclay and natural fibers. *Compos. Sci. Technol*. 68, 3344–3351 (2008).

15. Haq, M., Burgueño, R., Mohanty, A.K., Misra, M.: Bio-based unsaturated polyester/ layered silicate nanocomposites: Characterization and thermo-physical properties. *Compos. Part A Appl. Sci. Manuf*. 40, 540–547 (2009).

16. Malai Rajan, M.: The effect of hybrid filler loading on mechanical and physical properties of epoxy bio-composite. Bachelor of Engineering (Hons) (Mechanical Engineering. Universiti Teknologi Petronas, Tronoh, Perak, Malaysia (2012).

17. Mallick, P.K.: *Fibre-Reinforced Composites Materials, Manufacturing and Design*. CRC Press, Taylor & Francis Group, Boca Raton, USA (2007).

18. Thwe, M.M., Liao, K.: Characterization of bamboo–glass fiber reinforced polymer matrix hybrid composite. *J. Mater. Sci. Eng. Lett*. 19, 1873–1876 (2000).

19. Dixit, S., Verma, P.: The effect of hybridization on mechanical behaviour of coir/sisal/ jute fibres reinforced polyester composite material. *Res. J. Chem. Sci*. 2, 91–93 (2012).

20. Shanmugam, D., Thiruchitrambalam, M.: Static and dynamic mechanical properties of alkali treated unidirectional continuous Palmyra Palm Leaf Stalk Fiber/jute fiber reinforced hybrid polyester composites. *Mater. Des*. 50, 533–542 (2013). https://doi. org/10.1016/j.matdes.2013.03.048

21. Athijayamani, A., Thiruchitrambalam, M., Natarajan, U., Pazhanivel, B.: Effect of moisture absorption on the mechanical properties of randomly oriented natural fibers/ polyester hybrid composite. *Mater. Sci. Eng. A*. 517, 344–353 (2009). https://doi. org/10.1016/j.msea.2009.04.027

22. Alavudeen, A., Rajini, N., Karthikeyan, S., Thiruchitrambalam, M., Venkateshwaren, N.: Mechanical properties of banana/kenaf fiber-reinforced hybrid polyester composites: Effect of woven fabric and random orientation. *Mater. Des*. 66, 246–257 (2015). https://doi.org/10.1016/j.matdes.2014.10.067

23. Gupta, M.K., Srivastava, R.K.: A review on characterization of hybrid fibre reinforced polymer composite. *Gupta MK Am. J. Polym. Sci. Eng*. 4, 1–7 (2016).

24. Idicula, M., Malhotra, S.K., Joseph, K., Thomas, S.: Dynamic mechanical analysis of randomly oriented intimately mixed short banana/sisal hybrid fibre reinforced polyester composites. *Compos. Sci. Technol*. 65, 1077–1087 (2005). https://doi.org/10.1016/j. compscitech.2004.10.023

25. Venkateshwaran, N., ElayaPerumal, A., Alavudeen, A., Thiruchitrambalam, M.: Mechanical and water absorption behaviour of banana/sisal reinforced hybrid composites. *Mater. Des*. 32, 4017–4021 (2011). https://doi.org/10.1016/j.matdes.2011.03.002

26. Boopalan, M., Niranjanaa, M., Umapathy, M.J.: Study on the mechanical properties and thermal properties of jute and banana fiber reinforced epoxy hybrid composites. *Compos. Part B Eng.* 51, 54–57 (2013). https://doi.org/10.1016/j.compositesb.2013.02.033

27. Ramesh, M., Palanikumar, K., Reddy, K.H.: Mechanical property evaluation of sisal-jute-glass fiber reinforced polyester composites. *Compos. Part B Eng.* 48, 1–9 (2013). https://doi.org/10.1016/j.compositesb.2012.12.004

28. Sanjay, M.R., Arpitha, G.R., Yogesha, B.: Study on mechanical properties of natural - glass fibre reinforced polymer hybrid composites: A review. *Mater. Today Proc.* 2, 2959–2967 (2015). https://doi.org/10.1016/j.matpr.2015.07.264

29. Ahmed, K.S., Vijayarangan, S.: Tensile, flexural and interlaminar shear properties of woven jute and jute-glass fabric reinforced polyester composites. *J. Mater. Process. Technol.* 207, 330–335 (2008). https://doi.org/10.1016/j.jmatprotec.2008.06.038

30. Mishra, S., Mohanty, A.K., Drzal, L.T., Misra, M., Parija, S., Nayak, S.K., Tripathy, S.S.: Studies on mechanical performance of biofibre/glass reinforced polyester hybrid composites. *Compos. Sci. Technol.* 63, 1377–1385 (2003). https://doi.org/10.1016/S0266-3538(03)00084-8

31. Idicula, M., Boudenne, A., Umadevi, L., Ibos, L., Candau, Y., Thomas, S.: Thermophysical properties of natural fibre reinforced polyester composites. *Compos. Sci. Technol.* 66, 2719–2725 (2006). https://doi.org/10.1016/j.compscitech.2006.03.007

32. Sapuan, S.M., Aulia, H.S., Ilyas, R.A., Atiqah, A., Dele-Afolabi, T.T., Nurazzi, M.N., Supian, A.B.M., Atikah, M.S.N.: Mechanical properties of longitudinal basalt/woven-glass-fiber-reinforced unsaturated polyester-resin hybrid composites. *Polymers.* 12, 1–14 (2020). https://doi.org/10.3390/polym12102211

33. Sharma, V., Kundu, P. P.: Condensation polymers from natural oils. *Prog. Polym. Sci.* 33, 1199–1215 (2008). https://doi.org/10.1016/j.progpolymsci.2008.07.004

34. Mustapha, S. N. H.: *Mechanical, Thermal and Morphological Studies Of Epoxidized Palm Oil/Unsaturated Polyester / Montmorillonite Composites.* Universiti Teknologi Malaysia, Johor Bharu, Malaysia (2016).

35. Karak, N.: *Vegetable oil-Based Polymers.* Woodhead Publishing Limited, Sawston, Cambridge (2012).

36. Williams, G. I., Wool, R. P.: Composites from natural fibers and soy oil resins. *App. Compos. Mater.* 7(5–6), 421–432 (2000).

37. Aly, K. I., Sun, J., Kuckling, D., Younis, O.: Polyester resins based on soybean oil: Synthesis and characterization. *J. Polym. Res.* 27, 286 (2020).

38. Xuebin, W., Jincheng, W.: Studies on the preparation and properties of biodegradable polyester from soybean oil. *Green Process Syn.*, 8, 1–7 (2019).

39. Fakhari, A., Rahmat, A. R., Wahit, M. U., Kian, Y. S.: Synthesis of new bio-based thermoset resin from palm oil. *Adv. Mater. Res.* 931–932, 78–82 (2014). https://doi.org/10.4028/www.scientific.net/AMR.931-932.78

40. Clark, A. J., Ross, A. H., Bon, S. A. F.: Synthesis and properties of polyesters from waste grapeseed oil: Comparison with soybean and rapeseed oils. *J. Polym. Environ.* 25(1), 1–10 (2017). https://doi.org/10.1007/s10924-016-0883-3

41. Grishchuk, S., Leanza, R., Kirchner, P.: 'Greening' of unsaturated polyester resin based bulk molding compound with acrylated epoxidized soybean and linseed oils: Effect of urethane hybridization. *J. Reinf. Plastic Compos.* 30(17), 1455–1466 (2011). https://doi.org/10.1177/0731684411421541

42. Lu, J., Wool, R. P.: Additive toughening effects on new bio-based thermosetting resins from plant oils. *Compos. Sci. Technol.* 68(3–4), 1025–1033 (2008). https://doi.org/10.1016/j.compscitech.2007.07.009

43. Liu, C., Lei, W., Cai, Z., Chen, J., Hu, L., Dai, Y., Zhou, Y.: Use of tung oil as a reactive toughening agent in dicyclopentadiene-terminated unsaturated polyester resins. *Ind. Crops Prod.* 49, 412–418 (2013). https://doi.org/10.1016/j.indcrop.2013.05.023

44. Das, K., Ray, D., Banerjee, C., Bandyopadhyay, N. R., Mohanty, A. K., Misra, M.: Novel materials from unsaturated polyester resin/styrene/tung oil blends with high impact strengths and enhanced mechanical properties. *J. Appl. Polym. Sci.* 119, 2174–2182 (2010). https://doi.org/10.1002/app

45. Liu, C., Li, J., Lei, W., Zhou, Y.: Development of biobased unsaturated polyester resin containing highly functionalized castor oil. *Ind. Crops Prod.* 52, 329–337 (2014). https://doi.org/10.1016/j.indcrop.2013.11.010

46. Pfister, D. P., Larock, R. C.: Green composites from a conjugated linseed oil-based resin and wheat straw. *Compos. - A: Appl. Sci. Manuf.* 41(9), 1279–1288 (2010). https://doi.org/10.1016/j.compositesa.2010.05.012

47. Hosur, M., Maroju, H., Jeelani, S.: Comparison of effects of alkali treatment on flax fibre reinforced polyester and polyester-biopolymer blend resins. *Polym. Polym. Compos.* 23(4), 229–242 (2015). https://doi.org/10.1177/096739111502300404

48. Liu, W., Misra, M., Drzal, L. T., Mohanty, A. K.: Influence of processing methods and fiber length on physical properties of kenaf fiber reinforced soy based biocomposites. In *20th Technical Conference of the American Society for Composites 2005*, pp. 2111–2121 (2005).

49. García, G., Kieling, A. C., Bezazi, A., Boumediri, H.: Hybrid polyester composites reinforced with curauá fibres and nanoclays hybrid polyester composites reinforced with curauá fibres and nanoclays. *Fibers Polym.* 21(2), 399–406 (2020). https://doi.org/10.1007/s12221-020-9506-7

50. Baby, A., Nayak, S. Y., Heckadka, S. S., Purohit, S., Bhagat, K. K., Thomas, L. G.: Mechanical and morphological characterization of carbonized egg-shell fillers/Borassus fibre reinforced polyester hybrid composites. *Mater. Res. Exp.* 6(10), 105342 (2019).

51. Dhanola, A., Kumar, A., Kumar, A.: Influence of natural fillers on physico-mechanical properties of luffa cylindrica/polyester composites. *Mater. Today: Proc.*, 5, 17021–17029 (2018). https://doi.org/10.1016/j.matpr.2018.04.107

52. Mustapha, S. N. H., Aminuldin, A., Zakaria, S., Ramli, R. A., Salim, N., Roslan, R.: Mechanical and thermal properties of calcium carbonate filled kenaf reinforced unsaturated polyester/epoxidized palm oil composite. *AIP Conf. Proc.* (2018) https://doi.org/10.1063/1.5066831

53. Poorabdollah, M.: Diffusion and cure kinetic of sonicated nanoclay - reinforced unsaturated polyester resin. *J. Polym. Res.* 27(368), 1–15 (2020). https://doi.org/10.1007/s10965-020-02351-7

54. Hafiz, A. A. A.: *Synthesis and Characterization of EVA-Cloisite Clay Nanocomposites (Issue January)*. Master Thesis, The American University in Cairo, Egypt (2013).

55. Ganesan, K., Kailasanathan, C., Rajini, N., Ismail, S. O., Ayrilmis, N., Mohammad, F., Al-Lohedan, H. A., Tawfeek, A. M., Issa, Z. A., Aldhayan, D. M.: Assessment on hybrid jute/coir fibers reinforced polyester composite with hybrid fillers under different environmental conditions. *Constr. Build. Mater.* 301(April), 124117 (2021). https://doi.org/10.1016/j.conbuildmat.2021.124117

12 Natural Fiber/ Polyester-Based Hybrid Composites

Sabarish Radoor
King Mongkut's University of Technology
Jeonbuk National University

Aswathy Jayakumar
King Mongkut's University of Technology
Kyung Hee University

Jyothi Mannekote Shivanna
AMC Engineering College

Jasila Karayil
Government Engineering College

Jaewoo Lee
Jeonbuk National University

Jyotishkumar Parameswaranpillai
Alliance University

Suchart Siengchin
King Mongkut's University of Technology
Technische Universität Dresden

CONTENTS

DOI: 10.1201/9781003270980-12

12.1 INTRODUCTION

Present-day technologies such as aerospace, safe and quality transportation, and underwater solicitations demand a special amalgamation of material properties which can't be fulfilled by ceramics or metal alloys. The material properties could be high specific stiffness and specific strength, and less dense (Alothman et al., 2020; Thomas et al., 2021; Senthilkumar et al., 2015, 2022; Nasimudeen et al., 2021; Shahroze et al., 2021). Such combination of properties could be achieved by merging different materials in a thoughtful way and the resulting materials are termed as composites or hybrids. Recently, hybrids of carbon-reinforced polymers and glass fiber-reinforced polymers serve as alternatives to metals in several industrial applications (Fuqua et al., 2012; Joshi et al., 2004; Li et al., 2007; Herrera-Franco and Valadez-González, 2005). Broadly, these composites are categorized into two types, synthetic fiber-reinforced and the natural fiber-reinforced ones. Synthetic fibers are generally used as filler or reinforcing agents to develop polymer composite. The problem with synthetic fibers is its high energy combustion, non-renewability and non-biodegradability. Hence, there is a great interest in employing natural fibers as reinforcements in thermoplastic epoxy matrix. Various plant-based natural fibers such as coir, ramie, bamboo, sisal, jute and many more, and animal-based fibers such as wool, silk and many more have been explored as reinforcing materials in the composite's fabrication (Radoor et al., 2022).

The attractive features of natural fibers are low cost, biodegradability, abundance, renewability and excellent recyclability. Hence, natural fiber-based composite causes less environmental issues than synthetic fibers, and it was suggested as a good replacement for synthetic fiber (Chandrasekar et al., 2020; Senthilkumar et al., 2021). Nevertheless, the weak thermo-mechanical properties, poor compatibility with the polymer and high water absorption limits the applications of natural fibers-reinforced composite (Radoor et al., 2021; Parameswaranpillai et al., 2022; Chandrasekar et al., 2021). The compatibility of natural fiber with the matrix is enhanced by surface treatment such as alkali, permanganate and isocyanate. Hybrid composite is another solution for improving the properties of natural fiber-based composite (Chandrasekar et al., 2022a, b). Hybrid composite means two or more fibers in one matrix or two polymer blends and with one natural fiber reinforcement. The resulting composite shows unique properties such as high mechanical, thermal and water absorption, over individual fiber-reinforced composite (Sabarish et al., 2020; Radoor et al., 2020a, b).

12.2 PHYSICAL PROPERTIES OF NATURAL FIBER/ POLYESTER HYBRID COMPOSITES

Akil et al (2014) improved the durability of jute fiber by hybridizing it with synthetic glass fiber. In their work they monitored the effect of water aging on the mechanical properties of jute fiber and jute/glass fiber hybrid composite. It can be clearly seen

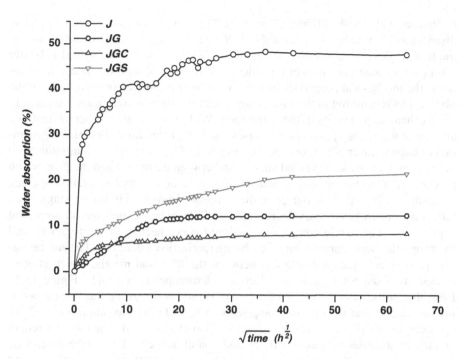

FIGURE 12.1 Water absorption curves of pultruded jute/glass-reinforced polyester hybrid composite. (Source: Reproduced with permission from Elsevier, License Number: 5230060516053.)

that water aging has detrimental effects on the mechanical properties of both composites. The water exposure causes the degradation of jute and thereby results in the mechanical damage. In the case of hybrid composite, the reduction in water uptake is low due to the fact that glass fiber acts as the barrier and thereby prevents contact between the fiber and water (Figure 12.1). Hence, hybrid composite shows superior tensile and flexural strength even after water aging. Polyester composite was developed through talc as filler. The talc-incorporated composites have less tendency for cracking and also can withstand bending force. Consequently, the flexural strength and Young's modulus of the composite increase with the increase in talc content. The authors therefore concluded that talc-incorporated polyester composites possess good mechanical, thermal and water absorption properties (Jahani and Ehsani, 2009).

Arumugam et al. (2020) studied the effect of chemical treatment and nanoparticles on the physical and mechanical properties of kenaf fiber/unsaturated polyester (UPE) composite. Their results suggested that inclusion of nanoparticle and chemically treatment has significantly improved the mechanical and thermal properties of kenaf fiber composite. The silver nanoparticle enhances the interfacial adhesion of the fiber and the polymer matrix. The mechanical bonding between the fiber and the polymer matrix is also improved by chemical treatment; the chemical treatment removes the lignin and cellulose from the fiber surface and makes the surface rough. The removal of lignin from the fiber surface is confirmed by monitoring the intensity of characteristic band of lignin (1736 cm^{-1}). The band at 1736 cm^{-1} was completely disappeared on treating the fiber with 5% NaOH. Athijayamani et al. (2009) studied the tensile, flexural and impact strength of roselle and sisal fiber hybrid polyester composite in both dry and wet conditions. In the case of both dry and wet conditions, the mechanical properties increase with increase in weight percentage of the fiber and its length. But in the case of wet condition, there is significant reduction in the mechanical properties of the composites. With increase in the fiber content, the amount of water absorption also increases due to high cellulose content and generation of large number of microvoids. The presence of moisture in the composite has a negative impact on the interfacial interaction between the matrix and the fiber, which is reflected in the morphological analysis. Presence of void and generation of crack is clearly visible in the micrograph of the composite immersed in water. Atiqah et al. (2014) developed kenaf glass fiber-reinforced UPE hybrid composite for structural applications. The hybrid composites displayed good mechanical performance and the properties were further improved by mercerization treatment. The mercerization process enhanced the adhesion between the fiber and matrix, which is quite evident from the SEM (scanning electron microscope) micrograph (Figure 12.2). Less fiber pulls out and absence of microvoid confirm the strong mechanical bonding in treated kenaf fiber hybrid composites. A hybrid biocomposite from UPE and epoxidized methyl soyate was developed by Haq et al. (2008). The tensile strength and tensile modulus increased with the addition of nanoclay. Due to the synergistic effect of hemp and the nanoclay, the hybrid composite displayed good stiffness and toughness. The author thus suggested hybrid biocomposites as suitable candidate for structural and transportation applications. Similarly, Dewan et al. (2013) observed an enhancement in the properties of developed hybrid composite of polyester reinforced with alkali/untreated jute and nanoclay. The combined effect of nanoclay and surface treatment make the hybrid composite induce thermo-mechanical stability to the composite. However, high percentage of clay leads to agglomeration and therefore to a reduction in the mechanical properties of the composite.

Prasanna Venkatesh et al. (2016) observed an enhancement in the tensile, flexural and impact properties of polyester composite reinforced with treated sisal/bamboo fiber. The mechanical properties increase with increase in fiber content and maximum mechanical properties was observed for fiber length 15 cm and fiber content of 20%. The untreated fiber has smooth surface, and during mechanical test the fiber was pulled out and deboned. This problem is minimized by fiber treatment. The fiber treatment reduces the moisture absorption capacity of the composite and improves the adhesion of the fiber with the polymer matrix. The good interfacial adhesion in the treated composite is confirmed from morphological and mechanical analyses. Rout et al. (2001) studies shows that an optimum percentage of alkali is required for improving the mechanical bonding of fiber with the matrix. They have varied the percentage of alkali from 2%–10%. The coir fiber modified with 2% and 5% alkali shows significant tensile and flexural strength. However, for 10% alkali-treated samples, a reduction in the mechanical properties was noted. High percentage of alkali thickens the cell wall of the fiber and thereby leads to poor adhesion between the fiber and matrix. Oil palm shell (OPS) is an agricultural waste, which could be employed as filler in polymer composite. Rosamah et al. (2016) utilized OPS

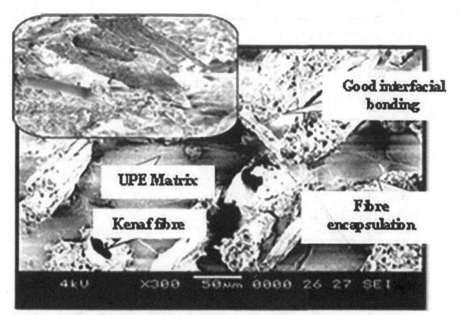

a) Fracture surface of treated S3(15/15v/v KG), fibre encapsulation and good interfacial bonding between kenaf fibres and UPE matrix

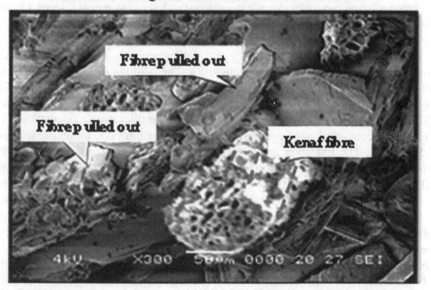

b) Fracture surface of untreated S3(15/15v/v KG), fibre pulled out between kenaf fibres and UPE matrix, matrix and non-uniform surface of UPE matrix

FIGURE 12.2 SEM images of kenaf/glass unsaturated polyester hybrid composite. (a) and (b) fracture surface of treated and untreated sample (Source: Reproduced with permission from Elsevier, License Number: 5230060994459.)

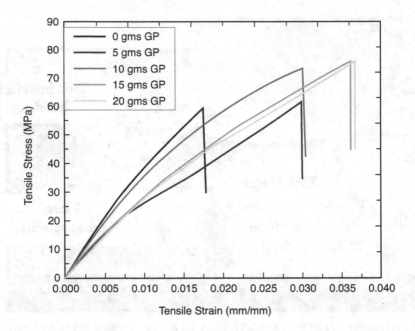

FIGURE 12.3 Tensile stress versus tensile strain of alkali-treated *Cordia dichotoma* fiber-reinforced polyester composite. (Source: Reproduced with permission from Elsevier, License Number: 5230071070742.)

nanoparticle as a filler in polyester composite. The addition of OPS nanoparticles improves the energy absorbing capability and surface area of the composite. The uniform distribution of nanoparticles in the polymer matrix reduces the free space within the hybrid composite and thereby enhances the filler matrix interaction. The 3% OPS nanoparticle-loaded composite exhibits maximum value for tensile strength and modulus. The reduction in the properties at high weight percentage of OPS nanoparticles is ascribed to the formation of agglomeration. Meanwhile, Reddy et al. (2020) added industrial waste granite powder into hybrid polyester composite. Here, in addition to granite powder, alkali-treated *Cordia dichotoma* fiber was also used as filler. The mechanical properties of the composites were investigated at different weight percentages of granite powder (5%–20%) by keeping *C. dichotoma* fiber content at fixed percentage (20 wt%). With increase in the granite powder, the tensile modulus and strength also increase. The granite powder is properly dispersed in the polymer matrix and enhances the adhesion between the constituent of the composite. However, beyond 15%, the granite powder has a negative effect on the mechanical properties (Figure 12.3).

High concentration of granite increases the viscosity of the system and causes exfoliation of the polymer matrix. Similar observation was noticed for impact strength. SEM morphological result revealed the well dispersion of granite filler on the surface of the *C. dichotoma* fiber. Arumuga Prabu and co-workers (2014) applied industrial waste, red mud (RM) to improve the properties of banana fiber-reinforced

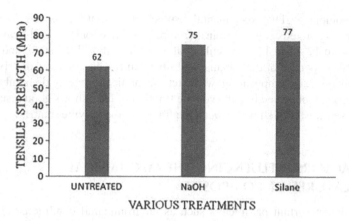

FIGURE 12.4   Effect of red mud on the tensile strength of untreated, silane and NaOH hybrid composites. (Source: Reproduced with permission from Elsevier, License Number: 5230070272022.)

FIGURE 12.5   (a–d) SEM images of tensile fractured samples of untreated and treated hybrid composite. (Source: Reproduced with permission from Elsevier, License Number: 5230070272022.)

polyester composite. The experimental results suggest that RM is an effective filter. The RM-incorporated composite can therefore be applied for high load-bearing application (Figure 12.4 and 12.5). Idicula et al. (2006) analyzed the static and dynamic mechanical properties of the banana/sisal fiber-reinforced polyester matrix. Results revealed that hybrid composite shows better mechanical property than unhybridized composite. They observed that 0.4 volume fraction of fiber shows good results due to the effective transfer of stress between the fiber and the polymer matrix.

12.3 FACTORS INFLUENCING THE MECHANICAL FEATURES OF COMPOSITES

Some of the important parameters such as environmental conditions, fiber layers stacking sequence, fiber treatment and fiber volume to weight ratio are known to influence the mechanical features of hybrid composites. The following sections will detail those parameters one by one.

12.3.1 EFFECT OF ENVIRONMENTAL CONDITIONS

The environmental conditions in which the composites are used have a greater influence on their mechanical features, due to which scientists have started exploring the mechanical conditions of prepared composites at varied environmental conditions. Natural fiber-reinforced composites possess internal strong water absorption property under moisture or wet environment. Hydrothermal degradation of jute-reinforced composites (JFRP) and glass-reinforced composite (GFRP) was explored by Liao et al. JFRP was able to absorb more amount of water and was faster than GFRP due to hydroscopic nature of jute fiber. Due to loss of resin particles, weight loss was observed and pectin dissolution occurred due to the micro-cracks attributing to hydrothermal water environments. There also occurred serious internal deterioration (Liao et al., 2012). Phani and Bose explored a similar work, wherein the hydrothermal aging was studied by acousto-ultrasonic technique. They found that incorporation of jute into GFRP reduced the degradation rate for the ones which were aged above 70 hours. Jute fiber and jute resin interfaces are expected to undergo faster degradation under moist conditions below 70 hours aging. Above 70 hours aging, the inflated jute fiber layer would absorb swelling strain from the resin, which acts as cushion. Additionally, jute fibers give added protection for the sandwiched glass layers (Phani and Bose, 1987). Composites made of orthophthalic polyester reinforced with curaua and E-glass fiber were exposed to distilled water and sea water and their mechanical properties were compared (Silva et al., 2008). Up to 330 hours, the diffusion rate for both the water remained same. However, after 330 hours, the diffusion rate for sea water reduced, attributing to the added NaCl ions on the fiber surface. The ones exposed to distilled water possessed lower tensile strength and tensile modulus than those exposed to sea water due to the greater absorption of water. Due to matrix plasticization, the flexural modulus for sea water-exposed samples degraded in a greater way. Water absorption ability and the swelling thickness for oil palm and jute-reinforced materials were tested. In the study by Abdul et al., the addition of jute fibers to the oil palm fibers at higher concentrations improved the mechanical characteristics. The work concluded the direct dependency of water adsorption ability and thickness swelling of composite on the density, existence of voids and the nature of fiber and matrix bonding of composite. Critical stress intensity factor, interlaminar shear strength and impact toughness for glass or textile fabric-reinforced hybrid were tested under ambient conditions and sea water. The considered parameters varied with the immersion time. Both the environments showed a different feature with respect to matrix cracking, growth of crack and fiber pull-out (Abdul Khalil et al., 2008).

12.3.2 Fiber Layers Stacking Sequence

The arrangement pattern of fiber layers in the hybrid composites is termed as stacking sequence. The symmetry in the sequence and the mechanical features of extremities and the core layer interestingly vary the impact, tensile and flexural strengths. A stacking sequence in jute- and glass-reinforced composites was varied by Ahmed and Vijayarangan (2008). Specimens with glass fibers at the extremities possessed a greater strength due to the stiffness and strength of glass fibers than those with jute fibers. Amico and group reported that hybrids with symmetrical stacking sequence are more capable when compared to the ones with asymmetrical stacking sequence owing to the structural integrity of the overall hybrid (Amico et al., 2008). Ortho-UPE was reinforced with long abaca and short bagasse fibers with altered sequences (Cao et al., 2006). Three abaca sheets were sandwiched between two bagasse mats; the orientations were parallel and at 90° cross-ply. Specimen with cross-ply orientation showed lesser tensile strength but exhibited greater load support, tested at both axial and transverse directions signifying its feasibility for biaxial loading. Woven coir (C) and glass (G) hybrids were made with different stacking sequences such as CGG, GGC, GGG and CGC (Jayabal et al., 2011). Maximum flexural, impact and tensile strengths were obtained for CGG than any other sequence. GGC sequence showed greater mechanical properties than GCG, and it also possessed greater breaking resistance than CGC sequence due to weaker coir fibers than glass fibers.

12.3.3 Fiber Treatment

An interfacial bonding and the compatibility between the matrix and fiber is an essential feature for the feasible performance. Chemical treatment is one of the best ways to reach the best interfacial bonding. The most preferred one is the alkali treatment, which aids in improving interfacial adhesion between matrix and fiber. Sisal and glass fiber-reinforced polyester was kept untreated, and treated with alkali and silane, and their mechanical characteristics were compared (John and Naidu, 2016). Alkali treatment removes the hemicellulose from natural fibers and helps in improved adhesion properties, and hence, hybrid treated with alkali showed maximum tensile strength than silane-treated and untreated hybrids. Silane treatment aids in increasing wetting property of sisal fiber resulting in better tensile strength than untreated hybrid. Jute fibers treated with alkali underwent acrylation and led to improved physical properties. Due to –OH groups esterification after alkali treatment, the composites showed reduced water uptake (Patel et al., 2008). Kapok and glass fiber-reinforced polyester was treated with alkali, and thus an increase in bond strength and interlaminar shear strength was observed (Venkata Reddy et al., 2008). The same group has also explored the effect of alkali treatment on bamboo fiber-reinforced hybrids (Reddy et al., 2010). Lignin and hemicellulose were removed by the alkali treatment to the coir/silk fiber composites, which increased fiber and matrix adhesion (Noorunnisa Khanam et al., 2009). Alavudeen and group have shown that banana/kenaf-reinforced hybrids showed better performance when the fibers were treated with 10% NaOH and 10% of sodium lauryl sulfate for 30 minutes (Alavudeen et al., 2015). Jute and E-glass fiber-reinforced UPE was prepared by Abdullah-Al et al. (2005). The dissimilar proportions of E-glass and jute (3:1) gave the better performance. The fiber surfaces when irradiated with UV radiation of different intensities have changed the mechanical properties. A Charpy impact strength of 40 kJ/m^2 was obtained for the jute/E-glass-reinforced fiber treated with UV radiation.

12.3.4 FIBER VOLUME

A measure of weight fraction or the volume of fibers in hybrid is termed as fiber volume. The weight or volume ratios of fiber in the matrix have affected the mechanical properties of hybrids both positively and negatively. Banana and glass fibers hybrids showed the linear variation of mechanical features with the glass fiber volume (Nunna et al., 2012). At higher ratio of glass fibers volume, the delamination of layers is observed and the failure of the hybrid occurred. 11% weight ration of glass fiber gave the optimal mechanical features. In the case of silk/glass hybrid composite has fixed weight ratio of 25%, silk-glass fiber of total weight ratio 25%, where the weight of glass varied from 0% to 50%. Each with 50% of weight ratio showed better performance (Priya and Rai, 2016). The glass fiber avoids the direct transformation of water tosilk, and it also transfers the efficient load transfer to silk. Wood flour hybridized with kenaf fiber was found to show greater properties with increase in kenaf fiber weight (Mirbagheri et al., 2007). In case of jute and glass fiber composite, 25% weight of jute showed better performance and beyond which the efficiency decreased due to the agglomerates formation (Abdullah-Al et al., 2005). These agglomerates would block the stress transfer and hence decreases the crack resistance and the mechanical properties. Mishra et al. fabricated pineapple leaf fiber- and sisal fiber-reinforced composites (Mishra et al., 2003). A very small amount of added glass fiber to the leaf and sisal fiber-reinforced matrix increased the mechanical properties of the formed hybrid. The group also explored the effect of alkali treatment and cyanoethylation to the sisal fibers on the mechanical features. Alkali treatment has improved the optimum properties the cyanoethylation increased the flexural strength of hybrids (Figure 12.6).

Overall, the mechanical features of hybrids increase linearly with the volume fraction of high strength fibers and it is applicable up to some ratios, beyond which the hybrids will fail due to agglomeration of reinforcing materials. Alkali treatment enhances the adhesion between fiber and matrix, leading to increased stiffness of the

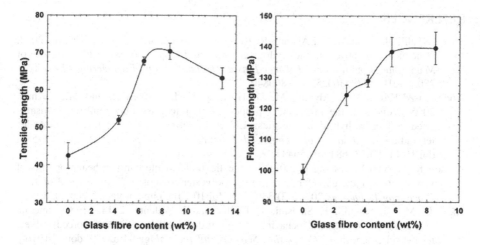

FIGURE 12.6 Effect of glass fiber content on tensile and flexural strength of sisal/glass hybrid polyester composite. (Source: Reproduced with permission from Elsevier, License Number: 5230080400566.)

composite. The deformation of hybrid occurs depending on the exposure time and temperature with respect to various environmental conditions. The performance and behavior of hybrids are also dependent on the extreme layer of the hybrid; the fiber of greater strength placed at the extreme layer would show better performance.

12.4   CONCLUSION

In this chapter, we presented an overview of reports on the enhancement of mechanical properties, water barrier, and thermal stability of natural fiber/polyester hybrid composites. It was observed that the effect of chemical treatment, coupling agent, fiber geometry, stacking sequence, fiber volume and concentration influences the overall properties of natural fiber hybrid composites. Hybrid composites exhibit superior physical properties such as tensile, impact, flexural and compressive strength compared to neat samples. The advantages of combination of natural and synthetic fibers are being eco-friendly, biodegradability, low cost and comparable mechanical properties, compared to glass fiber-reinforced composites. Hybrid composites have a wide range of applications such as automotive parts, construction and food packaging. The main challenges that we will face in the near future are to enhance the mechanical and water barrier properties of hybrid composites with low cost and to develop an eco-friendly strategy.

ACKNOWLEDGMENTS

The authors gratefully acknowledge the support received from King Mongkut's University of Technology North Bangkok (KMUTNB), Thailand, for the Post-Doctoral Program (Grant Nos. KMUTNB-64-Post-03 and KMUTNB-63-Post-03 to SR).

REFERENCES

Abdul Khalil, H. P. S., M. Siti Alwani, R. Ridzuan, H. Kamarudin, and A. Khairul. 2008. "Chemical composition, morphological characteristics, and cell wall structure of Malaysian oil palm fibers." *Polymer-Plastics Technology and Engineering* 47 (3):273–280. doi:10.1080/03602550701866840.

Abdullah-Al-Kafi, M. Z. Abedin, M. D. H. Beg, K. L. Pickering, and M.A. Khan. 2005. "Study on the mechanical properties of jute/glass fiber-reinforced unsaturated polyester hybrid composites: Effect of surface modification by ultraviolet radiation." *Journal of Reinforced Plastics and Composites* 25 (6):575–588. doi:10.1177/0731684405056437.

Ahmed, K. S., and S. Vijayarangan. 2008. "Tensile, flexural and interlaminar shear properties of woven jute and jute-glass fabric reinforced polyester composites." *Journal of Materials Processing Technology* 207 (1–3):330–335. doi:10.1016/j.jmatprotec.2008.06.038.

Akil, H. M., C. Santulli, F. Sarasini, J. Tirillò, and T. Valente. 2014. "Environmental effects on the mechanical behaviour of pultruded jute/glass fibre-reinforced polyester hybrid composites." *Composites Science and Technology* 94:62–70. doi:10.1016/j.compscitech.2014.01.017.

Alavudeen, A., N. Rajini, S. Karthikeyan, M. Thiruchitrambalam, and N. Venkateshwaren. 2015. "Mechanical properties of banana/kenaf fiber-reinforced hybrid polyester composites: Effect of woven fabric and random orientation." *Materials & Design (1980–2015)* 66:246–257. doi:10.1016/j.matdes.2014.10.067.

Alothman, O. Y., M. Jawaid, K. Senthilkumar, M. Chandrasekar, B. A. Alshammari, H. Fouad, M. Hashem, and S. Siengchin. 2020. "Thermal characterization of date palm/epoxy composites with fillers from different parts of the tree." *Journal of Materials Research and Technology* 9 (6): 15537–15546. doi:10.1016/j.jmrt.2020.11.020.

Amico, S. C., C. C. Angrizani, and M. L. Drummond. 2008. "Influence of the stacking sequence on the mechanical properties of glass/sisal hybrid composites." *Journal of Reinforced Plastics and Composites* 29 (2):179–189. doi:10.1177/0731684408096430.

Arumuga Prabu, V., M. Uthayakumar, V. Manikandan, N. Rajini, and P. Jeyaraj. 2014. "Influence of redmud on the mechanical, damping and chemical resistance properties of banana/polyester hybrid composites." *Materials & Design* 64:270–279. doi:10.1016/j.matdes.2014.07.020.

Arumugam, C., S. Arumugam, and S. Muthusamy. 2020. "Mechanical, thermal and morphological properties of unsaturated polyester/chemically treated woven kenaf fiber/AgNPs@PVA hybrid nanobiocomposites for automotive applications." *Journal of Materials Research and Technology* 9 (6):15298–15312. doi:10.1016/j.jmrt.2020.10.084.

Athijayamani, A., M. Thiruchitrambalam, U. Natarajan, and B. Pazhanivel. 2009. "Effect of moisture absorption on the mechanical properties of randomly oriented natural fibers/polyester hybrid composite." *Materials Science and Engineering: A* 517 (1–2):344–353. doi:10.1016/j.msea.2009.04.027.

Atiqah, A., M. A. Maleque, M. Jawaid, and M. Iqbal. 2014. "Development of kenaf-glass reinforced unsaturated polyester hybrid composite for structural applications." *Composites Part B: Engineering* 56:68–73. doi:10.1016/j.compositesb.2013.08.019.

Cao, Y., S. Shibata, and I. Fukumoto. 2006. "Mechanical properties of biodegradable composites reinforced with bagasse fibre before and after alkali treatments." *Composites Part A: Applied Science and Manufacturing* 37 (3):423–429. doi:10.1016/j.compositesa.2005.05.045.

Chandrasekar, M., I. Siva, T. Senthil Muthu Kumar, K. Senthilkumar, S. Siengchin, and N. Rajini. 2020. "Influence of fibre inter-ply orientation on the mechanical and free vibration properties of banana fibre reinforced polyester composite laminates." *Journal of Polymers and the Environment* 28 (11): 2789–2800. doi:10.1007/s10924-020-01814-8.

Chandrasekar, M., K. Senthilkumar, M. Jawaid, S. Alamery, H. Fouad, and M. Midani. 2022a. "Tensile, thermal and physical properties of Washightonia trunk fibres/pineapple fibre biophenolic hybrid composites." *Journal of Polymers and the Environment* 30 (10): 4427–4434. doi:10.1007/s10924-022-02524-z.

Chandrasekar, M., K. Senthilkumar, M. Jawaid, M. H. Mahmoud, H. Fouad, and M. Sain. 2022b. "Mechanical, morphological and dynamic mechanical analysis of pineapple leaf/Washingtonia trunk fibres based biophenolic hybrid composites." *Journal of Polymers and the Environment* 30 (10): 4157–4165. doi:10.1007/s10924-022-02482-6.

Chandrasekar, M., T. S. M. Kumar, K. Senthilkumar, S. Radoor, R. A. Ilyas, S. M. Sapuan, J. Naveen, and S. Siengchin. 2021. "Performance of natural fiber reinforced recycled thermoplastic polymer composites under aging conditions." In *Recycling of Plastics, Metals, and Their Composites*, Ilyas, A., Sapuan, S.M., and Bayraktar, E., Eds., Taylor & Francis, United Kingdom, pp. 127–139.

Dewan, M. W., M. K. Hossain, M. Hosur, and S. Jeelani. 2013. "Thermomechanical properties of alkali treated jute-polyester/nanoclay biocomposites fabricated by VARTM process." *Journal of Applied Polymer Science* 128 (6):4110–4123. doi:10.1002/app.38641.

Fuqua, M. A., S. Huo, and C. A. Ulven. 2012. "Natural fiber reinforced composites." *Polymer Reviews* 52 (3):259–320. doi:10.1080/15583724.2012.705409.

Haq, M., R. Burgueño, A. K. Mohanty, and M. Misra. 2008. "Hybrid bio-based composites from blends of unsaturated polyester and soybean oil reinforced with nanoclay and natural fibers." *Composites Science and Technology* 68 (15–16):3344–3351. doi:10.1016/j.compscitech.2008.09.007.

Herrera-Franco, P. J., and A. Valadez-González. 2005. "A study of the mechanical properties of short natural-fiber reinforced composites." *Composites Part B: Engineering* 36 (8):597–608. doi:10.1016/j.compositesb.2005.04.001.

Idicula, M., A. Boudenne, L. Umadevi, L. Ibos, Y. Candau, and S. Thomas. 2006. "Thermophysical properties of natural fibre reinforced polyester composites." *Composites Science and Technology* 66 (15):2719–2725. doi:10.1016/j.compscitech.2006.03.007.

Jahani, Y., and M. Ehsani. 2009. "The rheological modification of talc-filled polypropylene by epoxy-polyester hybrid resin and its effect on morphology, crystallinity, and mechanical properties." *Polymer Engineering & Science* 49 (3):619–629. doi:10.1002/pen.21294.

Jayabal, S., U. Natarajan, and S. Sathiyamurthy. 2011. "Effect of glass hybridization and staking sequence on mechanical behaviour of interply coir–glass hybrid laminate." *Bulletin of Materials Science* 34 (2):293–298. doi:10.1007/s12034-011-0081-9.

John, K., and S. Venkata Naidu. 2016. "Tensile properties of unsaturated polyester-based sisal fiber–glass fiber hybrid composites." *Journal of Reinforced Plastics and Composites* 23 (17):1815–1819. doi:10.1177/0731684404041147.

Joshi, S. V., L. T. Drzal, A. K. Mohanty, and S. Arora. 2004. "Are natural fiber composites environmentally superior to glass fiber reinforced composites?" *Composites Part A: Applied Science and Manufacturing* 35 (3):371–376. doi:10.1016/j.compositesa.2003.09.016.

Li, X., L. G. Tabil, and S. Panigrahi. 2007. "Chemical treatments of natural fiber for use in natural fiber-reinforced composites: A review." *Journal of Polymers and the Environment* 15 (1):25–33. doi:10.1007/s10924-006-0042-3.

Liao, M., Y. Yang, Y. Yu, and H. Hamada. 2012. "Hydrothermal ageing mechanism of natural fiber reinforced composite in hot water." In *ASME International Mechanical Engineering Congress and Exposition*, vol. 45196, pp. 1371–1378. American Society of Mechanical Engineers.

Mirbagheri, J., M. Tajvidi, J. C. Hermanson, and I. Ghasemi. 2007. "Tensile properties of wood flour/kenaf fiber polypropylene hybrid composites." *Journal of Applied Polymer Science* 105 (5):3054–3059. doi:10.1002/app.26363.

Mishra, S., A. K. Mohanty, L. T. Drzal, M. Misra, S. Parija, S. K. Nayak, and S. S. Tripathy. 2003. "Studies on mechanical performance of biofibre/glass reinforced polyester hybrid composites." *Composites Science and Technology* 63 (10):1377–1385. doi:10.1016/s0266-3538(03)00084-8.

Nasimudeen, N. A., S. Karounamourthy, J. Selvarathinam, S. Muthu Kumar Thiagamani, H. Pulikkalparambil, S. Krishnasamy, and C. Muthukumar. 2021. "Mechanical, absorption and swelling properties of vinyl ester based natural fibre hybrid composites." *Applied Science and Engineering Progress.* doi:10.14416/j.asep.2021.08.006.

Noorunnisa Khanam, P., G. Ramachandra Reddy, K. Raghu, and S. Venkata Naidu. 2009. "Tensile, flexural, and compressive properties of coir/silk fiber-reinforced hybrid composites." *Journal of Reinforced Plastics and Composites* 29 (14):2124–2127. doi:10.1177/0731684409345413.

Nunna, S., P. Ravi Chandra, S. Shrivastava, and A. K. Jalan. 2012. "A review on mechanical behavior of natural fiber based hybrid composites." *Journal of Reinforced Plastics and Composites* 31 (11):759–769. doi:10.1177/0731684412444325.

Parameswaranpillai, J., S. Krishnasamy, S. Siengchin, S. Radoor, R. Joy, J. J. George, C. Muthukumar, S. M. K. Thiagamani, N. V. Salim, and N. Hameed. 2022. "Thermal properties of the natural fiber-reinforced hybrid polymer composites: An overview." In *Natural Fiber-Reinforced Composites*, Senthilkumar, K., Muthu Kumar, T. S., Chandrasekar, M., Rajini, N., and Siengchin, S., Eds., Wiley, United States, pp. 31–51.

Patel, V. A., B. D. Bhuva, and P. H. Parsania. 2008. "Performance evaluation of treated—untreated jute—carbon and glass—carbon hybrid composites of bisphenol-c based mixed epoxy—phenolic resins." *Journal of Reinforced Plastics and Composites* 28 (20):2549–2556. doi:10.1177/0731684408093973.

Phani, K. K., and N. R. Bose. 1987. "Hydrothermal ageing of jute-glass fibre hybrid composites? an acousto-ultrasonic study." *Journal of Materials Science* 22 (6):1929–1933. doi:10.1007/bf01132918.

Prasanna Venkatesh, R., K. Ramanathan, and V. Srinivasa Raman. 2016. "Tensile, flexural, impact and water absorption properties of natural fibre reinforced polyester hybrid composites." *Fibres and Textiles in Eastern Europe* 24:90–94. doi:10.5604/12303666.1196617.

Priya, S. P., and S. K. Rai. 2016. "Mechanical performance of biofiber/glass-reinforced epoxy hybrid composites." *Journal of Industrial Textiles* 35 (3):217–226. doi:10.1177/1528083706055754.

Radoor, S., J. Karayil, A. Jayakumar, E. K. Radhakrishnan, L. Muthulakshmi, S. M. Rangappa, S. Siengchin, and J. Parameswaranpillai. 2020a. "Structure and surface morphology techniques for biopolymers." In *Biofibers and Biopolymers for Biocomposites*, Khan, A., Mavinkere Rangappa, S., Siengchin, S., and Asiri, A., Eds., Springer, Germany, pp. 35–70.

Radoor, S., J. Karayil, J. M. Shivanna, and S. Siengchin. 2021. "Water absorption and swelling behaviour of *wood plastic composites*." In Wood Polymer Composites, pp. 195–212.

Radoor, S., J. Karayil, R. Soman, A. Jayakumar, E. K. Radhakrishnan, J. Parameswaranpillai, and S. Siengchin. 2022. "Influence of nanoclay on the thermal properties of the natural fiber-based hybrid composites." In *Natural Fiber-Reinforced Composites*, pp. 239–254.

Radoor, S., J. Karayil, S. M. Rangappa, S. Siengchin, and J. Parameswaranpillai. 2020b. "A review on the extraction of pineapple, sisal and abaca fibers and their use as reinforcement in polymer matrix." *Express Polymer Letters* 14 (4):309–335. doi:10.3144/expresspolymlett.2020.27.

Reddy, B. M., R. M. Reddy, B. C. M. Reddy, P. V. Reddy, H. R. Rao, and Y. V. M. Reddy. 2020. "The effect of granite powder on mechanical, structural and water absorption characteristics of alkali treated *Cordia dichotoma* fiber reinforced polyester composite." *Polymer Testing* 91. doi:10.1016/j.polymertesting.2020.106782.

Reddy, E. V. S., A. V. Rajulu, K. H. Reddy, and G. R. Reddy. 2010. "Chemical resistance and tensile properties of glass and bamboo fibers reinforced polyester hybrid composites." *Journal of Reinforced Plastics and Composites* 29 (14):2119–2123. doi:10.1177/0731684409349520.

Rosamah, E., M. S. Hossain, H. P. S. Abdul Khalil, W. O. Wan Nadirah, R. Dungani, A. S. Nur Amiranajwa, N. L. M. Suraya, H. M. Fizree, and A. K. Mohd Omar. 2016. "Properties enhancement using oil palm shell nanoparticles of fibers reinforced polyester hybrid composites." *Advanced Composite Materials* 26 (3):259–272. doi:10.1080/09243046. 2016.1145875.

Rout, J., M. Misra, S. S. Tripathy, S. K. Nayak, and A. K. Mohanty. 2001. "The influence of fibre treatment on the performance of coir-polyester composites." *Composites Science and Technology* 61 (9):1303–1310. doi:10.1016/s0266-3538(01)00021-5.

Sabarish, R., S. Krishnasamy, S. Siengchin, K. Jasila, M. Vintu, M. S. Jyothi, C. Muthukumar, S. M. K. Thiagamani, and R. Nagarajan. 2020. "Plastics in fabric, textile and clothing." In *Reference Module in Materials Science and Materials Engineering*, Hashmi, M.S.J., Eds., Elsevier, Amsterdam, The Netherlands.

Senthilkumar, K., I. Siva, N. Rajini, and P. Jeyaraj. 2015. "Effect of fibre length and weight percentage on mechanical properties of short sisal/polyester composite." *International Journal of Computer Aided Engineering and Technology* 7 (1): 60. doi:10.1504/ IJCAET.2015.066168.

Senthilkumar, K., N. Saba, M. Chandrasekar, M. Jawaid, N. Rajini, S. Siengchin, N. Ayrilmis, F. Mohammad, and H. A. Al-Lohedan. 2021. "Compressive, dynamic and thermomechanical properties of cellulosic pineapple leaf fibre/polyester composites: Influence of alkali treatment on adhesion." *International Journal of Adhesion and Adhesives* 106 (April): 102823. doi:10.1016/j.ijadhadh.2021.102823.

Senthilkumar, K., S. Subramaniam, T. Ungtrakul, T. S. M. Kumar, M. Chandrasekar, N. Rajini, S. Siengchin, and J. Parameswaranpillai. 2022. "Dual cantilever creep and recovery behavior of sisal/hemp fibre reinforced hybrid biocomposites: Effects of layering sequence, accelerated weathering and temperature." *Journal of Industrial Textiles* 51 (2_suppl): 2372S–2390S. doi:10.1177/1528083720961416.

Shahroze, R. M., M. Chandrasekar, K. Senthilkumar, T. S. M. Kumar, M. R. Ishak, N. Rajini, S. Siengchin, and S. O. Ismail. 2021. "Mechanical, interfacial and thermal properties of silica aerogel-infused flax/epoxy composites." *International Polymer Processing* 36 (1):53–59. doi:10.1515/ipp-2020-3964.

Silva, R. V., E. M. F. Aquino, L. P. S. Rodrigues, and A. R. F. Barros. 2008. "Curaua/glass hybrid composite: The effect of water aging on the mechanical properties." *Journal of Reinforced Plastics and Composites* 28 (15):1857–1868. doi:10.1177/0731684408090373.

Thomas, S. K., J. Parameswaranpillai, S. Krishnasamy, P.M. S. Begum, D. Nandi, S. Siengchin, J. J. George, N. Hameed, N. V. Salim, and N. Sienkiewicz. 2021. "A comprehensive review on cellulose, chitin, and starch as fillers in natural rubber biocomposites." *Carbohydrate Polymer Technologies and Applications* 2 (December):100095. doi:10.1016/j.carpta.2021.100095.

Venkata Reddy, G., T. Shobha Rani, K. Chowdoji Rao, and S. Venkata Naidu. 2008. "Flexural, compressive, and interlaminar shear strength properties of kapok/glass composites." *Journal of Reinforced Plastics and Composites* 28 (14):1665–1677. doi:10.1177/0731684408090362.

13 Polyester-Based Bio-Composites for Marine Applications

Govindaraju Boopalakrishnan, Michael Johni Rexliene, Rajkumar Praveen, Viswanathan Balaji, Aravind Dhandapani, and Jayavel Sridhar
Madurai Kamaraj University

CONTENTS

DOI: 10.1201/9781003270980-13

13.1 INTRODUCTION

Concerns about the environment have boosted interest in renewable resource-based products. Polymeric materials, due to their wide collection of characteristics, are now an integral component of daily life. The efficient utilization of biodegradable polymers, which are being used more efficiently in a variety of disciplines, including medical, pharmacy, agro-chemistry, marine, and packaging industry, is growing rapidly (Alothman et al., 2020; Chandrasekar et al., 2020, 2022a). Plastics have numerous advantages over other traditional materials, including performance and adaptability, corrosion resistance, ease of processing, lightweight, durability, low cost, high productivity, environmental aspects, and so on, which determine their value to the society and ensure future industrial development. Biodegradable polymers can be considered ecologically benign and an appealing alternative to conventional polymers in terms of long-term sustainable development (Musiol et al., 2011).

Composites are heterogeneous materials that are made up of a mixture of (two or more) materials with distant physical and chemical characteristics. To attain greater mechanical performance in the desired plane, composite materials are combined with the high qualities of fiber and matrix components. Composites used in marine applications provide additional advantages such as flatness stealth and corrosion resistance to reliable and durable vessels (Cao and Grenestedt, 2004). In general, conditioning reinforcement fibers is to be done before composite makeup in aerospace and automotive sectors, which prefer to employ glass and carbon synthetic fibers for polymer strengthening (Chandrasekar et al., 2022b; Nasimudeen et al., 2021; Senthilkumar et al., 2015, 2021, 2022). Bio-composites are made up of one or more layers of natural fiber reinforcing with biopolymers (organic matrix). At the same time, biopolymers (natural biopolymers such as gelatin, corn zein, and soy protein; synthetic biopolymers such as poly(lactic acid) (PLA), poly(vinyl alcohol) (PVA), and other microbial polyesters) and reinforcements (cotton, hemp, flax, sisal, jute, and kenaf or recycled wood and paper) are renewable and degradable (Akampumuza et al., 2017) (Table 13.1).

Despite their higher cost, high-performance goods that are lightweight but robust are sufficient to endure rigorous packing conditions. One of the most significant drawbacks of using this vast family of composites is difficulty in applying two separate components at the same time. This makes end-of-life reuse and recycling extremely challenging. As a result, it is preferred to dispose of composite materials directly in a landfill or eliminate them.

In the mid-1960s, the hand lay-up process was used to produce mat and woven roving fiber glass. Polymeric composite materials are widely used in offshore structures, boats, submersibles, ships, and other marine structural applications. Sandwich construction methods were first utilized in the related industry in the 1970s, leading to the development of advanced fabrication technology and the usage of substitute resins such as vinyl ester and epoxy in the late 1970s and early 1980s. Polymer composites have tremendous potential for a broad range of uses, but their environmental effects and distance from a manufacturing industry limit their widespread use (Pickering, 2006). Plastic manufacturing, on the other hand, necessitates significantly non-renewable oil-based resources (Shahroze et al., 2021; Thomas et al., 2021).

TABLE 13.1

Composite Materials are Used for Various Purposes

Matrix Material		Reinforcement	
Polymer	Thermoplastic – nylon, polypropylene, polystyrene, polyethylene, polyether ether ketone (PEEK), polyphenylene sulfide (PPS), poly(p-phenylene oxide) (PPO), cellulose acetate Thermoset – polyester, vinyl ester, epoxy, phenolic, polyamide, polyurethane	Fiber-reinforced Continuous – unidirectional, bidirectional, and spatial Short-aligned and random	Natural Fibers – coir, cotton, bast, flax, hemp, jute, kenaf, soft or hard wood, grass, Bamboo, straw, leaf (pineapple), sisal Synthetic fibers – Carbon, Glass, Kevlar, etc.
Ceramics	Glass and cement	Particles reinforcement	Large and dispersion
Metal	Aluminum, copper, nickel, titanium	Molded sheets	Laminate and sandwich
		Reinforced matrix fibers	Cellulose, PHAs, PLAs, poly(esteramides) (PEAs), Poly(vinyl alcohol) (PVOH), PVA, or poly(vinyl alcohol) (PVAl), and thermoplastic starch

Petroleum-based synthetic polymers, such as polyolefin, are used in packing bottles and in molded goods employed in today's society. Worldwide, millions of tons of petrochemical plastics' disposal creates serious environmental issues; even production and consumption continue to rise. Furthermore, because these plastic wastes are resistant to microbial assault, they have become an unwelcome pollution on land, rivers, and seas. Nowadays, the usage of eco-friendly biodegradable polymers is the most practical way to replace the use of petroleum-based plastic components.

Regrettably, polymeric composite materials are sensitive to moisture and temperature, and prolonged exposure to such severe environments causes the material's mechanical qualities to deteriorate. For the prediction of structural service life, it is vital to understand the durability and life cycle of composites in terms of mechanical, chemical, and thermal characteristics (Hawileh et al., 2015; Yan et al., 2015). These biodegradable polymer compounds have the capacity to degrade completely in natural ecosystems (natural soil, lakes, active sludge, and seas). These eco-friendly polymers, on the other hand, can be chemically changed by the action of microbes or biological enzymes. According to some research, the biodegradability of bio-composites was the most essential aspect for many composites. A completely new approach on production and product must be provided, using present-day environmental assessment methods.

TABLE 13.2

Bio-Composites Classification (Mohanty et al., 2005)

Completely Degradable		Partly Degradable	Hybrid Bio-composites
Bio-fiber with biodegradable polymer or bio-fiber with petro-based biodegradable polymer		Bio-fiber with non-biodegradable polymer or synthetic fiber with biodegradable polymer	Fiber blending and matrix blending
Natural-based polymer	Petroleum-based polymer	Non-degradable petroleum-based polymer	
Polylactide, PLA	Aliphatic polyester	Polypropylene	Natural fibers such as jute and coir can be combined with polypropylene and viscose rayon
Thermoplastic starch	Aliphatic-aromatic polyester	Polyester	
Cellulose	Polyester amide	Polyethylene	
Polyhydroxy alkanoate (PHAs)	Poly-alklyene succinates	Polyvinyl alcohol	
	Polyvinyl alcohol		

13.2   DEGRADATION AND MARINE FOULING

The rate of degradation of bio-composite materials in a variety of natural habitats, not simply composting systems, is a significant quality for its long-term sustainability and reduction of the environmental impact associated with petrochemical-based plastics. Each year, several hundred thousand tons of plastic are estimated to be discarded directly or indirectly into the ocean, causing consequent negative impact on the pelagic and benthic marine environments (Gregory, 2009) (Table 13.2).

Fouling organisms must be protected aboard any boats and ship's vessel or device used in the marine applications, and fouling can be displaced using recently developed technologies. Toxic formulations such as cuprous oxide paints can raise the amounts of heavy metals, and other toxic substances in the marine ecology (e.g., tributyl tin is banned worldwide) are potential hazards. Self-polishing (exfoliating) surfaces produce polymer micro-debris, which concentrate poisons and are consumed by animals at the bottom of the food chain (Thompson et al., 2005). Low-surface energy coatings (e.g. silicone or polytetrafluoroethylene, PTFE) require extensive surface preparation to enable solid adhesion of the non-stick substance to the component.

13.3   RUSTING AND CAVITATION

A vapour bubble can form in a fast fluid flow with local pressures below the vapour pressure and sources for nucleation, and it will ascend until it reaches a higher pressure zone and collapse.. The term "cavitation" is accredited to Froude. Cavitation erosion of metal–matrix composites from a materials perspective is based on crystal displacement mechanics.

According to Hammond et al. (1993), GRP performs better than aluminum but not as well as nickel aluminum bronze (NiAl bronze) in cavitation erosion resistance.

According to Yamatogi et al. (2009), glass fiber has weak cavitation erosion resistance, whereas carbon fiber, unreinforced epoxy resin, and aramid fiber\ Nickel Aluminum Bronze (NAB) have strong cavitation erosion resistance. Some elastomeric coatings have very high erosion resistance, according to Kallas and Lichtman (1968); however, such systems may be damaged by coating substrate separation, resulting in early failure. Hydrofoils and stern gear are made of composites that work in a cavitating area, such as propellers (Anon, 2003) and rudders (Anon, 2012).

Due to the flow of an electric current, metals and semiconductors material may corrode. Electrochemical corrosion is not a concern because the majority of the constituent materials in fiber-reinforced laminates act as insulators. Carbon fibers should not be utilized with structural metals like light alloys in the presence of conducting fluid (seawater). To prevent the galvanic corrosion, cell formation and consequent loss of the alloy, and a thin glass fiber surface layer or polymer liners around bolt-holes, should be adequate.

13.4 SYNTHETIC COMPOSITES

When wood gets wet, it degrades and is attacked by marine boring creatures. In seawater, surface coatings protect steel, iron, and aluminum against rust and corrosion. As a result, one of the most essential factors in the acceptance of fiber-reinforced polymer and glass fiber-reinforced polyester for long-term usage in seawater has been their capacity to withstand this environment. Resins that cure at room temperature are compatible with reinforcing materials and are simple to laminate. Boat builders require resins with high mechanical and thermal qualities. Resins are simple to work with, when making composite parts.

In terms of mechanical characteristics and cost, vinyl esters are a cross between polyesters and epoxies. Because of the ester group's presence, they have a good affinity with polyesters, and they, like polyesters, require a catalyst and accelerator to cure at ambient temperature (Marsh, 2007). Vinyl esters have fewer reactive sites, so they may tolerate chemical corrosion well and have acceptable absorption and hydrolytic attack characteristics. As a result of these factors, they are a popular choice for boat hulls (Galanis, 2002; Marsh, 2007). A vinyl ester-based skin on a glass/polyester laminate can avoid hydrolysis-induced osmotic blistering. Even though epoxies are a more efficient barrier against humidity, boat builders and owners can adopt this protection instead of the more expensive epoxies.

Vinyl esters have a higher hardness than polyesters, making them more flexible. As a result, they can survive the effects of fatigue, which causes hulls and decks damage when utilized as a matrix in a fiber-reinforced polymer (FRP) laminate. Vinyl esters operate at a higher temperature of approximately 200°C. By using vinyl esters rather than polyester resins to create GRP components for their motor yachts, West Bay Son Ship of British Columbian was the first to advocate for this transition. Vinyl ester is increasingly being employed in the construction of larger ships and navy ships, because of its resilience.

Because of rigorous surface preparation required during treatment and the specific environmental conditions required to produce proper adhesion between fibers and resin, processing vinyl esters is more difficult and time-consuming when compared to polyesters. Furthermore, today's resins should be low in styrene to comply with more stringent industrial emission regulations (Hoge and Leach, 2016). In FRP constructions, polyester and epoxy resins are commonly utilized as matrices. Despite, this, there has been a substantial improvement with phenolic-based resins.

Phenolic resins can tolerate relatively high temperatures (about 200°C), and they are known for their fire resistance, while emitting minimal smoke or harmful fumes. Hand lay-up, hot-press, and infusion are all methods that can be used to process them. Mechanical properties, on the other hand, are 10%–20% lower than those of polyesters (Kimpara, 1991; Slater, 1994).

13.4.1 PLASTICIZING EFFECT

The plasticizing effect can be described. As the plasticizer concentration is raised, the fracture strength, elastic modulus, and viscosity of biopolymer–water mixtures decrease. When water molecules penetrate polymer composite materials, the plasticizing action is evident. Depending on whether the composite is made of a polymer matrix or natural fibers, it has a different impact on the composite's behavior. Effective plasticizers facilitate inter and intra-macromolecular contacts which may be shielded, segmental molecular motion must be facilitated, and internal friction in the biopolymer materials must be reduced. Plasticizers that are good solvents are usually good plasticizers. Some macromolecular associations and crystalline areas can be dissolved by a good solvent, while inter- and intra-molecular interactions are protected. The solvation shells reduce the activation energy of segmental motion by increasing the distance between chain segments. Many low-molecular-weight hydrophilic chemicals can operate as plasticizers for most biopolymers. Water, on the other hand, is the most effective.

13.4.2 SWELLING

When water is absorbed, it causes expansion strains in polymers and polymer composites that are analogous to thermal expansion. Water molecules diffuse between the polymer's macromolecular chains and fill the free volume, causing the distance between the macromolecules to grow. Swelling of polymers or composites occurs as the spacing between polymer chains widens. The steady-state or equilibrium state of the swelling process is the focus of the most swelling studies and characterizations. As the space between polymer chains grows, polymers or composites swell.

The stretch of osmotic pressure and a pull of material elasticity balance the equilibrium swelling of each network strand. Swelling measurements of filler-modified polymer composites are commonly used to quantify the polymer's cross-linking density. Araby et al. (2014), for example, evaluated the cross-linking density of graphene nanoplatelet-reinforced nano-composites produced to improve the electrical and thermal properties of styrene-butadiene by performing swelling investigations. The two most common methods for measuring swelling are gravimetrical and optical.

The mass ratio of the swollen network and solvent to the dry extracted material is taken into consideration in gravimetric measurements (Gevers et al., 2006).

13.4.3 HYDROLYSIS

In polymer materials, water diffusion can cause both physical and chemical interactions with polymer chains. When polymers are exposed to humidity or immersed in water, it causes the most common damaging process, called hydrolysis. It works on the basis of a chemical reaction between polymer chains and water molecules, commonly referred to as a chain cleavage mechanism. In the presence of water, this mechanism causes chain cleavage, resulting in the production of two broken chains; one has a hydrogen ion and the other with a hydroxide ion. In the presence of water, the esters, amides, and epoxide groups are the most likely to be hydrolyzed.

Aside from masts and propellers, a variety of secondary pieces, particularly in warships, are now made of polymer composites instead of metal. Discharge funnel, protective systems, shield, rudder, bulkhead, deck, hatches engine components, and heat exchangers are only some of the examples.

13.5 BIO-COMPOSITES

The development of eco-friendly composites offers a way to mitigate environmental issues caused by marine operations. Agricultural waste has emerged as a viable resource for the production of environmentally acceptable composites. This exciting field of study has various advantages, including biodegradability when combined with bio-based or natural polymers, light weight, low cost, and simple manufacturing (Kellersztein et al., 2019). Composite materials are made of two or even more materials that have been blended in such a way that the separate components can be easily distinguished.

Le Duigou et al. (2009) conducted a study on the aging of flax/PLA bio-composites in seawater. According to the results of this experiment, the absorption of water determined structural change, the degradation of fiber/matrix interface, different swelling at composite interfaces, and the degradation of matrix hydrolysis, and the degradation of fibers reduced the mechanical properties of the composites under the seawater aging mechanism.

The new material may be favored for a variety of reasons, including the fact that it is stronger, lighter, and less expensive than traditional materials. The majority of composites are made up of two materials: matrix and reinforcement. The reinforcement is more rigid and stronger than the matrix, giving the composite its upper-level qualities. The matrix keeps the reinforcement in an orderly pattern. Due to the uneven nature of reinforcement, the matrix also aids in the load transfer between reinforcements. These materials are also known as fiber-reinforced composite materials since they contain substantial fibers.

The most fundamental requirement for composites is the specificity of reinforcement. There are three types of reinforcements: particle, discontinuous fibers, and continuous fiber. Fibers are reinforcements that have one dimension that is at least one order of magnitude greater than the others. Even though there is a wide range of reinforcing options, the most frequent are glass, carbon, and aramid (commercially known as Kevlar). These composites are commonly referred to as "reinforced plastics" because matrix materials are frequently plastic. The most common plastic matrices are epoxy, polyester, and vinyl ester resins.

A composite material is a mixture of two or more chemically different and insoluble phases whose characteristics and structural performance outperform those of its constituents functioning alone. Nowadays, composites are now one of the most widely used types of engineered materials. Composites are employed for electrical, thermal, tribological, and environmental purposes in addition to their structural qualities. The type of reinforcement utilized in composite materials is frequently used to classify them. This reinforcement is encased in a matrix that keeps everything together. The reinforcement is used to make the composite stronger. The matrix material in glass fiber reinforced polymers (GFRP), for example, is polymer, and the reinforcing is glass fiber.

The development of natural biodegradable polymer composites as a replacement for petrochemical sources is an essential research subject at this level of renewable energy implementation. Cotton, flax, hemp, jute, Bamboo, and nettle fibers are largely constituted of cellulose from plants, whereas animal-origin fibers are made of protein (hair, silk, wool). Natural fibers come in a wide range of colors and textures, and their smooth surfaces aid with matrix adherence. In comparison to standard petroleum-based polymers, the market for biopolymers is currently quite tiny. Agropolymers (such as starch and cellulose) and biodegradable polyesters (obtained by microbiological production) are two types of biopolymers (Figures 13.1 and 13.2).

Biopolymers have the best prospects for application potential in the packaging market, which is also the largest market segment. Bio-plastics have the potential to grab up to 10% of the global polyolefin resin industry, which is utilized in packaging and agricultural products, and their production is estimated to be around 18 million t/year. Furthermore, agricultural fibers for use in the construction and automotive composites might have a market of 5,000 t/year. The price difference between bioplastics and synthetic plastics is expected to narrow due to continued advancements in production and processing technology, increases in base crude oil prices, close substitute energy prices, and government regulations encouraging greater use of renewable energy and waste materials (Fowler et al., 2006).

Bio-composites are materials that contain one or more biologically generated components. Due to a lack of sufficient strength, stiffness, and dimensional stability, these plastics are not appropriate for load-bearing applications. Fibers, on the other hand, offer tremendous strength and stiffness but, due to their fibrous structure, are challenging to utilize in load-bearing applications. Fibers provide strength and stiffness to fiber-reinforced composites, while the plastic matrix acts as an adhesive, binding the fibers in place so that structural composites can be created (Figure 13.3).

The principal physico-mechanical properties of natural fiber composites are that they are made of strong, light natural fibers that have a high strength-to-weight ratio.. However, because moisture absorption is often high and impact strength is low, reinforced materials' technical properties can be compromised (Kuciel et al., 2010). To begin with, the use of synthetic fibers such as glass and carbon fibers leads to significant structural lightening due to the massive specific weight difference between

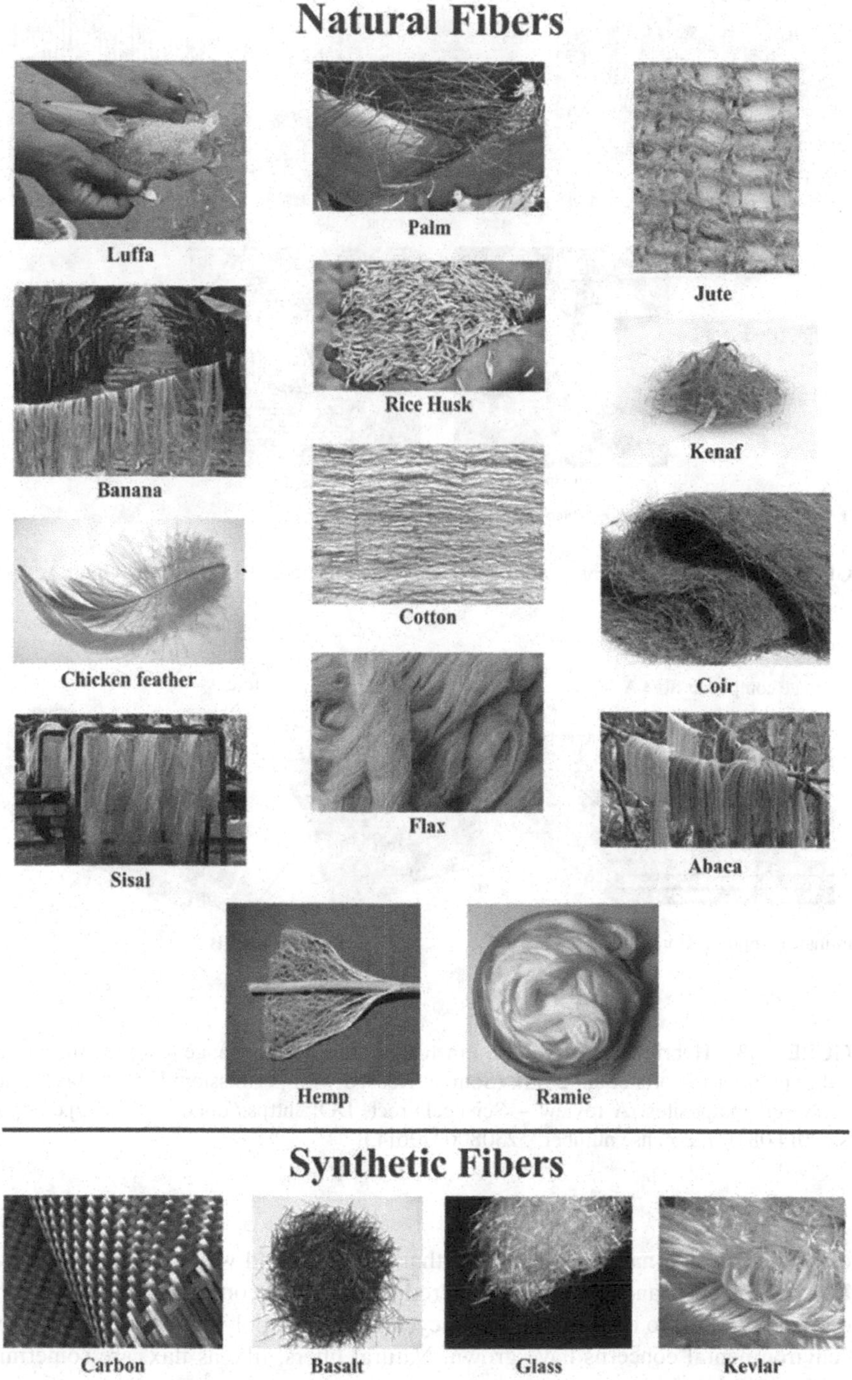

Natural Fibers

Synthetic Fibers

FIGURE 13.1 Natural and synthetic fibers of various sorts (Rajak et al., 2019). (Source: Open Access.)

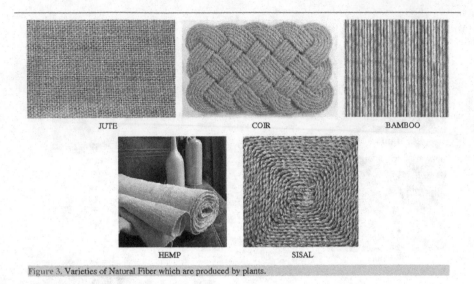

FIGURE 13.2 Plant-based natural fibers (Ashik and Sharma, 2015) (Source: Open Access.)

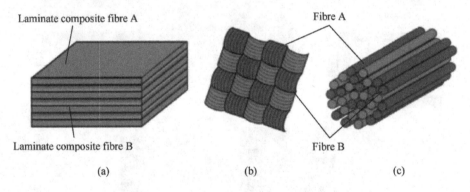

FIGURE 13.3 Hybrid structure: (a) laminate fiber sandwich arrangement; (b) inter-layer; (c) fiber by fiber (Swolfs et al., 2014). (Source: Reused with permission: Fibre hybridisation in polymer composites: A review – ScienceDirect. DOI: https://doi.org/10.1016/j.compositesa.2014.08.027, License number: 5230800470614.)

these materials and natural fibers (note that a linen strand weighs 1.4 g/cm³ versus 2.1 for a glass fiber) and a corresponding reduction in fuel consumption (and harmful emissions). A shift to bio-sourced and recyclable matrix polymers has been favored, as environmental concerns have grown. Natural fibers, like as flax, are sometimes used to reinforce them.

The hydrophilic character of the lignocellulose fibers, and thus the moist sensitivity of the composite material, is a possible issue with natural fiber-reinforced polymer matrix composites. Water absorption will be delayed by embedding hydrophilic fibers in a hydrophobic matrix, although the material will degrade over time due to diffusion and degradation. Moisture causes dimensional changes (swelling), changes in mechanical performance (plasticization, which results in higher strains to failure but lower moduli), and vulnerability to microbiological attack.

Acetylation is said to improve the resilience of composites when exposed to the environment. The matrix for composites containing various amounts of plant-derived precursor ingredients is increasingly made of bio-based and/or biodegradable thermoplastic polymers and thermosetting resins. The mechanical properties of the jute-epoxy composites were the best for all immersion times tested. Interface integrity and moisture resistance of jute fiber-matrix was better with epoxy resin (Costa and D'Almeida, 1999). Plant fibers can be acetylated to improve their composite's mechanical properties and hydrophobicity (Abdul Khalil et al., 2000). Jute and flax bast fibers, as well as coconut fiber (coir), oil palm empty fruit bunch, and oil palm frond were all to be considered for bio-composite production.

In marine manufacture, Short bio-fibers and wood dust are employed as fillers (thermoplastic blankets), as well as automobile, furniture, and train components. Long fibers such as hemp, bamboo, linen, and jute are employed as reinforcing materials to replace fiber glass. The fundamentals of fiber-reinforced plastics, which are a combination of polymeric resins and fibers, allow for the full or partial exploitation of components with a natural origin, such as flax fiber, hemp, bamboo, or jute. Instead of using materials of synthetic origin, from flax fiber, hemp, bamboo, and jute, aside from being less expensive than glass, they are also non-abrasive, non-toxic, and biodegradable. Unfortunately, these fibers are hydrophilic, and hence they should be protected from moisture absorption, which might have disastrous repercussions.

13.6 THE DIFFUSION PROPERTIES OF NATURAL FIBERS

Natural fibers are composites fabricated mostly of cellulose, hemicellulose, lignin, pectin, wax, and moisture (in the context of flax fibers) (Manfredi et al., 2006). The hemicellulose, which is the plant cell wall connected with the cellulose, is the fiber component responsible for moisture absorption (Methacanon et al., 2010). Higher hemicellulose content, as a result of the hollow-cavity shape of the fibers, leads to increased moisture sorption and biodegradation. The vulnerability of bio-composites to moisture, which causes dimensional changes and a loss of mechanical capabilities, drives researchers to investigate the use of a physico-chemical treatment to safeguard natural fibers.

When treated fibers are impregnated with a polymer matrix, the water resistance behavior changes. The coupling reaction between silane-treated fibers and an unsaturated polyester resin is explained using a co-polymerization process. In the presence of a peroxide initiator, the methacryl groups of the silane molecule might react with the double bonds of the unsaturated polyester. Oil palm fibers treated with methacryloxypropylmethoxysilane were found to provide significant protection against water uptake in unsaturated polyester composites.

13.6.1 THERMAL DEGRADATION

During service, composite materials destined for maritime applications may be subjected to thermal degradation. The term "thermal degradation" refers to a long-term change in material behavior caused by changes in temperature. The examination of thermal degradation is much more difficult to study for bio-composites, which have a matrix thermal behavior that differs significantly from natural fibers. When natural fibers are used as a composite material, distinct thermal behavior for the lignin, hemicellulose, and other components is introduced.

When natural fibers are exposed to heat, their non-homogeneous nature determines their behavior. The heat deterioration of natural fibers determined the viscoelastic behavior of wood. The temperature activates the viscoelastic qualities of wood up to a point where the temperature causes the wood to degrade.

13.6.2 MECHANICAL BEHAVIOR

The mechanical character of bio-fibers in bio-composites may also be estimated by using Halpin Tsai model (Le Duigou et al., 2014). The creation of hydrogen bonds between water molecules and cellulose fibers may induce deterioration in mechanical characteristics of bio-composites as moisture content rises. Natural fibers have a high concentration of hydroxyl groups (–OH) in their structure, which forms a wide network of –H bonds between the cellulose macromolecules with polymer. Water can cause dimensional variation in composites and break down the interfacial connection between fiber and matrix when it forms hydrogen bonds with cellulose in the fiber structure. One of the mechanisms involved in makeup of bio-composites is the permeability of natural fibers to polymer impregnation (mostly polar resins).

13.6.3 COUPLING EFFECT

In real situations, various damaging causes act simultaneously, and their effects on bio-composites are intertwined. Temperature, humidity, mechanical loadings, and even additional aging mechanisms are regularly coupled (Vauthier et al., 1998). In order to construct entirely biodegradable bio-composites without emitting poisonous or noxious components, benign materials should be expected to be used as a coupling agent. It's crucial to investigate the impact of fiber and coupling agent presence on the thermal stability of extruded polypropylene composites. In this scenario, the coupling agent is largely concentrated at the polymer/fiber interface, where the polymeric component interacts with the hydrophobic polymer matrix and the maleic anhydride interacts with hydrophilic fibers, promoting adhesion between the two phases.

13.6.4 MARINE USAGE

Several studies have shown that bio-composites have potential marine applications due to their superior mechanical properties and biodegradability. As a result of these characteristics, bio-composites have evolved as a feasible alternative to synthetic fiber-reinforced composites. As bio-composites are biodegradable, particular attention must be taken while integrating marine constructions. From small boats to submarines, composite materials have been widely utilized in marine structures.

13.6.5 FERRO-CEMENT

It's a cement-based composite material with one or more layers of mesh and tiny diameter rods covered with cement mortar. It was arguably the first composite material employed in the marine industry, and it was used to construct low-cost barges. Armature corrosion is a typical concern in chemically harsh marine environments, despite the fact that it is a low-cost composite. However, a few of ferro boats are still in operation today. It has a wide range of engineering qualities, including flexural strength, toughness, fatigue resistance, impact resistance, and fracture resistance.

13.7 GLASS-REINFORCED PLASTIC (FIBER GLASS)

Glass fibers were available after polyester resins were developed. Glass-reinforced plastic boats first appeared in the early 1950s and have remained a common marine composite construction approach to this day. It is strong, extremely light, and highly versatile, although fibers commonly used are GRP, and other fibers such as carbon, aramid, and basalt are also employed. The GRP composite's surface is comprised of diamond-hard aluminum oxide aggregate, which has excellent long-term wear resistance. GRP is extremely useful for its valuable properties in industries worldwide. Easy to shape, thermal insulation, anti-slip safety, weight ratio, high strength, fire retardancy, high energy absorption, chemical and corrosion resistance, good insulation to heat and sound are some of the properties of GRP.

13.8 ADHESIVE COMPOSITES

For connecting fiber composites to themselves and a variety of substrates, cutting-edge adhesive formulations are offered. To address the present requirement for composite bonding, these adaptable adhesives are offered as liquids, pastes, and films. Even when exposed to harsh environments, they are engineered to provide the best physical strength and endurance. "Hot molded" and "cold molded" boat building methods based on putting thin wood veneers over a frame were created as a result of wartime necessity (Table 13.3). To help speed up the process and lessen dependency on aluminum and steel, high-performance, urea-based adhesives have been employed successfully for molding naval hulls and in aircraft makeup. In bonded interfaces, impact/fracture toughness, tensile, compressive, creep, stress fracture, and shear properties are all outstanding. Thermal cycle resistance, cryogenic shock resistance, mechanical shock resistance, chemical resistance, low shrinkage, electrical insulation, and serviceability in extreme temperatures are just a few of the benefits.

TABLE 13.3

Composite Making Techniques

Open Molding	Closed Molding	Cast Polymer Molding	Additive Manufacture
• Hand lay-up • Spray up • Filament winding	• Vacuum bag molding • Vacuum infusion • Resin transfer molding • Compression molding • Pultrusion • Reinforced reaction injection molding • Centrifugal casting • Continuous lamination	• Gel-coated stone molding • Stone surface molding • Engineered stone molding	

13.9   ARAMID FIBER COMPOSITES

Aramid fibers are a type of synthetic organic polymer made by spinning a solid fiber from a liquid chemical combination, commonly utilized to reinforce sailboat parts such as the keel and bow sections. They also have better shock absorption properties, making them ideal for ocean sailing. All aramid fibers have tremendously strong and have a low density, resulting in a very high specific strength. Impact resistance is strong in all classes, and lower modulus grades are widely utilized in ballistic applications. The fibers are resistant to abrasion, chemical, and heat damage in addition to their great strength. However, if the fiber is exposed to ultraviolet (UV) radiation, it may eventually deteriorate.

13.10   CARBON FIBERS

It was rapidly being used in sailboats, super yacht furnishings, and highly ridged inner moldings because it provided stability and great effect while weighing less. Carbon fibers are an attractive material, and they are frequently substituted for the woven material's aesthetics, due to their superior material properties. Boats with honeycomb or foam hulls, structural mounts, keels, masts, poles, booms, and carbon winches, drums, and shafting are made up of carbon fiber-reinforced epoxy composites. In harsh sailing conditions, composites may aid to boost efficiency while minimizing the failure rate. Racing yachts use complex polymer composites more than any other marine structure to conserve weight and boost durability.

In comparable research, contemporary environmental problems like depletion and waste management must be considered on sailing yacht designs innovation. This leads to incorporation of natural fibers into the matrix, resulting in environmentally acceptable composite materials that might be used in marine applications. Furthermore, they have a high specific toughness and a small environmental imprint. For 2 years, the aging process of flax/PLA bio-composites was studied in natural saltwater. This research shows that bio-composites absorb a relatively substantial quantity of moisture, which is regulated by vegetal fibers (Le Duigou et al., 2014).

Controlling the aging of bio-composites in a marine habitat comprising both natural fiber and matrix is required. An additional coating layer of a comparable biopolymer on a bio-composite may improve weight loss through the interaction of fiber and matrix. The bio-composite's immersion mechanical and thermal characteristics demonstrate that the protective layers limit matrix hydrolysis, maintain composite qualities, and improve durability (Le Duigou et al., 2011). DuPont has created a marine composite Kevlar that is beneficial for providing an optimal blend of strength, rigidity, and lightweight qualities for a variety of nautical applications.[1]

This improves the greater speeds that may be obtained in patrol and service boats by boosting engine power. These Kevlar-based composites are lighter, and also harder, more damage-resistant, and perform much better during hydrodynamic fatigue loading.[2] Davies (2016) investigated the environmental destruction of composites used in highly loaded marine components, like tidal turbine blades or composite propellers; the use of composites is growing and necessitates a thorough understanding of the stress–seawater coupling. Rather than the theoretical framework, just a few experimental data sets are accessible, and the effort is being spent on completing a few tests that require particular test equipment. More research is critically needed in this area.

The FRP composite material was first utilized marine construction for boats shortly after World War II. As wood became more scarce and expensive, boat designers began to use FRP composites instead of timber, which was typically used in small nautical craft. Many boat builders and owners are moving away from wood since it is easily damaged by seawater and marine creatures, demanding expensive regular maintenance and repairs. The earliest attempts to make boat hulls and 12 small surf boats out of FRP composites were reported in 1947, for the United States Navy.

Despite the fact that sandwiched composites and modern FRP have carbon and aramid fibers along with vinyl ester or epoxy resin matrices, most marine ships are made of glass-reinforced polyester composites. They're frequently employed in structural applications that required excellent performance. The interest in marine renewable energy sources is rising, and composite components like tidal turbine blades are playing a crucial part in this new industry. Marine composites are being pushed into new aspects of quality and structure due to extremely high cyclic loads and long-term immersion.

13.11 FIBER-REINFORCED POLYMER (FRP) COMPOSITES

FRPs have been widely employed in the boat manufacturing and marine construction sectors for decades since they are the most durable, workable, and cost-effective option. These materials offer a higher strength-to-weight ratio (specific tensile strength) than traditional materials and are more resistant to material corrosion as well as marine environmental factors like UV, seawater, organisms, and fatigue loading (Guersel and Neser, 2012; Mouritz et al., 2001). It has become necessary to build marine structures in a cost-effective manner while also assessing their life cycle. Researchers in the field of marine technology can identify appropriate material solutions for the boatbuilding industry by investigating the behavior of FRPs in the marine environment. The intrinsic inhomogeneity of FRP and various distinctive damage mechanisms, including matrix/fiber fracture, fiber/matrix debonding, and inter-laminar delamination, as well as their intricate interactions (Marsh, 2010; Stewart, 2011).

FRPs have a number of additional benefits during the construction phase, such as the capacity to consolidate parts, lowering the number of parts, joints, fasteners, and manufacturing steps; offering significant weight savings and improved stability by lowering the center of gravity; and making it simpler to apply complex shapes, such as hulls, decks, and submarine fairings. (Guillermin, 2010). Fatigue characteristics are a key signal for the limit state in marine constructions susceptible to environmental cyclic loads, which must be considered by the designer concurrently nowadays. In comparison to metallic structural materials, the fatigue design approach for FRP is currently limited in terms of method availability and the inadequacy of available models to reliably characterize fatigue behavior.

13.12 GLASS-REINFORCED POLYMER-BASED COMPOSITES (GRP)

In addition to warships, high-performance materials are being utilized in civil applications. GRPs are used to make fishing boats, small boats, hovercraft, and catamarans all over the world. Yachts, sailboats, barges, and lifeboats are all examples of small boats that use GRP. GRP is the most widely used material, accounting for over 80% of the hulls on ships up to 20 m in length. GRP materials have significant weight and production cost advantages over steel, as well as low maintenance costs due to its extraordinary resistance to marine organisms and corrosion. The Dutch shipyards "Mulder and Rijke" have created a glass fiber/polyurethane composite lifeboat with improved flame resistance. The boat was engulfed in flames for 10 minutes in fire-resistance testing, with an outside temperature of 816°C recorded, but with an internal temperature of only 15.5°C (Tekalur et al., 2008).

A hybrid glass/carbon fiber composite was used in one of the early versions of the composite mast to improve both ballistic performance and rigidity. When compared to an aluminum shaft of the same size, it reduces the weight of 20% ± 50%. The composite shaft is more resistant to fatigue and corrosion, as well as vibration damping and air blast damage.

For excellent performance and safety, racing powerboats are highly composed of sophisticated and hybrid composites. Bennington-based Fothergill Composites Inc. has developed a safe cell cockpit made of carbon and aramid fibers with an aramid honeycomb core to safeguard the driver in all circumstances, especially in a high-speed collision. This construction can endure a 100-foot fall without sustaining serious damage.

The materials utilized in the composite are commercially accessible, and the formulation of the correct blend of fibers, resin, and laminate lay-up for marine applications provides the needed mechanical and environmental performance. Endurance testing in the marine habitat, as well as water uptake and fouling tests, are all part of the extensive development process. Glass or carbon fiber laminates are commonly used in composite propeller blades, although a thin layer of polyurethane, NAB, or stainless steel can be added to decrease impact damage. Although composite-based hubs have been realized, the FRP blades are normally bonded or connected to the metal hub of the propeller.

Lightweight composite propellers have less inertia, resulting in faster acceleration and deceleration rates, as well as increased speed and fuel economy. Hydroelastic tuning of the laminate might postpone the development of cavitation (Young, 2008). Stiffness, strength, and fatigue performance are all essential mechanical attributes in this application. The construction was designed to be the stiffest over the blade's length and robust enough to provide a large margin of safety over the design load. The composite was half as stiff as NAB in terms of material but had equivalent strength. The propeller's structural rigidity was restored through enhanced design. During the test, the metal insert at the root also improved in terms of fatigue performance. NAB flaws caused the failure, but the composite was unaffected.

Reinforcing a resin with reinforcing elements such as carbon fibers, fiber-glasses, or aramid fibers produces composite materials that are substantially stronger and have improved material qualities for a number of applications. They have a stronger strength-to-weight ratio than typical wood or steel frames, and produce a considerable easier finish with less skills. There has been considerable caution about selecting systems that suit current requirements without risking future generations' ability to meet their own.

13.13 CONCLUSION

Renewable bio-composites are used to make a variety of interior and exterior parts for autos, ships, sound-absorbing timber construction materials, and consumer products. Indeed, multiple studies have found that natural fiber reinforcement using as biopolymers has superior physical, thermal, and degradable qualities. Furthermore, incorporation of a compatibility agent may improve these qualities. However, as compared to other inorganic fillers, renewable bio-composites manufactured from bacterium cellulose, rice straw, rice husk, natural fiber, lignocellulose, cellulose, and paper sludge have a number of advantages. When compared to petroleum-based composites, bio-composites are the best and most environmentally beneficial option. Nonetheless, there remains a sense of traditionalism, if not outright refusal, to define polymer composite-based solutions for a variety of applications. This is owing to concerns about new ways to use old materials, as well as new and existing materials being used in novel applications. This suggests that more research is necessary to broaden the applicability and use of composite materials in marine industries.

In the case of thermoset composites, these materials are inappropriate that has restricted disposal alternatives at the end of their useful life (Singh et al., 2010). The usage of thermoplastic matrix composites, on the other hand, necessitates higher temperatures during manufacturing, potentially resulting in immediate global warming and climate change. Significant progress has been made in understanding the behavior of these materials and built buildings under mechanical, thermal, and fire-induced load scenarios over this time. Processing and manufacturing considerations have also received a lot of attention, resulting in the capacity to build extremely complicated, multi-material, massive, three-dimensional assemblies that can withstand tremendous loads. There are tons and tons of plastics and other waste being dumped into marine environments. If researchers focus on those recyclable materials for the making of composites, they may be used in the production of various parts in marine applications and solve environmental issues as well.

Bio-composites are gaining traction in both worldwide research and cutting-edge applications, which is becoming increasingly concerned about environmental preservation and, as a result, the eco-compatibility of composite materials. Many studies have looked into the physical properties of natural fibers like sisal, jute, and palm fibers, as well as the interfacial performances of hydrophilic natural fibers and hydrophobic polymer matrices due to insufficient interfacial bonding (Dufresne, 2008; Li and Mai, 2006). Natural fibers such as hemp, flax, and jute are mixed with polymeric matrices from both renewable and non-renewable sources to create materials that are increasingly more competitive than standard composites.

Despite the fact that their manufacturing needs additional care, such as a specific interface between the bio-fiber and the matrix, and more complex processing steps, there is a mineral and synthetic reinforcement's replacement with annually natural renewable fibers that has arisen as a result of public interest in environmental issues and awareness of future raw material exhaustion. Natural fibers in marine architecture have piqued the interest of many specialists and yacht designers. The current European directives on the disposal of toxic waste, on the other hand, make this industry quite fascinating and deserving of further investigation.

Bio-composites have been proven to exhibit mechanical qualities comparable to glass fiber-reinforced composites. One of the most serious concerns to be solved is the hydrophilicity of these materials, which drastically restricts their use and the amount of fiber produced. Natural materials in the marine industry are important because of their inherent qualities; they can solve a variety of problems and increase performance. Natural reinforcements for composites in the maritime industry aren't without their drawbacks. When selecting to employ vegetable fibers, one issue to consider is their suitability with the polymeric matrix that will sustain them. These procedures are currently being researched, and while many potential treatments were presented and tried, there are many components to be defined, as well as numerous alternatives to those that have already been tried.

Given the environmental benefits of adopting recyclable materials, extensive research is now being taken in this area to maximize longevity, regulate degradation, and mitigate inherent property losses. The development of next generation of composite materials, products, and processes should be guided by sustainability, industrial ecology, eco-efficiency, and eco-friendly manner under new legislation. Glass fiber-reinforced composites are being replaced with bio-composites (natural/ bio-fiber). Although there is still a long way to completely "green" composites, the road to innovation is teeming with new technologies and products.

REFERENCES

Abdul Khalil, H.P.S., Rozman, H.D., Ahmad, M.N., Ismail, H. (2000). Acetylated plant-fiber reinforced polyester composites: A study of mechanical, hygrothermal, and aging characteristics. *Polymer-Plastics Technology and Engineering*, 39(4):757–781.
Akampumuza, O., Wambua, P. M., Ahmed, A., Li, W and Qin, X. H (2017) Review of the applications of biocomposites in the automotive industry. *Polymer Composites*, 38(11), 2553–2569.

Alothman, O. Y., Jawaid, M., Senthilkumar, K., Chandrasekar, M., Alshammari, B. A., Fouad, H., Hashem, M., & Siengchin, S. (2020). Thermal characterization of date palm/epoxy composites with fillers from different parts of the tree. *Journal of Materials Research and Technology*, 9(6), 15537–15546. https://doi.org/10.1016/j.jmrt.2020.11.020

Anon. (2003) World's largest composite propeller successfully completes sea trials. Naval Architect 16.

Anon. (2012). The intelligent propeller made of carbon fiber. http://www.compositecarbonfiberprop.com/. Accessed 16:37 on 19 August 2013.

Araby, S., Meng, Q., Zhang, L., Kang, H., Majewski, P., Tang, Y., & Ma, J. (2014). Electrically and thermally conductive elastomer/graphene nanocomposites by solution mixing. *Polymer*, 55(1), 201–210.

Ashik, K. P., & Sharma, R. S. (2015). A review on mechanical properties of natural fiber reinforced hybrid polymer composites. *Journal of Minerals and Materials Characterization and Engineering*, 3(5), 420.

Cao. J and Grenestedt J. L. (2004). Design and testing of joints for composite sandwich/steel hybrid ship hulls. *Composites Part A: Applied Science and Manufacturing*, 35, 1091–1105.

Chandrasekar, M., Senthilkumar, K., Jawaid, M., Alamery, S., Fouad, H., & Midani, M. (2022a). Tensile, thermal and physical properties of washightonia trunk fibres/pineapple fibre biophenolic hybrid composites. *Journal of Polymers and the Environment*, 30(10), 4427–4434. https://doi.org/10.1007/s10924-022-02524-z

Chandrasekar, M., Senthilkumar, K., Jawaid, M., Mahmoud, M. H., Fouad, H., & Sain, M. (2022b). Mechanical, morphological and dynamic mechanical analysis of pineapple leaf/washingtonia trunk fibres based biophenolic hybrid composites. *Journal of Polymers and the Environment*, 30(10), 4157–4165. https://doi.org/10.1007/s10924-022-02482-6

Chandrasekar, M., Siva, I., Kumar, T. S. M., Senthilkumar, K., Siengchin, S., & Rajini, N. (2020). Influence of fibre inter-ply orientation on the mechanical and free vibration properties of banana fibre reinforced polyester composite laminates. *Journal of Polymers and the Environment*, 28(11), 2789–2800. https://doi.org/10.1007/s10924-020-01814-8

Costa, F.H.M.M and D'Almeida, J.R.M. (1999). Effect of water absorption on the mechanical properties of sisal and jute fiber composites. *Polymer-Plastics Technology and Engineering*, 38(5):1081–1094.

Davies, P (2016) Environmental degradation of composites for marine structures: New materials and new applications. *Philosophical Transactions of the Royal Society A: Mathematical, Physical and Engineering Sciences*, 374(2071), 20150272.

Dufresne, A. (2008). Cellulose-based composites and nano-composites. In: Belgacem, M. N. and Gandini, A. (eds.) *Monomers, Polymers and Composites from Renewable Resources* (pp. 401–418). Elsevier, Oxford.

Fowler, P. A., Hughes, J. M and Elias, R. M. (2006). Bio composites: Technology, environmental credentials and market forces. *Journal of the Science of Food and Agriculture*, 86(12), 1781–1789.

Galanis, K. (2002). Hull construction with composite materials for ships over 100 m in length; Massachusetts Institute of Technology: Massachusetts, MA, USA. (Doctoral dissertation, Massachusetts Institute of Technology).

Gevers, L. E., Vankelecom, I. F., & Jacobs, P. A. (2006). Solvent-resistant nanofiltration with filled polydimethylsiloxane (PDMS) membranes. *Journal of Membrane Science*, 278(1–2), 199–204.

Gregory, M. R. (2009). Environmental implications of plastic debris in marine settings—entanglement, ingestion, smothering, hangers-on, hitch-hiking and alien invasions. *Philosophical Transactions of the Royal Society B: Biological Sciences*, 364(1526), 2013–2025.

Guersel, K. T and Neser, G. (2012). Fatigue properties of fiberglass bolted, bonded joints in marine structures. *Sea Technology*, 53(11), 37–41.

Guillermin, O. (2010). Composites put wind in the sails of all kinds of vessels. *Reinforced Plastics*, 54(4), 28–31.

Hammond, D.A., Amateau, M.F., Queeney, R.A. (1993) Cavitation erosion performance of fiber reinforced composites. *Journal of Composite Materials* 27(16), 1522–1544.

Hawileh, R. A., Abu-Obeidah, A., Abdalla, J. A and Al-Tamimi, A. (2015). Temperature effect on the mechanical properties of carbon, glass and carbon–glass FRP laminates. *Construction and Building Materials*, 75, 342–348.

Hoge, J and Leach, C. (2016). Epoxy resin infused boat hulls. *Reinforced Plastics*, 60(4), 221–223.

Kallas, D.H and Lichtman, J.Z. (1968) Cavitation erosion. In: Rosato DV, Schwartz RT (eds.) *Environmental Effects on Polymeric Materials* (Vol. 1, pp. 223–280), Wiley-Interscience, London-Sydney-New York.

Kellersztein, I., Shani, U., Zilber, I and Dotan, A. (2019). Sustainable composites from agricultural waste: The use of steam explosion and surface modification to potentialize the use of wheat straw fibers for wood plastic composite industry. *Polymer Composites*, 40(S1), E53–E61.

Kimpara, I. (1991). Use of advanced composite materials in marine vehicles. *Marine Structures*, 4(2), 117–127.

Kuciel, S., Kuźniar, P., & Liber-Kneć, A. (2010). Polymer biocomposites with renewable sources. *Archives of Foundry Engineering*, 10(3), 53–56.

Le Duigou, A., Bourmaud, A., Davies, P and Baley, C. (2014). Long term immersion in natural seawater of Flax/PLA biocomposite. *Ocean Engineering*, 90, 140–148.

Le Duigou, A., Davies, P and Baley, C. (2009). Seawater ageing of flax/poly (lactic acid) bio-composites. *Polymer Degradation and Stability*, 94(7), 1151–1162.

Le Duigou, A., Deux, J. M., Davies, P and Baley, C. (2011). Protection of flax/PLLA bio composites from seawater ageing by external layers of PLLA. *International Journal of Polymer Science*, 2011, 235805.

Li, Y and Mai, Y. W. (2006). Interfacial characteristics of sisal fiber and polymeric matrices. *The Journal of Adhesion*, 82(5), 527–554.

Manfredi, L. B., Rodríguez, E. S., Wladyka-Przybylak, M and Vázquez, A. (2006). Thermal degradation and fire resistance of unsaturated polyester, modified acrylic resins and their composites with natural fibres. *Polymer Degradation and Stability*, 91(2), 255–261.

Marsh, G. (2007). Vinyl ester-the midway boat building resin. *Reinforced Plastics*, 51(8), 20–23.

Marsh, G. (2010). Marine composites drawbacks and successes. *Reinforced Plastics*, 54(4), 18–22.

Methacanon, P., Weerawatsophon, U., Sumransin, N., Prahsarn, C and Bergado, D. T. (2010). Properties and potential application of the selected natural fibers as limited life geotextiles. *Carbohydrate Polymers*, 82(4), 1090–1096.

Mohanty, A. K., Misra, M., and Drzal, L. T. (eds.). (2005). *Natural Fibers, Biopolymers, and Bio-Composites*. CRC Press, UK.

Mouritz, A. P., Gellert, E., Burchill, P and Challis, K. (2001). Review of advanced composite structures for naval ships and submarines. *Composite Structures*, 53(1), 21–42.

Musiol, M. T., Rydz, J., Sikorska, W. J., Rychter, P. R and Kowalczuk, M. M. (2011). A preliminary study of the degradation of selected commercial packaging materials in compost and aqueous environments. *Polish Journal of Chemical Technology*, 13(1), 55–57.

Nasimudeen, N. A., Karounamourthy, S., Selvarathinam, J., Kumar Thiagamani, S. M., Pulikkalparambil, H., Krishnasamy, S., & Muthukumar, C. (2021). Mechanical, absorption and swelling properties of vinyl ester based natural fibre hybrid composites. *Applied Science and Engineering Progress*. https://doi.org/10.14416/j.asep.2021.08.006

Pickering, S. J. (2006). Recycling technologies for thermoset composite materials—current status. *Composites Part A: Applied Science and Manufacturing*, 37(8), 1206–1215. https://doi.org/10.1016/j.compositesa.2005.05.030

Rajak, D. K., Pagar, D. D., Menezes, P. L., & Linul, E. (2019). Fiber-reinforced polymer composites: Manufacturing, properties, and applications. *Polymers*, 11(10), 1667.

Senthilkumar, K., Saba, N., Chandrasekar, M., Jawaid, M., Rajini, N., Siengchin, S., Ayrilmis, N., Mohammad, F., & Al-Lohedan, H. A. (2021). Compressive, dynamic and thermomechanical properties of cellulosic pineapple leaf fibre/polyester composites: Influence of alkali treatment on adhesion. *International Journal of Adhesion and Adhesives*, 106, 102823. https://doi.org/10.1016/j.ijadhadh.2021.102823

Senthilkumar, K., Siva, I., Rajini, N., & Jeyaraj, P. (2015). Effect of fibre length and weight percentage on mechanical properties of short sisal/polyester composite. *International Journal of Computer Aided Engineering and Technology*, 7(1), 60. https://doi.org/10.1504/IJCAET.2015.066168

Senthilkumar, K., Subramaniam, S., Ungtrakul, T., Kumar, T. S. M., Chandrasekar, M., Rajini, N., Siengchin, S., & Parameswaranpillai, J. (2022). Dual cantilever creep and recovery behavior of sisal/hemp fibre reinforced hybrid biocomposites: Effects of layering sequence, accelerated weathering and temperature. *Journal of Industrial Textiles*, 51(2_ suppl), 2372S–2390S. https://doi.org/10.1177/1528083720961416

Shahroze, R. M., Chandrasekar, M., Senthilkumar, K., Senthil Muthu Kumar, T., Ishak, M. R., Rajini, N., Siengchin, S., & Ismail, S. O. (2021). Mechanical, interfacial and thermal properties of silica aerogel-infused flax/epoxy composites. *International Polymer Processing*, 36(1), 53–59. https://doi.org/10.1515/ipp-2020-3964

Singh, M., Summerscales, J. and Wittamore, K. (2010) Disposal of composite boats and other marine composites. In: Goodship, V. (ed.) *Management, Recycling and Reuse of Waste Composites* (pp. 495–519). Woodhead Publishing, Cambridge.

Slater, J. E. (1994). Selection of a blast-resistant GRP composite panel design for naval ship structures. *Marine Structures*, 7(2–5), 417–440.

Stewart, R. (2011). Better boat building—trend to closed-mould processing continues. *Reinforced Plastics*, 55(6), 30–36.

Swolfs, Y., Gorbatikh, L., & Verpoest, I. (2014). Fiber hybridization in polymer composites: A review. *Composites Part A: Applied Science and Manufacturing*, 67, 181–200.

Tekalur, S.A.; Shivakumar, K and Shukla, A. (2008) Mechanical behavior and damage evolution in E-glass vinyl ester and carbon composites subjected to static and blast loads. *Composites Part B: Engineering*, 39, 57–65.

Thomas, S. K., Parameswaranpillai, J., Krishnasamy, S., Begum, P. M. S., Nandi, D., Siengchin, S., George, J. J., Hameed, N., Salim, N. V., & Sienkiewicz, N. (2021). A comprehensive review on cellulose, chitin, and starch as fillers in natural rubber biocomposites. *Carbohydrate Polymer Technologies and Applications*, 2, 100095. https://doi.org/10.1016/j.carpta.2021.100095

Thompson, R., Moore, C., Andrady, A., Gregory, M., Takada, H and Weisberg, S. (2005) Letter: New directions in plastic debris. *Science*, 310 (5751), 1117.

Vauthier, E., Abry, J. C., Bailliez, T and Chateauminois, A. (1998). Interactions between hygrothermal ageing and fatigue damage in unidirectional glass/epoxy composites. *Composites Science and Technology*, 58(5), 687–692.

Yamatogi, T., Murayama, H., Uzawa, K., Kageyama, K and Watanabe, N. (2009) Study on cavitation erosion of composite materials for marine propeller. In: *17th International Conference on Composite Materials Edinburgh, Scotland*.

Yan, L., Chouw, N and Jayaraman, K. (2015). Effect of UV and water spraying on the mechanical properties of flax fabric reinforced polymer composites used for civil engineering applications. *Materials & Design*, 71, 17–25.

Young, Y.L. (2008) Fluid–structure interaction analysis of flexible composite marine propellers. *Journal of Fluids and Structures*, 24(6), 799–818.

NOTES

1 See https://www.dupont.com/products-and-services/fabrics-fibers-nonwovens/fibers/uses-and-applications/composites.html
2 See http://www.dupont.com/products-and-services/fabrics-fibers-nonwovens/fibers/uses-and-applications/composites.html

14 Polyester-Based Biocomposites for Building and Construction Applications

Saurabh Tayde, Ajinkya Satdive, Bhagwan
Toksha, and Aniruddha Chatterjee
Maharashtra Institute of Technology

Shravanti Joshi
Marathwada Institute of Technology

CONTENTS

14.1 INTRODUCTION

Polyester resin belongs to the class of polymer materials, commonly referred to as 'laminating resin' or 'fibreglass resin' (Cook, 2001). In general terms, it is referred to as 'unsaturated polyester resin' (UPR). Polymer materials comprise foams, epoxies, polyesters and silicon resins largely used in the construction and building industry.

The ease in usage and being cheaper than epoxies/urethanes make it an economical component used in a wide range of industrial components (Chandrasekar et al., 2020; Senthilkumar et al., 2021; Alothman et al., 2020). This material offers incredible physical and chemical properties, making its performance superior and a choice of the construction sector (Plank, 2005). The main characteristics of these materials involve gain in strength at a faster rate, very high abrasion/chemical/corrosion/heat resistance and excellent impact and compressive strength. The lower styrene emission rates and application-ready viscosity responses allow the mass-scale usage and consumption of these materials (Crawford and Throne, 2002; Vanderlaan and Forster, 1988). Polyesters are consumed in the construction sector in anchoring grouts, resin mortars and concrete additives, coatings, electrical insulation and lighting sealants, bridge building high-tension and strong ropes and power belts. Characteristic-based classification of the polyester material forms the following classes:(1) flexible polyester, (2) chemical-resistant polyester, (3) general purpose and speciality polyester, (4) low-styrene emission polyester and (5) drinking water-compliant polyester. Construction and buildings are one of the highest consuming segments of polyester materials along with the electrical and transport sectors. The expected yearly growth of the polyester resin market from 2019 to 2027 is 6.3%, creating a market size of more than USD 19 billion by 2027 (Unsaturated Polyester Resin Market Size | Industry Report, 2027, n.d.).

There is a tremendous growth expected in the construction industry which warrants the huge demand for construction resources (García-Gonzále et al., 2020; Ślosarczyk et al., 2020). The anchoring grouts are used in drilled or formed holes in various construction instances such as concrete, masonry or natural rock. These anchoring grouts are made of polyester and are used for securing bolts, bars, tendons or dowels (Thomas et al., 2021; Senthilkumar et al., 2015, 2022; Nasimudeen et al., 2021). Anchoring grouts made from polyester composites feature rapid installation, rapid strength gain and a high degree of corrosion protection (Lim et al., 2009). Polyester-based coatings are used in the construction industry to provide a barrier for concrete and steel substrates, particularly in mineral acid resistance. These coatings are also useful in tank linings, floors and sewage treatment plant pipeline systems (Bouzit et al., 2022). The next application of polyesters in the construction industry is concreting. The concretes realized from polyesters are applied as a repair material similar to epoxy mortars. Composites are developed with phases of materials from biological sources, glass cloth or glass flakes (Shahroze et al. 2021; Chandrasekar et al., 2022a). The challenge of using this material is the limited pot-life, thermal contraction and volume shrinkage during the cure. Owing to its rapid strength gain capacities, the polyester concrete is used in bridge decks for galvanic corrosion protection, subsequent cracking and spalling of the concrete, for airport runway maintenance and the installation of runway lights, and as a filler in cable slots cut across asphalt runways and also to fill over PVC cable ducts installed in concrete runways. Loop and lightning sealants made from polyester composites are used in road traffic signal controllers and inductance loops, airport runway lighting and automatic barriers or gates sealing horizontal cable slots in concrete and asphalt pavements. Features of rapid cure to form a tough resilient seal with a flexible degree of movement, without the need of a primer, adhesion in dry or damp conditions to concrete, and resistance to various ambient attacks make them suitable for such applications (Al-Haydari et al., 2021; Halil Akın and Polat, 2022; Pothan et al., 2003; Tabatabai et al., 2018). Cement is one of the highest consumed key components in the construction industry. The production of cement is not environmentally friendly and accounts for nearly 10% of global carbon emissions (Marrero et al., 2017). With time, the problems associated with the concrete such as non-eco-friendly, harmful environmental potential, low durability, low post-cracking capacity, brightness and limited fatigue life are rising (Noori et al., 2021). The substitute materials with appropriate mechanical, physical and thermal performance characteristics are required to replace cement simultaneously controlling the final product cost.

Agro-originated biopolymers, i.e., starch, are generally cheaper, biodegradable and renewable materials with potential as a substitute for non-biodegradable synthetic polymers (Avérous, 2007; Ślosarczyk et al., 2020; Verma et al., 2022). The inherent limitations such as poor mechanical properties compared to conventional thermoplastics, long post-processing ageing and high hygroscopicity hinder the possibility of using these materials. The multiphase composite materials are considered to overcome these drawbacks (Okonkwo et al., 2020; Reddy et al., 2013; Xie et al., 2013). Biodegradable co-polyesters such as poly(butylene adipate-co-terephthalate) exhibit good mechanical performance and higher elongation at break than synthetic polymers explored in these situations (Fukushima et al., 2012; Mohanty and Nayak, 2012). The resultant biocomposite materials with a high interface area leading to strong nanofiller–polymer interactions, higher dispersion, and improved mechanical, thermal and barrier properties exhibit improved behaviour along with the biodegradability eliminating the possibility of any (eco)toxic effects (Ali et al., 2021; Mini et al., 2022; Rivera-Gómez and Galán-Marín, 2017). Lignocellulosic fibres obtained from natural sources, i.e., hemp fibre, are studied as a potential material in civil construction applications as polymer composite reinforcement in substitution of energy-intensive and non-recyclable synthetic fibres (Islam and Ahmed, 2018; Sanjay et al., 2018). The qualification criteria such as mechanical properties, tensile strength and modulus create a possibility of creating stronger thermoset polymeric matrices using polyester (Callister and Rethwisch, 2018).

The scientific community is particularly working on various aspects of the construction industry, as it is consuming a high amount of total energy in the world. The research frontiers include finding possible solutions towards enhanced thermal insulation of building envelope, optimizing the building's energy performance, acoustical design and sustainable green concrete, reducing the use of conventional Portland cement and withstanding the adverse environmental conditions. In this scenario, the present chapter begins with a brief description of the polyester materials, useful properties and various applications of polyesters in the construction industry and market volume as an introduction to the topic. Then chapter proceed towards the discussion of the latest updates and reflections about the fibres used in construction and buildings, various blends of polymer resins, characteristics of natural fibres and mechanical properties of polyester-based biocomposites used in the construction and building industry. Finally, in order to stimulate ideas among readers about the new applications of polyester materials in building products, this chapter concludes with a discussion of future prospects.

14.2 POLYESTER-BASED BIOCOMPOSITES AS A CONSTRUCTION MATERIAL

Polyester-based biocomposites are gaining importance nowadays for building and construction applications. Polyester-based biocomposites possess good strength and toughness, as they have low density and exhibit good specific properties. The composites which are used for building and construction applications are supposed to have good functional and structural stability during the end-use applications and resistance to various environmental factors (Conzatti et al., 2012). A trend of green building is focusing on the use of biocomposite as the main material. Polyester-based biocomposites are replacing many synthetic composite materials because of the negative impact of synthetic composites on the environment (Chandrasekar et al., 2022b). For the building and construction applications, various parts such as doors, windows, fencing and roofing are biocomposites. The polymer biocomposites are petroleum-based synthetic polymers (polyester, phenolics, epoxy, polyethylene, polypropylene, etc.) reinforced with natural fibres. Biocomposites can be biopolymers (e.g., polylactic acid) reinforced with natural fibres. Also, it can be biopolymers reinforced with synthetic fibres (e.g., kevlar fibre, glass fibre, etc.). The polyester-based biocomposites used for building and construction applications can be classified as structural and non-structural. Polyester-based biocomposites are gaining worldwide attention because of their low weight and manufacturing cost.

UPR is the liquid polymer matrix synthesized by condensation polymerization using dicarboxylic acid (phthalic acid/anhydride) reacting with polyhydric alcohol and unsaturated dicarboxylic acid. The viscosity of UPR is reduced with the help of a styrene monomer. The styrene monomer reacts with the double bond of UPR to form a cross-linked structure. Cobalt octate/cobalt is used as a curing agent along with methyl ethyl ketone peroxide (MEKP). Cobalt octate act as an accelerator while MEKP work as an initiator. The ratio between them and their percentage played an important role in deciding curing time with the exothermic behaviour of the thermosetting polyesters. Cobalt can be used in the range of 0.05%–1% (Mohd Nurazzi et al., 2017; Kuppusamy and Neogi, 2014). The excess use of styrene monomer in the UPR may lead to phase separation between them. Thus, 30%–40% of styrene monomer can be used commercially with UPR (Hsu and Lee, 1993). The use of biobased polyester composite in the building and construction industry is playing an important role in replacing concrete and metal materials (Keller, 2003).

14.3 BLENDS OF POLYESTER RESIN

As the individual polyester polymer matrix could not provide the mechanical and other properties required for the specific application, polyester can be blended with another polymer or copolymer, which remains together physically to exhibit the desired mechanical and other functional properties. After curing, the glass transition temperature of unsaturated polyester increases, and due to high crosslinking it shows brittleness, which is a major reason to limit it for engineering and may be for building and construction applications. Therefore, using a blend of polyester resin is a suitable way to achieve the performance more cheaply (Cherian and Thachil, 2001).

14.4   UPR-EPOXY BLEND

Reactive blending is used to prepare a blend of UPR-epoxy resin. Both the resins are compatible with each other as they have good miscibility. Their blends have improved thermal, mechanical and weather resistance properties. Bioepoxies in the form of epoxidized oil can be used to blend with UPR. The blending of epoxidized oil in UPR reduces its crosslinking density and ultimately the glass transition temperature. It has been observed that modulus decreases but the impact properties increase in this type of blending (Mustapha et al., 2014).

14.5   UPR-PHENOLIC RESIN BLEND

This blend can be used when fire resistance is required for building and construction applications. Phenolic resins possess good flammability resistance, whereas UPR is highly flammable and generates toxic fumes on combustion. Therefore, the blend of UPR and phenol-formaldehyde exhibits good fire resistance properties with low smoke generation on burning. After curing the phenolic resins form a thermally stable cross-link structure which helps to improve its heat and fire resistance. Their blend also shows compatibility with a natural fibre like kenaf, which is a polyester-based biocomposite with good thermal and somewhat low mechanical properties (Mohd Mahadar et al., 2016).

14.6   UPR-NATURAL RUBBER BLEND

It has been observed from the studies that the blend of UPR with natural rubber shows improved impact properties but reduced modulus in specific conditions (Hameed, 2012). Natural rubber is available at a low cost, possessing good mechanical and abrasion resistance properties with poor resistance to oil. Mechanical mixing can be used to blend UPR with natural rubber. As the natural rubber is elastomeric and imparts impact, fracture and toughness resistance to its blend with UPR.

14.7   UPR-VINYL ESTER (VE) BLEND

The UPR-VE blend can be prepared by blending them mechanically, followed by curing at atmospheric temperature conditions using MEKP. Their blend shows good thermal and physical properties (Ardhyananta et al., 2017). UPR-VE blend is a low cost material that can be used to produce composites for building and construction applications with enhanced tensile and stiffness properties.

14.8   FIBRES USED IN CONSTRUCTION AND
BUILDING COMPOSITES

Looking forward to the negative effect of non-degradable polymer matrix composites on the environment, various countries are emphasizing to use of biocomposites for building, construction, automotive and other applications (Gupta et al., 2011). The use of natural fibres as a reinforcing agent makes the polymer matrix composite

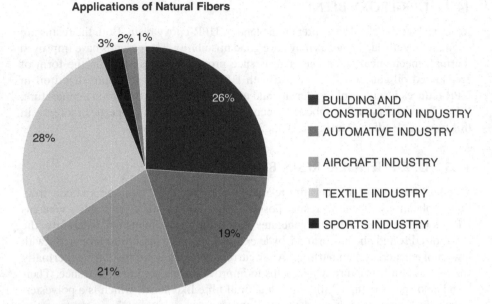

Applications of Natural Fibers

- BUILDING AND CONSTRUCTION INDUSTRY
- AUTOMATIVE INDUSTRY
- AIRCRAFT INDUSTRY
- TEXTILE INDUSTRY
- SPORTS INDUSTRY

FIGURE 14.1   Applications of natural fibres in different sectors. (Shireesha and Nandipati, 2019.)

a biocomposite. Natural fibres are derived from bio-resources and are biodegradable, cheap, eco-friendly, non-toxic, high specific strength and lightweight (Lee et al., 2012; Gupta et al., 2015). But they have poor mechanical and thermal resistance and high water absorption, brittleness and flammability properties (Bajpai et al., 2013). The natural fibre-based polyester composite exhibits a moderate strength-to-weight ratio and good impact resistance properties, while the use of synthetic fibres such as glass, carbon, Kevlar and boron as a reinforcement makes them expensive. Also, synthetic fibres are non-degradable by biological means, which causes pollution.

Plant fibres are renewable and cellulosic. Various natural fibres such as sisal, kenaf, hemp, jute, coir, bamboo, sugarcane and banana. are used as a reinforcing agent in UPR for building and construction applications.

From Figure 14.1, it is found that the natural fibres of about 26% are used for building and construction applications.

Figure 14.2 shows the development of natural fibre used in polymer matrix composites. The fibres obtained from the plant or animal resources are known as bio-fibres. They may be short or long fibres. Fibres having a length-to-diameter ratio greater than 100 are known as long fibres, while those less than 100 are called short fibres. The plant fibres are obtained from the plant stem or leaf. Cellulose, hemicellulose and lignin are the important constituents of plant fibres. Cellulose is nothing but a natural polymer and is crystalline in nature containing hydroxyl groups. It has a degree of polymerization of about 1000. The bio-fibres have advantages in terms of environment, biological, production, economic and general aspects. They show CO_2 neutrality and can degrade biologically. Bio-fibres do not have any issue related

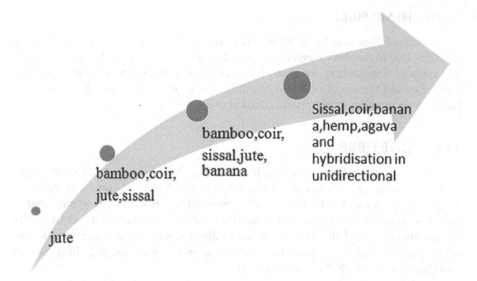

FIGURE 14.2   Natural fibre development for composite applications. (Shireesha and Nandipati, 2019.)

to skin itching while handling them as compared to glass fibres. Also, they are low cost and renewable as compared to glass fibres and other synthetic fibres. The direct use of plant fibres as reinforcement in the polyester matrix may have poor interfacial properties. Hence, the plant fibres need alkali treatment to have proper adhesion with the polymer matrix. The coupling agents such as silane can also be used to improve the fibre–matrix interfacial properties and ultimately the mechanical properties of biocomposites. The chemical treatment helps to improve the surface tension, surface properties and removal of undesired components from plant fibres. As the plant fibres contain cellulose and hydroxyl groups, they are hygroscopic in nature, which may be a challenge while processing them and utilizing them for end-use application. Various factors such as type of fibre, fibre composition, fibre orientation, properties of fibre and type of polymer matrices affect the properties of biocomposite. Along with the advantageous properties, biocomposites have some disadvantages as well. Dimensional instability, moisture absorption, limited biological resistance, limited strength, low UV and fire resistance are some of the major disadvantages of biocomposites obtained from natural fibres. Some plant fibres are discussed in the following subsections.

14.9   COIR FIBRE

It is obtained from coconut and is available easily at low cost. It is non-toxic and mainly contains cellulose 36%–43%, hemicellulose 0.15%–0.25% and lignin 41%–45%. It has poor adhesion with the polymer matrix because of its high fibrillar angle and lignin content. Therefore, it is treated with alkali for proper adhesion with the polymer matrix (Geethamma et al., 1998).

14.10   HEMP FIBRE

It is obtained from the hemp plant. They are cellulosic fibres and possess good strength. Due to the absorption of moisture, its tensile and flexural strength decreases and reduces fibre–matrix adhesion. It requires alkali treatment to modify surface properties and the use of maleated coupling agents to enhance interfacial adhesion between fibres and polymer matrix (Dhakal et al., 2007).

14.11   JUTE FIBRE

It is also a cellulosic fibre containing microfibrils. Due to the presence of cellulose, it absorbs moisture and can cause an increase in the density of composite material. They are cheaper and lightweight. Due to poor adhesion with the polymer matrix, they also need chemical treatment. The alkali treatment improves the crystallinity of jute fibres, which helps to enhance their adhesion with the polymer matrix. The alkali treatment helps to improve the mechanical properties of the jute fibres (Shaikh and Channiwala, 2010; Wang et al., 2008).

14.12   NATURAL FIBRE CHARACTERISTICS

Almost 30% of natural fibres are produced per year and used as a constituent in various applications such as building materials, automotive, packaging, sports equipment, paper-making, clothing, etc.(Jawaid and Khalil, 2015; Asim et al., 2015). Due to its density and eco-friendly nature over traditional composites, natural fibres have become a material of interest for the industries (Asim et al., 2015). While selecting the natural fibres for building and construction applications, one of the most important properties, i.e., mechanical characteristics, should always be considered. These characteristics differ in natural fibres and mainly depend on the type of natural fibre, chemical composition, growth conditions and structural strength. As compared with synthetic fibres, natural fibres exhibit poor mechanical properties. Nevertheless, the main advantages include low health risk, low density, low cost, non-abrasiveness, non-irritating to skin, renewability, recyclability and biodegradability (Malkapuram et al., 2008; Ku et al., 2011; Asim et al., 2021; Arumugam et al., 2020; Odesanya et al., 2021). When compared to synthetic fibres, their processing is more environmentally friendly, resulting in improved working conditions and a lower risk of dermatological or respiratory disorders. Natural fibres' most intriguing feature is their favourable environmental impact. Natural fibres have a lower density (1.6 g/cm^3) than glass fibre (2.5 g/cm^3), allowing for lighter composites to be produced. Natural fibres have a number of disadvantages over synthetic fibres, including non-uniformity, a wide range of diameters and mechanical qualities (Bismarck et al., 2005). As a result, developing high-performance natural fibre/polymer composites is a crucial problem to be solved in order to increase natural fibre adoption as a quality alternative to conventional reinforcing fibres (John and Thomas, 2008; Akil et al., 2011). Fibres of various types, spanning from long-continuous to short-discontinuous, have been employed to reinforce composites. The properties of natural fibre-reinforced polymer composites are determined by each component in the composite, namely the reinforcing fibres and resin matrix (Md Shah et al., 2019).

Natural fibres often have a high amount of hydroxyl groups, making them polar and hydrophilic in nature (Wambua et al., 2003). Most plastics, on the other hand, are hydrophobic by nature. As a result, one of the major disadvantages of natural fibres/polymers composites is the incompatibility of hydrophilic natural fibres with hydrophobic thermoplastic matrices, which results in unfavourable composite properties (Asim et al., 2021). Natural fibres' polar nature causes natural fibre-based composites to absorb a lot of moisture, causing fibre swelling and voids in the fibre–matrix interphase. Moisture will result in a porous product if it is not removed from natural fibres prior to compounding by drying. Mechanical qualities may deteriorate and dimensional stability may be lost as a result of excessive moisture absorption (Alvarez et al., 2004). To strengthen the adhesion between fibre and matrix while reducing their ability to absorb moisture, it is essential to modify the fibre surface using chemical modifications (Malkapuram, Kumar, and Singh Negi, 2008; Li et al., 2020). The fibre content/amount of filler can also influence the characteristics of natural fibre-reinforced polymer composites. As a result, the impact of fibre content on the characteristics of natural fibre-reinforced composites is quite important. The processing parameters utilized are another essential component that has a considerable impact on the composites' properties and interfacial properties (Ku et al., 2011). When natural fibres are heated over 200°C, they degrade, limiting the types of plastics that can be utilized as a matrix (Bismarck et al., 2005; Akil et al., 2011). As a result, the best processing processes and parameters must be carefully chosen to produce the best composite products.

14.13   CONCLUSIONS AND FUTURE TRENDS

Polyester-based biocomposites have found a vital place in construction applications, and a wide range of products is now being used, with significant sales turnover. Moreover, this trend will further increase with the expansion of the construction industry. The availability challenges with conventional materials technology shift will alter the demands of lignosulfonate, lignite, concrete and gypsum products. This vacuum will possibly be filled by various polymer derivatives with an expected requirement for more sophisticated building products. Reason supplementing to this situation is rapid deployment of modern tools and technology such as machine plastering in the place of the conventional building methods. The shift from a labour-intensive strategy to an automated approach to construction using prefabricated building materials created the need for various novel materials. Polyester-based biocomposites have a perceived advantage over synthetic materials. Being convinced of 'bio' needs to be confirmed and practised as being environmentally more preferable. The conventional materials used in the construction industry are currently under scanner for potential health risks for their products. Therefore, polyester-based biocomposites are possibly non-toxic, emission-free alternatives. Before mass-scale use of these materials, there is a need for a set of standards to qualify recycling criteria, minimizing the need to dispose of building materials. The main requirement of the construction industry is the admixtures allowing the economic, high-quality materials which are compatible with modern tools and technology, without sacrificing environmental aspects. It is conclusive to assume that polyester-based biocomposites will make a major contribution to this target.

REFERENCES

Akil, Hazizan Md, Mohd Firdaus Omar, Adlan Akram Mohamad. Mazuki, Suhana Binti Safiee, Z. A.M. Ishak, and Azhar Abu Bakar. 2011. "Kenaf Fiber Reinforced Composites: A Review." *Materials & Design* 32 (8–9): 4107–21. https://doi.org/10.1016/J. MATDES.2011.04.008.

Al-Haydari, Israa Saeed, Ghadah Ghassan Masood, Safaa Adnan Mohamad, and Hussein Mohammed Noor Khudhur. 2021. "Stress–Strain Behavior of Sustainable Polyester Concrete with Different Types of Recycled Aggregate." *Materials Today: Proceedings* 46 (January): 5160–66. https://doi.org/10.1016/J.MATPR.2021.01.591.

Ali, Mohammed S., A. A. Al-Shukri, M. R. Maghami, and Chandima Gomes. 2021. "Nano and Bio-Composites and Their Applications: A Review." *IOP Conference Series: Materials Science and Engineering* 1067 (1): 012093. https://doi. org/10.1088/1757-899X/1067/1/012093.

Alothman, Othman Y., Mohammad Jawaid, Senthilkumar Krishnasamy, Chandrasekar Muthukumar, Basheer A. Alshammari, Hassan Fouad, Mohamed Hashem, and Suchart Siengchin. 2020. "Thermal Characterization of Date Palm/Epoxy Composites with Fillers from Different Parts of the Tree." *Journal of Materials Research and Technology* 9 (6): 15537–46. https://doi.org/10.1016/j.jmrt.2020.11.020.

Alvarez, Vera Alejandra, A. N. Fraga, and A. Vázquez. 2004. "Effects of the Moisture and Fiber Content on the Mechanical Properties of Biodegradable Polymer–Sisal Fiber Biocomposites." *Journal of Applied Polymer Science* 91 (6): 4007–16. https://doi. org/10.1002/APP.13561.

Ardhyananta, H., F. D. Puspadewa, Sigit Tri Wicaksono, Widyastuti, A. T. Wibisono, Budi Agung Kurniawan, H. Ismail, and A. V. Salsac. 2017. "Mechanical and Thermal Properties of Unsaturated Polyester/Vinyl Ester Blends Cured at Room Temperature." *IOP Conference Series: Materials Science and Engineering* 202 (1): 012088. https://doi. org/10.1088/1757-899X/202/1/012088.

Arumugam, Chinnappa, Senthilkumar Arumugam, and Sarojadevi Muthusamy. 2020. "Mechanical, Thermal and Morphological Properties of Unsaturated Polyester/ Chemically Treated Woven Kenaf Fiber/AgNPs@PVA Hybrid Nanobiocomposites for Automotive Applications." *Journal of Materials Research and Technology* 9 (6): 15298– 312. https://doi.org/10.1016/J.JMRT.2020.10.084.

Asim, Mohammad, Khalina Abdan, M. Jawaid, Mohammed Nasir, Zahra Dashtizadeh, M. R. Ishak, and M. Enamul Hoque. 2015. "A Review on Pineapple Leaves Fibre and Its Composites" *International Journal of Polymer Science* 2015: 950567.

Asim, Mohammad, Mohammad Jawaid, Hassan Fouad, and Othman Y. Alothman. 2021. "Effect of Surface Modified Date Palm Fibre Loading on Mechanical, Thermal Properties of Date Palm Reinforced Phenolic Composites." *Composite Structures* 267 (July): 113913. https://doi.org/10.1016/J.COMPSTRUCT.2021.113913.

Avérous, Luc. 2007. "Biodegradable Multiphase Systems Based on Plasticized Starch: A Review." *Journal of Macromolecular Science, Part C: Polymer Reviews* 44 (3): 231–74. https://doi.org/10.1081/MC-200029326.

Bajpai, Pramendra Kumar, Inderdeep Singh, and Jitendra Madaan. 2013. "Tribological Behavior of Natural Fiber Reinforced PLA Composites." *Wear* 297 (1–2): 829–40. https://doi.org/10.1016/J.WEAR.2012.10.019.

Bismarck, Alexander, Supriya Mishra, and Thomas Lampke. 2005. "Plant Fibers as Reinforcement for Green Composites." In *Natural Fibers, Biopolymers, and Biocomposites*, edited by Amar Mohanty, Manjusri Misra, and Lawrence Drzal, 1st ed., pp. 52–128. CRC Press. https://doi.org/10.1201/9780203508206-10.

Bouzit, S., F. Merli, E. Belloni, R. Akhrraz, S. Asri Ssar, M. Sonebi, S. Amziane, C. Buratti, and M. Taha. 2022. "Investigation of Thermo-Acoustic and Mechanical Performance of Gypsum-Plaster and Polyester Fibers Based Materials for Building Envelope." *Materials Today: Proceedings* 58 (January): 1578–81. https://doi.org/10.1016/J. MATPR.2022.03.560.

Callister, William D., and David G. Rethwisch, eds. 2018. *Materials Science and Engineering : An Introduction*, 10th ed. Wiley, New York.

Chandrasekar, M., K. Senthilkumar, M. Jawaid, Salman Alamery, Hassan Fouad, and Mohamad Midani. 2022a. "Tensile, Thermal and Physical Properties of Washightonia Trunk Fibres/Pineapple Fibre Biophenolic Hybrid Composites." *Journal of Polymers and the Environment* 30 (10): 4427–434. https://doi.org/10.1007/s10924-022-02524-z.

Chandrasekar, M., K. Senthilkumar, M. Jawaid, Mohamed H. Mahmoud, H. Fouad, and Mohini Sain. 2022b. "Mechanical, Morphological and Dynamic Mechanical Analysis of Pineapple Leaf/Washingtonia Trunk Fibres Based Biophenolic Hybrid Composites." *Journal of Polymers and the Environment* 30 (10): 4157–165. https://doi.org/10.1007/ s10924-022-02482-6.

Chandrasekar, M., I. Siva, T. Senthil Muthu Kumar, K. Senthilkumar, Suchart Siengchin, and N. Rajini. 2020. "Influence of Fibre Inter-Ply Orientation on the Mechanical and Free Vibration Properties of Banana Fibre Reinforced Polyester Composite Laminates." *Journal of Polymers and the Environment* 28 (11): 2789–800. https://doi.org/10.1007/ s10924-020-01814-8.

Cherian, Benny, and Eby Thomas Thachil. 2001. "Toughening Studies of an Unsaturated Polyester Resin Using Maleated Elastomers." *Progress in Rubber and Plastics Technology* 17 (4): 205–24. https://doi.org/10.1177/147776060101700401.

Conzatti, Lucia, Francesco Giunco, Paola Stagnaro, Massimo Capobianco, Maila Castellano, and Enrico Marsano. 2012. "Polyester-Based Biocomposites Containing Wool Fibres." *Composites Part A: Applied Science and Manufacturing* 43 (7): 1113–19. https://doi. org/10.1016/j.compositesa.2012.02.019.

Cook, Gordon. 2001. "Polyester Fibres." In *Handbook of Textile Fibres*, 328–91. Woodhead Publishing. https://doi.org/10.1533/9781855734852.328.

Crawford, Roy J., and James L. Throne. 2002. "Rotational Molding Polymers." In *Rotational Molding Technology*, pp. 19–68. William Andrew Publishing. https://doi.org/10.1016/ B978-188420785-3.50004-6.

Dhakal, H. N., Z. Y. Zhang, and M. O.W. Richardson. 2007. "Effect of Water Absorption on the Mechanical Properties of Hemp Fibre Reinforced Unsaturated Polyester Composites." *Composites Science and Technology* 67 (7–8): 1674–83. https://doi.org/10.1016/J. COMPSCITECH.2006.06.019.

Fukushima, Kikku, Meng Hsiu Wu, Sergio Bocchini, Amaliya Rasyida, and Ming Chien Yang. 2012. "PBAT Based Nanocomposites for Medical and Industrial Applications." *Materials Science and Engineering: C* 32 (6): 1331–51. https://doi.org/10.1016/J. MSEC.2012.04.005.

García-Gonzále, Julia, Paulo C. Lemos, Alice S. Pereira, Julia Ma Morán-Del Pozo, M. Ignacio Guerra-Romero, Andrés Juan-Valdés, and Paulina Faria. 2020. "Biodegradable Polymers on Cementitious Materials." In *XV International Conference on Durability of Building Materials and Components (DBMC 2020)*, pp. 99–104. https://doi.org/10.23967/ DBMC.2020.017.

Geethamma, V. G., K. Thomas Mathew, R. Lakshminarayanan, and Sabu Thomas. 1998. "Composite of Short Coir Fibres and Natural Rubber: Effect of Chemical Modification, Loading and Orientation of Fibre." *Polymer* 39 (6–7): 1483–91. https://doi.org/10.1016/ S0032-3861(97)00422-9.

Gupta, Anu, Ajit Kumar, Amar Patnaik, and Sandhyarani Biswas. 2011. "Effect of Different Parameters on Mechanical and Erosion Wear Behavior of Bamboo Fiber Reinforced Epoxy Composites." *International Journal of Polymer Science* 2011. https://doi.org/10.1155/2011/592906.

Gupta, M K, R K Srivastava, and Himanshu Bisaria. 2015. "Potential of Jute Fibre Reinforced Polymer Composites: A Review." *International Journal of Fiber and Textile Research* 5 (3): 30–38. http://www.urpjournals.com.

Halil Akın, Muhammed, and Rıza Polat. 2022. "The Effect of Vehicle Waste Tires on the Mechanical, Hardness and Stress–Strain Properties of Polyester-Based Polymer Concretes." *Construction and Building Materials* 325 (March): 126741. https://doi.org/10.1016/J.CONBUILDMAT.2022.126741.

Hameed, Awham M. 2012. "Effect of Water Absorption on Some Mechanical Properties of Unsaturated Polyester Resin/Natural Rubber Blends." *Jordan Journal of Physics* 5 (3): 119–27.

Hsu, C. P., and L. James Lee. 1993. "Free-Radical Crosslinking Copolymerization of Styrene/Unsaturated Polyester Resins: 3. Kinetics-Gelation Mechanism." *Polymer* 34 (21): 4516–23. https://doi.org/10.1016/0032-3861(93)90158-7.

Islam, Mohammad S., and Syed Ju Ahmed. 2018. "Influence of Jute Fiber on Concrete Properties." *Construction and Building Materials* 189 (November): 768–76. https://doi.org/10.1016/J.CONBUILDMAT.2018.09.048.

Jawaid, M., and H. P. S. Abdul Khalil. 2015. "Cellulosic/Synthetic Fibre Reinforced Polymer Hybrid Composites : A Review." *Carbohydrate Polymers* 86 (1): 1–18. https://doi.org/10.1016/j.carbpol.2011.04.043.

John, Maya Jacob, and Sabu Thomas. 2008. "Biofibres and Biocomposites." *Carbohydrate Polymers* 71 (3): 343–64. https://doi.org/10.1016/j.carbpol.2007.05.040.

Keller, Thomas. 2003. *Use of Fibre Reinforced Polymers in Bridge Construction*. International Association for Bridge and Structural Engineering (IABSE). https://doi.org/10.2749/SED007.

Ku, H., H. Wang, N. Pattarachaiyakoop, and M. Trada. 2011. "A Review on the Tensile Properties of Natural Fiber Reinforced Polymer Composites." *Composites Part B: Engineering* 42 (4): 856–73. https://doi.org/10.1016/J.COMPOSITESB.2011.01.010.

Kuppusamy, R. R.P., and Swati Neogi. 2014. "Influence of Curing Agents on Gelation and Exotherm Behaviour of an Unsaturated Polyester Resin." *Bulletin of Materials Science* 36 (7): 1217–24. https://doi.org/10.1007/S12034-013-0591-8.

Lee, Koon Yang, Puja Bharadia, Jonny J. Blaker, and Alexander Bismarck. 2012. "Short Sisal Fibre Reinforced Bacterial Cellulose Polylactide Nanocomposites Using Hairy Sisal Fibres as Reinforcement." *Composites Part A: Applied Science and Manufacturing* 43 (11): 2065–74. https://doi.org/10.1016/J.COMPOSITESA.2012.06.013.

Li, Mi, Yunqiao Pu, Valerie M. Thomas, Chang Geun Yoo, Soydan Ozcan, Yulin Deng, Kim Nelson, and Arthur J. Ragauskas. 2020. "Recent Advancements of Plant-Based Natural Fiber–Reinforced Composites and Their Applications." *Composites Part B: Engineering* 200 (November): 108254. https://doi.org/10.1016/J.COMPOSITESB.2020.108254.

Lim, Siong Kang, Mohd Warid Hussin, Fadhadli Zakaria, and Tung Chai Ling. 2009. "GGBFS as Potential Filler in Polyester Grout: Flexural Strength and Toughness." *Construction and Building Materials* 23 (5): 2007–15. https://doi.org/10.1016/J.CONBUILDMAT.2008.08.030.

Malkapuram, Ramakrishna, Vivek Kumar, and Yuvraj Singh Negi. 2008. "Recent Development in Natural Fiber Reinforced Polypropylene Composites" *Journal of Reinforced Plastics and Composites* 28 (10): 1169–89. https://doi.org/10.1177/0731684407087759.

Marrero, R. E., H. L. Soto, F. R. Benitez, C. Medina, and O. M. Suarez. 2017. "Study of High-Strength Concrete Reinforced with Bamboo Fibers." *Advanced Materials - TechConnect Briefs* 2: 301–4.

Md Shah, Ain U., Mohamed T.H. Sultan, and Mohammad Jawaid. 2019. "Sandwich-Structured Bamboo Powder/Glass Fibre-Reinforced Epoxy Hybrid Composites – Mechanical Performance in Static and Dynamic Evaluations" *Journal of Sandwich Structures & Materials* 23 (1): 47–64. https://doi.org/10.1177/1099636218822740.

Mini, K. M., Dhanya Sathyan, and K. Jayanarayanan. 2022. "Biofiber Composites in Building and Construction." In *Advances in Bio-Based Fiber: Moving Towards a Green Society*, pp. 335–65. Woodhead Publishing. https://doi.org/10.1016/B978-0-12-824543-9.00019-0.

Mohanty, S., and S. K. Nayak. 2012. "Biodegradable Nanocomposites of Poly(Butylene Adipate-Co-Terephthalate) (PBAT) and Organically Modified Layered Silicates." *Journal of Polymers and the Environment* 20 (1): 195–207. https://doi.org/10.1007/S10924-011-0408-Z.

Mohd Mahadar, Marliana, Azman Hassan, Nor Yuziah Mohd Yunus, H. P. S. Abdul Khalil, and Mohamad Haafiz Mohamad Kassim. 2016. "Thermal Properties and Mechanical Performance of Unsaturated Polyester/Phenolic Blends Reinforced by Kenaf Fiber." *Advanced Materials Research* 1134 (December): 61–65. https://doi.org/10.4028/WWW.SCIENTIFIC.NET/AMR.1134.61.

Mohd Nurazzi, N., A. Khalina, S. M. Sapuan, A. M. Dayang Laila, and M. Rahmah. 2017. "Curing Behaviour of Unsaturated Polyester Resin and Interfacial Shear Stress of Sugar Palm Fibre." *Journal of Mechanical Engineering and Sciences* 11 (2): 2650–64. https://doi.org/10.15282/JMES.11.2.2017.8.0242.

Mustapha, S. N.H., A. R. Rahmat, A. Arsad, and S. K. Yeong. 2014. "Novel Bio-Based Resins from Blends of Functionalised Palm Oil and Unsaturated Polyester Resin." *Materials Research Innovations* 18 (December): S6-326–S6-330. https://doi.org/10.1179/143289 1714Z.000000000978.

Nasimudeen, Nadeem Ahmed, Sharwine Karounamourthy, Joshua Selvarathinam, Senthil Muthu Kumar Thiagamani, Harikrishnan Pulikkalparambil, Senthilkumar Krishnasamy, and Chandrasekar Muthukumar. 2021. "Mechanical, Absorption and Swelling Properties of Vinyl Ester Based Natural Fibre Hybrid Composites." *Applied Science and Engineering Progress*. https://doi.org/10.14416/j.asep.2021.08.006.

Noori, Adel, Yubin Lu, Pooya Saffari, Jinguan Liu, and Jinfu Ke. 2021. "The Effect of Mercerization on Thermal and Mechanical Properties of Bamboo Fibers as a Biocomposite Material: A Review." *Construction and Building Materials* 279 (April): 122519. https://doi.org/10.1016/J.CONBUILDMAT.2021.122519.

Odesanya, Kazeem Olabisi, Roslina Ahmad, Mohammad Jawaid, Sedat Bingol, Ganiyat Olusola Adebayo, and Yew Hoong Wong. 2021. "Natural Fibre-Reinforced Composite for Ballistic Applications: A Review." *Journal of Polymers and the Environment* 29 (12): 3795–3812. https://doi.org/10.1007/S10924-021-02169-4.

Okonkwo, E. G., C. N. Anabaraonye, C. C. Daniel-Mkpume, S. V. Egoigwe, P. E. Okeke, F. G. Whyte, and A. O. Okoani. 2020. "Mechanical and Thermomechanical Properties of Clay-Bambara Nut Shell Polyester Bio-Composite." *The International Journal of Advanced Manufacturing Technology* 108 (7): 2483–96. https://doi.org/10.1007/S00170-020-05570-W.

Plank, Johann. 2005. "Applications of Biopolymers in Construction Engineering." In *Biopolymers Online*. John Wiley & Sons, Ltd. https://doi.org/10.1002/3527600035. BPOLA002.

Pothan, Laly A., Zachariah Oommen, and Sabu Thomas. 2003. "Dynamic Mechanical Analysis of Banana Fiber Reinforced Polyester Composites." *Composites Science and Technology* 63 (2): 283–93. https://doi.org/10.1016/S0266-3538(02)00254-3.

Reddy, Murali M., Singaravelu Vivekanandhan, Manjusri Misra, Sujata K. Bhatia, and Amar K. Mohanty. 2013. "Biobased Plastics and Bionanocomposites: Current Status and Future Opportunities." *Progress in Polymer Science* 38 (10–11): 1653–89. https://doi.org/10.1016/j.progpolymsci.2013.05.006.

Rivera-Gómez, C., and C. Galán-Marín. 2017. "Biodegradable Fiber-Reinforced Polymer Composites for Construction Applications." In *Natural Fiber-Reinforced Biodegradable and Bioresorbable Polymer Composites*, pp. 51–72. Woodhead Publishing. https://doi.org/10.1016/B978-0-08-100656-6.00004-2.

Sanjay, M. R., P. Madhu, Mohammad Jawaid, P. Senthamaraikannan, S. Senthil, and S. Pradeep. 2018. "Characterization and Properties of Natural Fiber Polymer Composites: A Comprehensive Review." *Journal of Cleaner Production* 172 (January): 566–81. https://doi.org/10.1016/J.JCLEPRO.2017.10.101.

Senthilkumar, K., I. Siva, N. Rajini, and P. Jeyaraj. 2015. "Effect of Fibre Length and Weight Percentage on Mechanical Properties of Short Sisal/Polyester Composite." *International Journal of Computer Aided Engineering and Technology* 7 (1): 60. https://doi.org/10.1504/IJCAET.2015.066168.

Senthilkumar, K., N. Saba, M. Chandrasekar, M. Jawaid, N. Rajini, Suchart Siengchin, Nadir Ayrilmis, Faruq Mohammad, and Hamad A. Al-Lohedan. 2021. "Compressive, Dynamic and Thermo-Mechanical Properties of Cellulosic Pineapple Leaf Fibre/ Polyester Composites: Influence of Alkali Treatment on Adhesion." *International Journal of Adhesion and Adhesives* 106 (April): 102823. https://doi.org/10.1016/j.ijadhadh.2021.102823.

Senthilkumar, K., S. Subramaniam, Thitinun Ungtrakul, T. Senthil Muthu Kumar, M. Chandrasekar, N. Rajini, Suchart Siengchin, and Jyotishkumar Parameswaranpillai. 2022. "Dual Cantilever Creep and Recovery Behavior of Sisal/Hemp Fibre Reinforced Hybrid Biocomposites: Effects of Layering Sequence, Accelerated Weathering and Temperature." *Journal of Industrial Textiles* 51 (2_suppl): 2372S–2390S. https://doi.org/10.1177/1528083720961416.

Shahroze, R. M., M. Chandrasekar, K. Senthilkumar, T. Senthil Muthu Kumar, M. R. Ishak, N. Rajini, Suchart Siengchin, and S. O. Ismail. 2021. "Mechanical, Interfacial and Thermal Properties of Silica Aerogel-Infused Flax/Epoxy Composites." *International Polymer Processing* 36 (1): 53–59. https://doi.org/10.1515/ipp-2020-3964.

Shaikh, A A, and S A Channiwala. 2010. "To Study the Characteristics of Jute Polyester Composite for Randomly Distributed Fiber Reinforcement." In *Proceedings of the World Congress on Engineering*. London, UK.

Shireesha, Y, and Govind Nandipati. 2019. "State of Art Review on Natural Fibers." *Materials Today: Proceedings* 18: 15–24. https://doi.org/10.1016/j.matpr.2019.06.272.

Ślosarczyk, Agnieszka, Izabela Klapiszewska, Patryk Jedrzejczak, Lukasz Klapiszewski, and Teofil Jesionowski. 2020. "Biopolymer-Based Hybrids as Effective Admixtures for Cement Composites." *Polymers* 12 (5): 1180. https://doi.org/10.3390/POLYM12051180.

Tabatabai, Habib, Morteza Janbaz, and Azam Nabizadeh. 2018. "Mechanical and Thermo-Gravimetric Properties of Unsaturated Polyester Resin Blended with FGD Gypsum." *Construction and Building Materials* 163 (February): 438–45. https://doi.org/10.1016/J.CONBUILDMAT.2017.12.041.

Thomas, Seena K., Jyotishkumar Parameswaranpillai, Senthilkumar Krishnasamy, P.M. Sabura Begum, Debabrata Nandi, Suchart Siengchin, Jinu Jacob George, Nishar Hameed, Nisa V. Salim, and Natalia Sienkiewicz. 2021. "A Comprehensive Review on Cellulose, Chitin, and Starch as Fillers in Natural Rubber Biocomposites." *Carbohydrate Polymer Technologies and Applications* 2 (December): 100095. https://doi.org/10.1016/j.carpta.2021.100095.

Unsaturated Polyester Resin Market Size|Industry Report, 2027. n.d. Accessed June 13, 2022. https://www.grandviewresearch.com/industry-analysis/unsaturated-polyester-resin-upr-market.

Vanderlaan, Douglas, and Wolfgang Forster. 1988. US4918120A - Low Styrene Emission Unsaturated Polyester Resins - Google Patents, Issued 1988. https://patents.google.com/patent/US4918120A/en.

Verma, Deepak, Vaishally Dogra, Arun Kumar Chaudhary, and Ravikant Mordia. 2022. "Advanced Biopolymer-Based Composites: Construction and Structural Applications." In *Sustainable Biopolymer Composites: Biocompatibility, Self-Healing, Modeling, Repair and Recyclability: A Volume in Woodhead Publishing Series in Composites Science and Engineering*, pp. 113–28. https://doi.org/10.1016/B978-0-12-822291-1.00010-5.

Wambua, Paul, Jan Ivens, and Ignaas Verpoest. 2003. "Natural Fibres: Can They Replace Glass in Fibre Reinforced Plastics?" *Composites Science and Technology* 63 (9): 1259–64. https://doi.org/10.1016/S0266-3538(03)00096-4.

Wang, Wei-ming, Zai-sheng Cai, and Jian-yong Yu. 2008. "Study on the Chemical Modification Process of Jute Fiber." *Journal of Engineered Fibers and Fabrics* 3 (2). https://doi.org/10.1177/155892500800300203.

Xie, Fengwei, Eric Pollet, Peter J. Halley, and Luc Avérous. 2013. "Starch-Based Nano-Biocomposites." *Progress in Polymer Science* 38 (10–11): 1590–1628. https://doi.org/10.1016/j.progpolymsci.2013.05.002.

15 Polyester-Based Biocomposites for Food Packaging Applications

Ana Luiza Machado Terra, Ana Claudia Araujo de Almeida, Bruna da Silva Vaz, Jorge Alberto Vieira Costa, Michele Greque de Morais, and Juliana Botelho Moreira
Federal University of Rio Grande

CONTENTS

15.1 INTRODUCTION

The principal functions of food packaging are to ensure the safety and quality of food and prolong the shelf life of products (Moustafa et al., 2019). Polymers that contain the ester functional group in their main chain are called polyesters. These polymers can be natural or synthetic, aliphatic, semi-aromatic, or aromatic. In addition, polyesters are adaptable and diverse and can be rigid or soft, opaque or transparent thermoset, or thermoplastic. This characteristic allows them to transform into various shapes and sizes of films or containers. As essential properties of food packaging, they include a barrier to gases, aromatic compounds and water vapors, mechanical strength, and thermal stability. Polyesters have these predefined properties, so they have a high demand for these packaging materials. Main examples include polycaprolactone (PCL), poly-3-hydroxybutyrate (PHB), poly(glycolic acid) (PGA), poly(lactic acid) (PLA), polybutylene adipate-co-terephthalate (PBAT), and polyethylene terephthalate (PET) (Hamid & Samy, 2021; Kausar, 2019).

Most polyesters are not biodegradable and are referred to as persistent pollutants in many environmental niches. As a result, there is growing interest in developing sustainable materials to alleviate environmental pollution due to these polymers. Thus, environmental pollution caused by polyesters has encouraged the development of sustainable materials to reduce the effects of plastic pollution (Hamid & Samy, 2021). In this context, biopolymers are potential substitutes for petroleum-based polyesters due to their biodegradability, compatibility, and renewability properties. Biopolymers can be classified as natural biopolymers, chemically synthesized biodegradable polymers, and polymers of microbial origin (Moreira et al., 2022; Vinod et al., 2020). Besides, biopolymers allow the incorporation of a variety of additives such as antifungals, antioxidants, antimicrobials, and dyes (Maftoonazad & Ramaswamy, 2018).

The growing consumer demand for safe food has encouraged research in new technologies such as intelligent and active food packaging systems. In the first system, biologically active compounds are added to the packaging material to ensure food preservation. In the second, food quality information is obtained with an indicator external or internal to the packaging (Moustafa et al., 2019; Priyadarshi et al., 2021). Currently, food packaging combines various materials and technologies to take advantage of functional characteristics or design properties and ensure product quality and freshness during distribution and storage (Hamid & Samy, 2021).

A polyester composite consists of material composed of a polymer matrix with reinforcing fillers integrated into this matrix, resulting in synergistic chemical, physical, and mechanical properties (Wang et al., 2017). Biocomposites are materials obtained from natural and renewable sources. These biomaterials combine the optimized functional properties of composites with the environmental appeal of biopolymers (Abdulkhani et al., 2020). However, the use of polyester-based biocomposites in food packaging is limited to industrial application due to mechanical properties and relatively weak barriers (Tyagi et al., 2021).

Polyester-based bionanocomposites consist of a biopolymer matrix reinforced with nanofillers (Arora et al., 2018). Nanofillers can increase the surface area of the biocomposite material. Nanomaterials can improve the functionality of polyester-based food packaging, improving the barrier, mechanical, optical, and thermal properties (Priyadarshi et al., 2021). Despite the advantages, there is great concern regarding the migration of nanomaterials to food through the packaging material. Moreover, there is a growth in interest to develop biocompatible, safe, and non-toxic nanostructures. Therefore, more research is needed to investigate the toxic potential and immunogenic effects of these nano-bionanocomposites (Bajpai et al., 2018; Priyadarshi et al., 2021). Based on this, the chapter addresses the potential of polyester-based biocomposites for the development of food packaging materials, focusing on strategies to improve their properties through nanotechnological approaches.

15.2 MAIN POLYESTERS APPLIED IN FOOD PACKAGING

PCL is a polyester extracted from fossil sources, with ring-opening polymerization of ε-caprolactone and an average molecular weight between 3000 and 90,000 g/mol (Li et al., 2021; Ma et al., 2020). The tensile strength of bulk PCL is about 25–43 MPa and the modulus of elasticity is about 330–360 MPa. The mechanical characteristics and adjusting physicochemical properties contribute to PCL resistance to physical, chemical, and mechanical aggressions without significantly losing its properties (Dwivedi et al., 2019). PCL is compatible with other polymers to form polymeric blends and overcomes mechanical limitations. In polymer blends, PCL would act as the majority phase. The other polymer, preferably with a more rigid character, would function as the dispersed phase (Schmatz et al., 2019; Terra et al., 2021a, b). Another crucial property of PCL is its high solubility in organic solvents due to the hydrophobic methylene fractions in its repeating units (Dwivedi et al., 2019). Furthermore, PCL is non-toxic and compatible with human cells and tissues and is considered safe by the Food and Drug Administration (FDA) to develop food packaging materials (Malik, 2020).

PLA polymer is a linear aliphatic polyester obtained by fermentation from natural sources (Jem & Tan, 2020; Sanusi et al., 2020). The solubility properties of amorphous PLA are high in organic solvents such as chlorinated and acetonitrile (Buhecha et al., 2019). PLA is also biodegradable and biocompatible with human cells and are Generally Recognized as Safe certified by the FDA (Gigante et al., 2019; Malek et al., 2021). When processed PLA is exposed to natural environments with high temperatures and humidity, it is gradually hydrolyzed to form carbon dioxide (CO_2) and water (Kalita et al., 2021). The barrier properties to water vapor, gases, and organic substances of PLA are poor (Jin et al., 2018; Sari et al., 2021). Thus, its processability is restricted, making it necessary to formulate polymeric blends (Khatsee et al., 2018; Mohandesnezhad et al., 2020; Moreira et al., 2018a).

PGA is a linear aliphatic polyester, which has high crystallinity and mechanical strength, good gas barrier properties, and low deformation resistance. PGA also has low solubility in organic solvents; however, it is soluble in fluorinated solvents such as hexafluoroisopropanol and water (Samantaray et al., 2020). Due mainly to its high hydrophilicity, the formation of polymeric blends with this polyester becomes an alternative (Jem & Tan, 2020; Wang et al., 2021a, b). In addition to being miscible in the formation of polymer blends, PGA is also compostable and has high biodegradation rates. The biodegradation rate is high (~1 month at 30°C) when compared to other biodegradable polyesters such as PCL (2–3 years) (Gentile et al., 2014; Jem & Tan, 2020; Samantaray et al., 2020). The biodegradation of PGA occurs through slow chemical hydrolysis in an aqueous environment, which may have a better adjunct efficiency to enzymatic catalysis (Samantaray et al., 2020). PGA is also considered non-toxic and biocompatible with human cells and tissues. This biopolymer can be used as a matrix for human tissue regeneration, drug carrier, and packaging materials (Guido et al., 2021).

PHB is a short-chain biological polyester belonging to the family of polyhydroxyalkanoate. PHB is an aliphatic monomer produced by bacteria that use glucose, sucrose, and lactose as carbon sources to obtain this polymer (Atakav et al., 2021). In addition, studies show that microalgae (Kamravamanesh et al., 2019; Silva et al., 2018) also synthesize PHB as an energy reserve. PHB has high crystallinity, rigidity, and resistance to deformation. The high crystallinity of PHB also contributes to its low water solubility and good gas barrier (Santos et al., 2017). Another property of PHB is its biodegradability, which can be either by aerobic or anaerobic biological means. However, this biopolymer is miscible in organic solvents such as chloroform and has low processability due to low thermal stability. Thus, plasticizers are used to overcome this limitation (Moreira et al., 2022). In addition, nanotechnological approaches, such as the fabrication of nanocomposites and nanofibers, have been used to contribute to the thermal properties of PHB (Garcia-Garcia et al., 2018; Jayakumar et al., 2020; Kuntzler et al., 2018). PHB can be applied both in the medical field and in the food industry, as it is FDA-approved, non-toxic, and biocompatible with human cells and tissues (Chen & Zhang, 2017; Moreira et al., 2022).

PBAT is a synthetic aliphatic-aromatic thermoplastic copolyester based on fossil resources (Pavon et al., 2020; Rodrigues et al., 2016). The excellent biodegradability characteristic of PBAT is related to the aliphatic unit in the molecule chain (Jian et al., 2020; Rodrigues et al., 2016). PBAT has good thermal stability and mechanical properties similar to some polyethylenes. This polyester also has more flexibility than other biodegradable polyesters such as PLA and PHB, making it interesting for packaging applications (Fukushima et al., 2012; Jian et al., 2020). However, the production costs of PBAT are high (Pavon et al., 2020). In this way, the addition of low-cost materials (natural fibers) is an alternative to improve the economic viability of processing this polymer. In addition, the reinforcement with biomaterials contributes to some polymer properties, maintaining the material's biodegradability. Melt blending (extrusion or injection molding) is considered one of the most commonly applied techniques for developing PBAT-based composites (Ferreira et al., 2019).

PET is the most common semi-crystalline and transparent thermoplastic in the polyester family (Singh et al., 2021). The polymeric film developed with PET for food packaging material features lightweight, high tensile strength, excellent chemical resistance, elasticity, and thermal stability ($-60°C$ to $220°C$) (Robertson, 2014). Despite the various applications of PET, the inappropriate disposal of this polyester in the environment harms the marine ecosystem (Singh et al., 2021). On the other hand, PET is one of the most recycled polymers worldwide. In this sense, research has investigated the recycling of post-consumer PET waste and the properties of composite materials produced from recycled PET (Ragaert et al., 2017; Raheem et al., 2019; Nasir et al., 2022; Singh et al., 2021).

15.3 MECHANICAL AND BARRIER PROPERTIES OF POLYESTER-BASED BIOCOMPOSITES

Knowledge of the mechanical properties of materials used in food packaging is crucial to protect food against physical damage and control mass transfers that act in the degradation of products (Berthet et al., 2015). To be applied in packaging, the material must have some characteristics, including tensile strength, elongation at break, and modulus of elasticity (Ahmad et al., 2016). The mechanical properties of composites are related to the constituent reinforcement and matrix' properties. The transfer of forces applied to the material depends on relatively strong interfacial bonds, as weak interfaces can influence the integrity of the composite (De Rosa et al., 2009). Studies demonstrate that the nanostructures addition to the polymer matrix develops bionanocomposites with improved mechanical and barrier properties for application in food packaging (Table 15.1).

The barrier properties of packaging influence food quality during its shelf life (Berthet et al., 2015). Oxygen (O_2) is one of the main ones responsible for degradation

TABLE 15.1
Studies on Polyester-Based Bionanocomposites

Polymer	Type of Reinforcement	Method	Principal Results	Reference
PLA	Cellulose nanowhisker (NWC)	Solvent casting	The incorporation of 1% NWC improved the mechanical properties, including tensile strength (21.22 MPa) and tensile modulus (11.35 MPa).	Sucinda et al. (2021)
PLA	Nanoclay added to the antimicrobial agents cinnamaldehyde (Ci) and thymol (Thy)	Supercritical impregnation with CO_2	Young's modulus reduction (close to 86% and 77% for thymol-impregnated bionanocomposites and nanoclay-added films, respectively).	Villegas et al. (2017)
PLA/PHB	Cellulose nanocrystals (CNC)	Electrospinning	The addition of 1% by weight of CNC increased the values of modulus and tensile strength (E = 230 MPa and TS = 16 MPa).	Arrieta et al. (2016)
PLA/Poly-(3-hydroxybutyrate-co-3-hydroxyvalerate) (PHBV)	Nanocrystalline cellulose (NCC)	Solvent casting	The addition of 0.25 NCC promoted higher flexural strength (MPa) and flexural modulus (GPa) for the bionanocomposite developed in the polymeric proportion of (70/30 PLA/PHBV).	Dasan et al. (2017)
PCL	Microfibrillated cellulose (MFC), without and with the addition of zinc oxide nanoparticles (ZnO)	Extrusion process	The nanocomposites showed better mechanical properties compared to the PCL matrix. ZnO-MFC composite addition to the polymer matrix provided a higher Young's modulus (169.5 MPa) in relation to the PCL polymer (88.5 MPa).	Reis et al. (2021)

(Continued)

TABLE 15.1 (*Continued*)
Studies on Polyester-Based Bionanocomposites

Polymer	Type of Reinforcement	Method	Principal Results	Reference
Polycaprolactone (PCL) and poly (L-lactic acid)	Phycocyanin/polyvinyl alcohol nanoparticles (PC-PVAn)	Electrospinning	The incorporation of PC-PVAn provided higher tensile strength to the PCL nanofibers and PLA nanofibers (66 and 31 kPa) compared to the pure polymers (62 and 19 kPa) in static collector.	Schmatz et al. (2019)
PBAT/PLA	Nano-polyhedral oligomeric silsesquioxane (POSS(epoxy)8)	Twin-screw extrusion	The addition of POSS (epoxy 8) increased the mechanical performance of the films in the longitudinal and transverse directions. Higher tensile strength was obtained in the materials with 1% by weight of POSS (epoxy 8) (30 MPa longitudinal strength and 20 MPa transverse strength).	Qiu et al. (2021)
PLA	Titanium dioxide (TiO_2)	*Electrospinning* and casting	Maximum tensile strengths of 2.71 ± 0.11 MPa and 14.49 ± 0.13 MPa for the nanofibers and films, respectively, with a TiO_2 content of 0.75% by weight.	Feng et al. (2019)
PLA	TiO_2 nanoparticles	Casting method	The tensile strength of the PLA/TiO_2 composites increased by 63.3 MPa at a TiO_2 content of 0.6% by weight.	Li et al. (2020)
PHB	Silver nanoparticles (AgNPs)	Biological synthesis by *Bacillus megaterium*	Increased mechanical properties of PHB-AgNps compared to pure PHB, such as elongation at break, tensile strength, and Young's modulus, by 1.305%, 35.42 and 1.058 N/mm², respectively.	Jayakumar et al. (2020)

reactions that occur in food. The presence of this gas can promote the oxidation of several compounds that compromise the sensory and nutritional characteristics of foods. Oxygen also favors the emergence of microorganisms and acts in the maturation process of fresh products (Jafarzadeh & Jafari, 2021). CO_2 influences the inhibition of the respiration rate of microorganisms. Thus, CO_2 is considered a bacteriostatic and fungistatic agent. The rate of transmission of this gas or permeability should be a criterion for choosing the used material in food packaging (Berthet et al., 2015). The control of moisture permeability in packaging is also of great importance to control the growth of microorganisms that promote aerobic spoilage and preserve the sensory characteristics of the food (Jafarzadeh & Jafari, 2021).

Fortunati et al. (2012) evaluated the barrier properties of PLA nanocomposites added with cellulose nanocrystals, developed by the solvent casting method. The authors investigated the incorporation of pure (CNC) and surfactant-modified cellulose nanocrystals (s-CNC) into the polymer matrix. Films containing 1% by weight s-CNC showed a 34% reduction in water permeability. On the other hand, bionanocomposites with 5 wt% surfactant-modified CNC (PLA/5s-CNC) had the highest reduction in oxygen transfer rate.

Díez-Pascual and Díez-Vicente (2014) incorporated different concentrations of Zn nanoparticles into PHB-based biocomposites using the solution casting technique. There was a reduction in water absorption (66%) and water vapor permeability (38%) with increasing ZnO content (5% by weight). Oxygen permeability also decreased with increasing ZnO concentration (~53% decrease compared to pure PHB). Vasile et al. (2017) evaluated the influence of plasticization and the addition of nanoparticles on the barrier properties of PLA films. The materials were plasticized with Cu-doped ZnO, and Ag nanoparticle composites (ZnO:Cu/Ag) were added by the melt blending technique. The addition of 0.5 ZnO:Cu/Ag in the nanobiocomposites promoted a reduction in the water vapor transmission rate (11.35 $g/m^2/day$) and CO_2 and O_2 transmission rates (230 and 97 $mL/m^2/day$, respectively).

Manikandan et al. (2020) developed PHB nanocomposites with graphene nanoplatelets (Gr-NPs) (0%–1.3% by weight) for application in food packaging. Compared to pure PHB, the nanocomposite containing 0.7% Gr-NPs promoted 2× increase in tensile strength. Moreover, UV and visible light penetration was reduced by 3× with the addition of Gr-NPs. The nanocomposite application enhanced the shelf life (4×) of foods sensitive to moisture and oxygen (chips and dairy products).

From this perspective, bionanocomposites reinforcements have shown positive results in the mechanical and barrier properties of food packaging. Moreover, with the bionanocomposites application, the environmental appeal of packaging will be met by using materials of biological origin.

15.4 POLYESTER-BASED BIOCOMPOSITES IN ACTIVE FOOD PACKAGING

Active packaging systems involve the use of antimicrobial, antioxidant, scavengers, emitter, and absorber agents and their interaction with food or the environment (Figure 15.1). Antimicrobial packaging works by preventing or delaying the

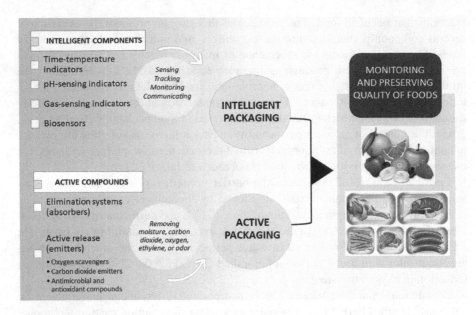

FIGURE 15.1 Characteristics, classification, and application of smart and active packaging materials. (Sani et al., 2021.)

growth of microorganisms in food products. Therefore, active packaging helps to maintain the quality of foods (Moreira et al., 2018b). The migration of antimicrobial compounds to foods can cause reactions with lipids and proteins, leading to the loss of their activity. Thus, the development of packaging with the addition of bionano-composites allows the controlled release of these agents, controlling the incidence of contaminating microorganisms in the food (Talegaonkar et al., 2017). These bion-anomaterials can be developed with polymeric nanostructures, including nanopar-ticles, nanofibers, nanoclays, nanocrystals, or nanowhiskers (Sharma et al., 2020). Kaffashi et al. (2016) developed PCL bionanocomposites with different concentra-tions of triclosan-loaded PLA nanoparticles (30% by weight), produced through a melt mixing process. The authors found that PLA nanoparticles contributed to reduc-ing the rapid release of triclosan from the bionanocomposites and thus, extending the antibacterial property of the samples for up to 2 years. Villegas et al. (2017) evaluated the antimicrobial capacity of PLA films with the addition of C30B nanoclay (5.0% w/w). Cinnamaldehyde and thymol were incorporated into the nanocomposites by supercritical impregnation using carbon dioxide. According to the study, PLA films without the active compound addition showed growth of microorganisms with values around 2.1×10^7 and 6.0×10^7 CFU/mL for *Staphylococcus aureus* and *Escherichia coli*, respectively.

In another study, Lyu et al. (2019) evaluated the antioxidant capacity of PCL films developed with grapefruit seed extract. The concentration of 5% of the active compound provided high inhibition of *Listeria monocytogenes* (decreased bacterial count by 5.8 log). Moreover, bionanomaterials developed by electrospinning from functional or renewable polymers allow encapsulating of active agents. In this way, nanofibers can be applied as reinforcement to improve the physical properties of composites and develop active packaging with antimicrobial properties (Maftoonazad & Ramaswamy, 2018). In this sense, Kuntzler et al. (2018) evaluated the antimicrobial capacity of PHB nanofibers containing phenolic compounds from the microalga *Spirulina* sp. LEB 18. The addition of 1% of the active compound in nanomaterials developed with 35% of PHB showed a greater area of inhibition (7.5 ± 0.4 mm) for *S. aureus*. Thus, active compounds attributed antimicrobial capacity to the nanofibers to develop active food packaging.

Feng et al. (2019) evaluated the antimicrobial activity of nanofibers and PLA/TiO$_2$ composite films through *electrospinning* and casting processes, respectively. In the study, materials containing 0.75% by weight of TiO$_2$ nanoparticles showed an increase in antimicrobial capacity. The nanofibers and films developed showed areas of antimicrobial inhibition for *E. coli* (4.86 ± 0.50 and 3.69 ± 0.40 mm) and *S. aureus* (5.98 ± 0.77 and 4.63 ± 0.45 mm), respectively. Thus, both materials can be applied in food packaging to control contaminating microorganisms present in food.

15.5 POLYESTER-BASED BIOCOMPOSITES FOR MONITORING FOOD QUALITY

Intelligent packaging is a system that detects, tracks, and communicates to the consumer possible changes in food (Figure 15.1). These packaging systems are crucial to guarantee the quality of the products, allowing decision-making about food safety (Kalpana et al., 2019). Besides, intelligent packaging help prevents food poisoning by users, reduces food waste, indicates the integrity of the seal, and acts against the counterfeiting of products (Moreira et al., 2018a, b).

The intelligent packaging system can be classified into indicators, sensors, and data support (Tyagi et al., 2021). The indicators directly show the quality of the packaged product due to its interactions/reactions with compounds generated during the growth of microorganisms. The sensors detect the presence of microorganisms, interacting with components inside the package (food macromolecules and gases), and environmental factors. Regarding data support, barcodes and Radio Frequency Identification tags are included (Kalpana et al., 2019; Ramos et al., 2018).

Using nanotechnological compounds in food packaging is a promising technology for quality control and food safety (Duncan, 2011). From this perspective, bionanocomposites are materials with great potential for intelligent biodegradable coatings due to their ability to monitor the deterioration of food products (Abreu et al., 2012). Studies have addressed the electrospun nanofibers' potential for application in active and intelligent food packaging. Moreira et al. (2018a) developed pH-indicating membranes from PLA/poly(ethylene oxide) (PEO) fibers containing phycocyanin. Variations in pH (from 3 to 4 and 5 to 6) caused a color change (ΔE) of the compound added to the nanomaterials (ΔE around 18). Thus, nanofibers loaded with phycocyanin are promising for use in intelligent food packaging, as the pigment color changes with pH variation. In this same segment, Terra et al. (2021a) also developed pH indicator membranes from polymeric nanofibers for application in food packaging. In the study, PCL/PEO nanofibers were developed with phycocyanin concentrations of 0.5%, 1%, and 2% (w/v). Nanomaterials containing 2% of the pigment were exposed to pH buffers ranging from 3 to 4 and 5–6 and showed a color change $\Delta E \geq 8.5$. These data indicate that this color change is perceptible to the human eye. The authors also found no membranes with phycocyanin additions returned to the initial staining color, confirming that the indicators could not tamper. In another study, Terra et al. (2021b) produced PCL/PEO nanofibers with curcumin, quercetin, and phycocyanin for application as pH–time indicators in food packaging. Color changes detectable by the human eye ($\Delta E \geq 5$) were evidenced in nanofibers containing only phycocyanin (2%, w/v) in the pH range 3–6, showing that the developed indicator is promising for food quality control.

15.6 BIOSAFETY OF BIONANOCOMPOSITES IN FOODS

Nanomaterials have a high surface area and small size that facilitate migration and absorption by consumers' bodies (Li & Huang, 2008). Research is still insufficient to measure the risks related to nanomaterials in the human body (Youssef & El-Sayed, 2018). The safety of biocomposites in food packaging is still not fully understood (Abdul-Khalil et al., 2019). The effects of nanomaterials on human health and their application in the food sector should be better exposed (Sharma et al., 2020). There are still differences between countries regarding regulations and legislative measures regarding the application of nanotechnology in food packaging. This aspect makes it difficult to guarantee product safety, the environment, and the marketing of new products that benefit consumers (Ramos et al., 2018).

In this sense, the main limitation of nanotechnology is the migration speed of nanocomposites from food packaging and the consequences for human health, animals, and the environment (Dimitrijevic et al., 2015). Regarding food intake by consumers, the migration of nanomaterials must be evaluated from the mouth to absorption in the gastrointestinal system (Silvestre et al., 2011). The speed at which nanomaterials migrate into products depends on the food' physicochemical properties and coating material characteristics. In this context, the diffusivity is related to the polymer characteristics (concentration, molecular weight, and viscosity) and the food composition. Moreover, factors such as temperature, pH, and contact time of the packaging with the product also influence this phenomenon (Huang et al., 2015).

Therefore, the use of nanocomposites in food packaging is essential to improve the characteristics of packaging materials. However, research should deepen the investigation into the impact of these particles on the human body. In addition, there should be uniformity of regulations related to nanomaterials applied to food around the world to promote safety in the application and consumption of these products.

15.7 CONCLUSIONS AND FUTURE PROSPECTS

Biocomposites represent a new class of materials for creating food packaging. These biomaterials have promising potential to improve the barrier, mechanical, thermal, and functional properties, and biodegradability of polyester composites. Polyester-based biocomposites are versatile materials that allow the development of packages for application in various food products. Governments and industries are investing in the development of biocomposites through research to increase the application efficiency of these materials. In addition, research for environmentally friendly packaging has gained prominence among companies seeking more sustainable processes to reduce environmental pollution and resource depletion. Therefore, biocomposite market is expected to increase considerably with technological innovations related to the identification/application of new nanocomposite materials in food packages.

Active and intelligent packages added with polyester-based bionanocomposites have been investigated to ensure food quality, integrity, and safety without altering the taste and physical characteristics of the product. However, despite the advantages of bionanocomposites as packaging materials for food, safety and health concerns arise regarding nanomaterials accumulation in the human body and the environment. Therefore, detailed protocols must report the bionanomaterials migration from packaging to food and their toxicity. In addition, regulatory policies must be considered during the manufacturing and processing of nanocomposite-based active and intelligent packages.

ACKNOWLEDGMENTS

This study was financed in part by the Coordenação de Aperfeiçoamento de Pessoal de Nível Superior – Brasil (CAPES) – Finance Code 001. This research was developed within the scope of the Capes-PrInt Program (Process # 88887.310848/2018-00).

REFERENCES

Abdul-Khalil, H. P. S., Saurabh, C. K., Syakir, M. I., Nurul-Fazita, M. R., Banerjee, A., Fizree, H. M., Rizal, S., and P. M. D. Tahir. 2019. Barrier properties of biocomposites/ hybrid films. In *Mechanical and Physical Testing of Biocomposites, Fibre-Reinforced Composites and Hybrid Composites*, eds. J. Mohammad., T. Mohamed and S. Naheed, 241–258. Cambridge: Woodhead Publishing.

Abdulkhani, A., Echresh, Z., and M. Allahdadi. 2020. Effect of nanofibers on the structure and properties of biocomposites. In *Fiber-Reinforced Nanocomposites: Fundamentals and Applications*, eds, B. Han., T. A. Nguyen and K. S. Bhat, pp. 321–357. Amsterdam: Elsevier.

Abreu, D. A. P., Cruz, J. M. and P. P. Losada. 2012. Active and intelligent packaging for the food industry. *Food Rev. Int.* 28:146–187.

Ahmad, M., Nirmal, N. P., Danish, M., Chuprom, J., and S. Jafarzedeh. 2016. Characterisation of composite films fabricated from collagen/ chitosan and collagen/soy protein isolate for food packaging applications. *RSC Adv.* 6:2191–2204.

Arora, B., Bhatia, R., and P. Attri. 2018. Bionanocomposites: Green materials for a sustainable future. In *New Polymer Nanocomposites for Environmental Remediation*, eds, C. M. Hussain and A. K. Mishra, pp. 699–712. Amsterdam: Elsevier.

Arrieta, M. P., López, J., López, D., Kennya, J. M., and L. Peponi. 2016. Biodegradable electrospun bionanocomposite fibers based on plasticized PLA–PHB blends reinforced with cellulose nanocrystals. *Ind. Crops Prod.* 93:290–301.

Atakav, Y., Pinar, O., and D. Kazan. 2021. Investigation of the physiology of the obligate alkaliphilic *Bacillus marmarensis* GMBE 72T considering its alkaline adaptation mechanism for poly(3-hydroxybutyrate) synthesis. *Microorganisms* 9(2):462.

Bajpai, V. K., Kamle, M., Shukla, S., Mahato, D. K., Chandra, P., Hwang, S. K., Kumar, P., Huh, Y. S., and Y-K. Han. 2018. Prospects of using nanotechnology for food preservation, safety, and security. *J. Food Drug Anal.* 26:1201–1214.

Berthet, M. A., Angellier-Coussy, H., Guillard, V., and N. Gontard. 2015. Vegetal fiber-based biocomposites: Which stakes for food packaging applications? *J. Appl. Polym. Sci.* 133:42528.

Buhecha, M. D., Lansley, A. B., Somavarapu, S., and A. S. Pannala. 2019. Development and characterization of PLA nanoparticles for pulmonary drug delivery: Co-encapsulation of theophylline and budesonide, a hydrophilic and lipophilic drug. *J. Drug Delivery Sci. Technol.* 53:101128.

Chen, G.-Q., and J. Zhang. 2017. Microbial polyhydroxyalkanoates as medical implant biomaterials. *Artif. Cells Nanomed. Biotechnol.* 46(1):1–18.

Dasan, Y.K., Bhat, A. H., and A. Faiz. 2017. Polymer blend of PLA/PHBV based bionanocomposites reinforced with nanocrystalline cellulose for potential application as packaging material. *Carbohydr. Polym.* 157:1323–1332.

De Rosa, I. M., Santulli, C., and F. Sarasini. 2009. Acoustic emission for monitoring the mechanical behaviour of natural fibre composites: A literature review. *Compos. Part A Appl. Sci. Manuf.* 40:1456–1469.

Díez-Pascual, A. M., and A. L. Díez-Vicente. 2014. Poly(3-hydroxybutyrate)/ZnO bionanocomposites with improved mechanical, barrier and antibacterial Properties. *Int. J. Mol. Sci.* 15:10950–10973.

Dimitrijevic, M., Karabasil, N., Boskovic, M., Teodorovic, V., Vasilev, D., Djordjevic, V., Kilibarda, N., and N. Cobanovic. 2015. Safety aspects of nanotechnology applications in food packaging. *Procedia Food. Sci.* 5:57–60.

Duncan, T.V. 2011. Applications of nanotechnology in food packaging and food safety: Barrier materials, antimicrobials and sensors. *J. Colloid. Interface Sci.* 363:1–24.

Dwivedi, R., Kumar, S., Pandey, R., Mahajan, A., Nandana, D., Katti, D. S., and D. Mehrotra. 2019. Polycaprolactone as biomaterial for bone scaffolds: Review of literature. *J. Oral Biol. Craniofac. Res.* 10(1):381–388.

Feng, S., Zhang, F., Ahmed, S., and Y. Liu. 2019. Physico-mechanical and antibacterial properties of PLA/TiO$_2$ composite materials synthesized via electrospinning and solution casting processes. *Coatings* 9(8):525.

Ferreira, F. V., Cividanes, L. S., Gouveia, R. F., and L. M. F. Lona. 2019. An overview on properties and applications of poly(butylene adipate-co-terephthalate)–PBAT based composites. *Polym. Eng. Sci.* 59:E7–E15.

Fortunati, E., Peltzer, M., Armentano, I., Torre, L., Jiménez, A., and J. M. Kenny. 2012. Effects of modified cellulose nanocrystals on the barrier and migration properties of PLA nanobiocomposites. *Carbohydr. Polym.* 90:948–956.

Fukushima, K., Wu, M.-H., Bocchini, S., Rasyida, A., and M.-C. Yang. (2012). PBAT based nanocomposites for medical and industrial applications. *Mater. Sci. Eng. C.* 32:1331–1351.

Garcia-Garcia, D., Garcia-Sanoguera, D., Fombuena, V., Lopez-Martinez, J., and R. Balart. 2018. Improvement of mechanical and thermal properties of poly(3-hydroxybutyrate) (PHB) blends with surface-modified halloysite nanotubes (HNT). *Appl. Clay Sci.* 162:487–498.

Gentile, P., Chiono, V., Carmagnola, I., and P. V. Hatton. 2014. An overview of poly(lactic-co-glycolic) acid (PLGA)-based biomaterials for bone tissue engineering. *Int. J. Mol. Sci.* 15(3):3640–3659.

Gigante, V., Coltelli, M.-B., Vannozzi, A., Panariello, L., Fusco, A., Trombi, L., Donnarumma, G., Danti, S., and A. Lazzeri. 2019. Flat die extruded biocompatible poly(lactic acid) (PLA)/poly(butylene succinate) (PBS) Based Films. *Polymers* 11:1857.

Guido, C., Testini, M., D'Amone, S., Cortese, B., Grano, M., Gigli, G., and I. E. Palamà. 2021. Capsid-like biodegradable poly-glycolic acid nanoparticles for a long-time release of nucleic acid molecules. *Mater. Adv.* 2(1):310–321.

Hamid, L., and I. Samy. 2021. Fabricating natural biocomposites for food packaging. In *Fiber-Reinforced Plastics*, eds. L. Hamid, I. S. Fahin, 1–14. London: IntechOpen.

Huang, J. Y., Li, X., and W. Zhou. 2015. Safety assessment of nanocomposite for food packaging application. *Trends Food Sci. Technol.* 45:187–199.

Jafarzadeh, S., and S. M. Jafari. 2021. Impact of metal nanoparticles on the mechanical, barrier, optical and thermal properties of biodegradable food packaging materials. *Crit. Rev. Food Sci. Nutr.* 61:2640–2658.

Jayakumar, A., Prabhu, K., Shah, L., and P. Radha. 2020. Biologically and environmentally benign approach for PHB-silver nanocomposite synthesis and its characterization. *Polym. Test.* 81:106197.

Jem, K. J., and B. Tan. 2020. The development and challenges of poly (lactic acid) and poly (glycolic acid). *Adv. Ind. Eng. Polym. Res.* 3(2):60–70.

Jian, J., Xiangbin, Z., and H. Xianbo. 2020. An overview on synthesis, properties and applications of poly(butylene-adipate-co-terephthalate) –PBAT. Adv. *Ind. Eng. Polym. Res.* 3:19–26.

Jin, F.-L., Hu, R.-R., and S.-J. Park. 2018. Improvement of thermal behaviors of biodegradable poly(lactic acid) polymer: A review. *Compos. B. Eng.* 164:287–296.

Kaffashi B., Davodi, S., and E. Oliaei. 2016. Poly(e-caprolactone)/triclosan loaded polylactic acid nanoparticles composite: A long-term antibacterial bionanocomposite with sustained release. *Int. J. Pharm.* 508:10–21.

Kalita, N. K., Sarmah, A., Bhasney, S. M., Kalamdhad, A., and V. Katiyar. 2021. Demonstrating an ideal compostable plastic using biodegradability kinetics of poly(lactic acid) (PLA) based green biocomposite films under aerobic composting conditions. *Environ. Challenges* 3:100030.

Kalpana, S., Priyadarshini, S. R., Leena, M. M., Moses, J. A., and C. Anandharamakrishnan. 2019. Intelligent packaging: Trends and applications in food systems. *Trends Food Sci. Technol.* 93:145–157.

Kamravamanesh, D., Kiesenhofer, D., Fluch, S., Lackner, M., and C. Herwig. 2019. Scale-up challenges and requirement of technology-transfer for cyanobacterial poly (3-hydroxybutyrate) production in industrial scale. *Int. J. Biobased Plast.* 1(1):60–71.

Kausar, A. 2019. Review of fundamentals and applications of polyester nanocomposites filled with carbonaceous nanofillers. *J. Plast. Film Sheeting* 35:22–44.

Khatsee, S., Dananarong, D., Punyodom, W., and P. Worajittiphon. 2018. Electrospinning polymer blend of PLA and PBAT: Electrospinnability–solubility map and effect of polymer solution parameters toward application as antibiotic-carrier mats. *J. Appl. Polym. Sci.* 135:46486.

Kuntzler, S. G., de Almeida, A. C. A., Costa, J. A. V., and M. G. Morais. 2018. Polyhydroxybutyrate and phenolic compounds microalgae electrospun nanofibers: A novel nanomaterial with antibacterial activity. *Int. J. Biol. Macromol.* 113:1008–1014.

Li, M., Pu, Y., Chen, F., and A. J. Ragauskas. 2021. Synthesis and characterization of lignin-grafted-poly(e-caprolactone) from different biomass sources. *N. Biotechnol.* 60:189–199.

Li, S., Chen, G., Qiang, S., Yin, Z, Zhang, Z., and Y. Chen. 2020. Synthesis and evaluation of highly dispersible and efficient photocatalytic TiO$_2$/poly lactic acid nanocomposite films via sol-gel and casting processes. *Int. J. Food Microbiol.* 331:108763.

Li, S. D., and L. Huang. 2008. Pharmacokinetics and biodistribution of nanoparticles. *Mol. Pharm.* 5:473–680.

Lyu, J. S., Lee, J-S., and J. Han. 2019. Development of a biodegradable polycaprolactone film incorporated with an antimicrobial agent via an extrusion process. *Sci Rep.* 9:20236.

Ma, Q., Shi, K., Su, T., and Z. Wang. 2020. Biodegradation of polycaprolactone (PCL) with different molecular weights by candida antarctica lipase. *J. Polym. Environ.* 28:2947–2955.

Maftoonazad, N., and H. Ramaswamy. 2018. Novel techniques in food processing: Bionanocomposites. *Curr. Opin. Food Sci.* 23:49–56.

Malek, N. S. A., Faizuwan, M., Khusaimi, Z., Bonnia, N. N., Rusop, M., and N. A. Asli. (2021). Preparation and characterization of biodegradable polylactic acid (PLA) film for food packaging application: A review. *J. Phys. Conf. Ser.* 1593:338–341.

Malik, N. 2020. Thermally exfoliated graphene oxide reinforced polycaprolactone-based bactericidal nanocomposites for food packaging applications. *Mater. Technol.* 37(5):345–354.

Manikandan, N., Pakshirajan, K., and G. Pugazhenthi. 2020. Preparation and characterization of environmentally safe and highly biodegradable microbial polyhydroxybutyrate (PHB) based graphene nanocomposites for potential food packaging applications. *Int. J. Biol. Macromol.* 154:866–877.

Mohandesnezhad, S., Pilehvar-Soltanahmadi, Y., Alizadeh, E., Goodarzi, A., Davaran, S., Khatamian, M., Zarghami, N., Samiei, M., Aghazadeh, M., and A. Akbarzadeh. 2020. In vitro evaluation of Zeolite-nHA blended PCL/PLA nanofibers for dental tissue engineering. *Mater. Chem. Phys.* 252:123152.

Moreira, J. B., Kuntzler, S. G., Vaz, B. S., Silva, C. K., Costa, J. A. V., and M. G. Morais. 2022. Polyhydroxybutyrate (PHB)-based blends and composites. In *Biodegradable Polymers, Blends and Composites*, eds. S. M. Rangappa, J. Parameswaranpillai, S. Siengchin and M. Ramesh, pp. 389–413. Cambridge: Elsevier.

Moreira, J. B., Morais, M.G., Morais, E.G., Vaz, B.S., and J.A.V. Costa. 2018b. Electrospun polymeric nanofibers in food packaging. In *Impact of Nanoscience in the Food Industry*, eds. A. M. Grumezescu and A. M. Holban, pp. 387–417. London: Academic Press.

Moreira, J. B. M., Terra, A. L. M., Costa, J. A. V., and M. G. Morais. 2018a. Development of pH indicator from PLA/PEO ultrafine fibers containing pigment of microalgae origin. *Int. J. Biol. Macromol.* 118:1855–1862.

Moustafa, H., Youssef, A. M., Darwish, N. A., and A. I. Abou-Kandil. 2019. Eco-friendly polymer composites for green packaging: Future vision and challenges. *Compos. B Eng.* 172:16–25.

Nasir, N. H. M., Usman, F., Usman, F., Saggaf, A. and Saloma. 2022. Development of composite material from Recycled Polyethylene Terephthalate and fly ash: Four decades progress review. *Curr. Opin. Green Sustain. Chem.* 5:100280.

Pavon, C., Aldas, M., Rosa-Ramírez, H. R., López-Martínez, J., and M. P. Arrieta. 2020. Improvement of PBAT Processability and mechanical performance by blending with pine resin derivatives for injection moulding rigid packaging with enhanced hydrophobicity. *Polymers* 12:2891.

Priyadarshi, R., Roy, S., Ghosh, T., Biswas, D., and J-W, Rhim. 2021. Antimicrobial nanofillers reinforced biopolymer composite films for active food packaging applications - A review. *Sustain. Mater. Technol.* 32:e00353.

Qiu, S., Zhou, Y., Waterhouse, G. I. N., Gong, R., Xie, J., Zhang, K., and Xu, J. 2021. Optimizing interfacial adhesion in PBAT/PLA nanocomposite for biodegradable packaging films. *Food Chem.* 334:127487.

Ragaert, K., Delva, L., and K. Van Geem. 2017. Mechanical and chemical recycling of solid plastic waste. *Waste Manag.* 69:24–58.

Raheem, A.B., Noor. Z.Z., Hassan, A., Abd Hamid, M.K., Samsudin, S.A., Sabeen, A.H. 2019. Current developments in chemical recycling of post-consumer polyethylene terephthalate wastes for new materials production: A review. *J. Clean Prod.* 225:1052–1064.

Ramos, O. L., Pereira, R. N., Cerqueira, M. A., Martins, J. R., Teixeira, J. A., Malcata, X., and A. A. Vicente. 2018. Bio-based nanocomposites for food packaging and their effect in food quality and safety. In *Food Packaging and Preservation*, eds. A. Grumezescu., A. M. Holban, pp. 271–306. London: Academic Press.

Reis, R. S., Souza, D. H, S., Marques, M. F. V., Luz, F. S., and S. N. Monteiro. 2021. Novel bionanocomposite of polycaprolactone reinforced with steam-exploded microfibrillated cellulose modified with ZnO. *J. Mater. Sci. Technol.* 13:1324–1335.

Robertson, G. L. 2014. Food Packaging. In *Encyclopedia of Agriculture and Food Systems*, ed. N. K. V. Alfen, pp. 232–249. London: Academic Press.

Rodrigues, B. V. M., Silva, A. S., Melo, G.F.S., Vasconscellos, L. M. R., Marciano, F. R., and A. O. Lobo. 2016. Influence of low contents of superhydrophilic MWCNT on the properties and cell viability of electrospun poly (butylene adipate-co-terephthalate) fibers. *Mater Sci. Eng. C Mater. Biol. Appl.* 59:782–791.

Samantaray, P. K., Little, A., Haddleton, D., McNally, T., Tan, B., Sun, Z., Huang, W., Ji, Y., and C. Wan. 2020. Poly (glycolic acid) (PGA): A versatile building block expanding high performance and sustainable bioplastic applications. *Green Chem.* 22:4055.

Sani, M. A., Azizi-Lalabai, M., Tavassoli, M., Mohammadi, K., and D. J. McClements. 2021. Recent advances in the development of smart and active biodegradable packaging materials. *Nanomaterials* 11:1331.

Santos, A. J., Valentina, L. V. O. D., Schulz, A. A. H., and M. A. T. Duarte. 2017. From obtaining to degradation of PHB: Material properties. Part I. *Ingeniería y Ciencia* 13:269–298.

Sanusi, O. M., Benelfellah, A., Bikiaris, D. N., and N. Aït Hocine. 2020. Effect of rigid nanoparticles and preparation techniques on the performances of poly(lactic acid) nanocomposites: A review. *Polym. Adv. Technol.* 32(1):1–17.

Sari, N. H., Suteja, S., Sapuan, S. M, and R. A. Ilyas. 2021. Properties and food packaging application of poly-(lactic) acid. In *Bio-Based Packaging: Material, Environmental and Economic Aspects*, eds. S. M. Sapuan, and R. A. Ilyas, pp. 245–263. Hoboken: JohnWiley & Sons Ltd.

Schmatz, D. A., Costa, J. A. V., and M. G. Morais. 2019. A novel nanocomposite for food packaging developed by electrospinning and electrospraying. *Food Packag. Shelf Life* 20:100314.

Sharma, R., Jafari, S. M., and S. Sharma. 2020. Antimicrobial bio-nanocomposites and their potential applications in food packaging. *Food Control* 112:107086.

Silva, C. K., Costa, J. A. V., and M. G. Morais. 2018. Polyhydroxybutyrate (PHB) synthesis by *Spirulina* sp. LEB 18 using biopolymer extraction waste. *Appl. Biochem. Biotechnol.* 185(3):822–833.

Silvestre, C., Duraccio, D., and S. Cimmino. 2011. Food packaging based on polymer nanomaterials. *Prog. Polym. Sci.* 36:1766–1782.

Singh, A. K., Bedi, R., and B. S. Kaith. 2021. Composite materials based on recycled polyethylene terephthalate and their properties – A comprehensive review. *Compos. B. Eng.* 219:108928.

Sucinda, E. F., Abdul Majid, M. S., Ridzuan M. J. M., Cheng, E. M., Alshahrani, H. A., and N. Mamat. 2021. Development and characterisation of packaging film from Napier cellulose nanowhisker reinforced polylactic acid (PLA) bionanocomposites. *Int. J. Biol. Macromol.* 187:43–53.

Talegaonkar, S., Sharma, H., Pandey, S., Mishra, P. K., and R. Wimmer. 2017. Bionanocomposites: Smart biodegradable packaging material for food preservation. In *Food Packaging: Nanotechnology in the Agri-Food Industry*, ed. A. Grumezescu, pp. 79–104. Lodon: Academic Press.

Terra, A. L. M., Moreira, J. B., Costa, J. A. V., and M. G. Morais. 2021a. Development of pH indicators from nanofibers containing microalgal pigment for monitoring of food quality. *Food Biosci.* 44:101387.

Terra, A. L. M., Moreira, J. B., Costa, J. A. V., and M. G. Morais. 2021b. Development of time-pH indicator nanofibers from natural pigments: An emerging processing technology to monitor the quality of foods. *LWT - Food Sci. Technol.* 142:111020.

Tyagi, P., Salem, K. S., Hubbe, M. A., and L. Pal. 2021. Advances in barrier coatings and film technologies for achieving sustainable packaging of food products – A review. *Trends Food Sci. Technol.* 115:461–485.

Vasile, C., Rapa, M., Stefan, M., Stan, M., Macavei, S., Darie-Nita, R. N., Barbu-Tudoran, L., Vodnar, D. C., Popa, E. E., Stefan, R., Borodi, R., Borodi, G., and M. Brebu. 2017. New PLA/ZnO:Cu/Ag bionanocomposites for food packaging. *Polym. Lett.* 11:531–544.

Villegas, C., Arrieta, M. P., Rojas, A., Torres, A., Faba, S., Toledo, M. J., Gutierrez, M. A., Romero, J., Galotto, M. J., and X. Valenzuela. 2017. PLA/organoclay bionanocomposites impregnated with thymol and cinnamaldehyde by supercritical impregnation for active and sustainable food packaging. *Compos. B. Eng.* 176:107336.

Vinod, A., Sanjay, M. R., Suchart, S., and P. Jyotishkumar. 2020. Renewable and sustainable biobased materials: An assessment on biofibers, biofilms, biopolymers and biocomposites. *J. Clean. Prod.* 258:120978.

Wang, G., Yu, D., Kelkar, A.D., and L. Zhang. 2017. Electrospun nanofiber: Emerging reinforcing filler in polymer matrix composite materials. *Prog. Polym. Sci.* 75:73–107.

Wang, K., Shen, J., Ma, Z., Zhang, Y., Xu, N., and S. Pang. 2021a. Preparation and properties of poly (ethylene glycol-co-cyclohexane-1,4-dimethanol terephthalate)/polyglycolic acid (PETG/PGA) blends. *Polymers* 13(3):452.

Wang, R., Sun, X., Chen, L., and W. Liang. 2021b. Morphological and mechanical properties of biodegradable poly(glycolic acid)/poly(butylene adipate-co-terephthalate) blends with in situ compatibilization. *RSC Adv.* 11:1241–1249.

Youssef, A. M., and S. M. El-Sayed. 2018. Bionanocomposites materials for food packaging applications: Concepts and future outlook. *Carbohydr. Polym.* 193:19–27.

16 An Experimental and Numerical Investigation of Bio-Based Polyurethane Foam for Acoustical Applications

Durgam Muralidharan Nivedhitha, Subramanian Jeyanthi, Selvaraj Vinoth Kumar, and Amol Manoj
Vellore Institute of Technology

CONTENTS

DOI: 10.1201/9781003270980-16

LIST OF SYMBOLS AND ABBREVIATIONS

Symbols

F	Frequency
F_l	Lower frequency
F_s	Sampling rate
F_u	Upper frequency
H_I	Transfer function at incident wave
H_R	Transfer function at reflected wave
k	Constant
k_0	Complex wave number
Ø	Porosity
P_1	Pressure at microphone 1
P_2	Pressure at microphone 2
P_i	Incident pressure
P_r	Reflected pressure
r	Reflection coefficient
rho	Density of air
S	Microphone spacing
Zair	Characteristic impedance of air
Z_s	Surface impedance
Z_{s-n}	Normalised surface impedance
α	Absorption coefficient
Λ	Viscous characteristic length
λ	Wavelength
Λ′	Thermal characteristic length
σ	Flow resistivity
Υ	Adiabatic exponent of air
ώ	Angular frequency
Γ́	Propagation constant

Abbreviations

FPF	Flexible polyurethane foam
HXDI	Hydro-generated xylene diisocyanate
PU	Polyurethane
RPF	Rigid polyurethane foam
TMXDI	Meta-tetra methyl xylene diisocyanate
XDI	Xylene diisocyanate

FIGURE 16.1 Sound insulation material in an automobile.

16.1 BACKGROUND AND MOTIVATION

Sound absorption is a technique used in various fields to control noise pollution. A unique material is used as the absorbent material, a combination of solid and liquid phases, which is most often a porous material. Many examples can be found in the automation, building and industrial sectors, where certain materials play an essential role in controlling the noise pollution. Different environmental and economic constraints prompted designers and manufacturers to find a solution to increase their usage and potential. The efficiency of material combinations is considered. The bio-based material polyurethane (PU) foam is developed and analysed to predict its efficiency (Figure 16.1).

In the present scenario, the PU foam obtained from the polyol and isocyanate-based mixer is used as an absorbent material in different areas. This type of foam, known as gasoline-based foam, affects human activities such as breathing and causes throat and lung problems. This motivated to spark a particular interest in developing bio-based PU foam, where polyol is derived from natural vegetable oils such as castor oil. Luffa fibre is a product obtained from the *Luffa cylindrica* plant, which is incorporated with bio-based PU foam in different percentages to form a composite material that contributes to enhanced sound absorption properties.

16.2 STATE OF THE ART IN ACOUSTICS ABSORPTION

16.2.1 DEVELOPMENT OF ACOUSTIC ABSORBING MATERIAL

Sound-absorbing material is widely used in various fields such as automobiles, industrial environments and buildings. Efforts have been made accordingly concerning our work area. In cars, the material is used for sound absorption; it should be thin and light. In industry, materials are preferably made lightweight and cheap, as these two parameters have major influence for increasing the efficiency of the automobile. Therefore, improving the absorption performance of these materials is of great importance. PU foam is an absorbing material that meets all the above requirements and is enhanced and evolved by incorporating natural fibres in different percentages, resulting in a composite material. The main advantage of adding fibre particles is strengthening the porous medium, so flow resistance is an essential improved factor that provides a good value for the sound absorption coefficient. Another important motivating factor was a particular interest in chemically minimising the impact of PU foam on human health. Bio-based polyol is blended with isocyanate to produce eco-friendly PU foam.

16.2.2 FINITE ELEMENT-BASED NUMERICAL MODEL

Various calculation methods are adopted to describe the acoustic behaviour of absorbent or porous materials at multiple frequencies. Delany–BezMiki and Joley–Hanson–Champoux–Allard are the dconventional methods adopted to develop the drag factor model. Atala et al. [1] have developed a Finite Element Design based on a numerical model which explains the acoustic absorption coefficient of porous material composed of porous reinforcements with different properties. One end of the impedance generates the incident plane wave, and a porous medium defines the other end. This domain is designed using the fundamentals of BIOT theory, which solves the solution for both pressure and displacements of the porous matrix. By adopting BIOT theory, many unknown properties and characteristics of absorbent material can be determined. As we consider five significant properties of any porous material: porosity, tortuosity, flow resistivity, viscous characteristics, and thermal characteristics length, BIOTS theory gives a theoretical explanation as these parameters appear in the BIOT general model.

16.3 PROBLEM DEFINITION

Recent developments in PU foams relate to composite formation incorporated with various fibres such as glass fib, jute fibre and textile. This area is still being developed to exploit its full potential. The disadvantage of PU foam is its chemical formation of polyol and isocyanate derived from gasoline chemicals, which can harm human health. Therefore, to reduce the impact of these synthetic foams, the development of bio-based foam from vegetable oils to produce PU FOAM is being carried out. Luffa fruit, obtained from the *L. cylindrica* plant, is used as fibre to improve sound absorption properties.

16.4 OBJECTIVES

The main object of this present research is to fabricate a bio-based PU foam. Generally, PU is formed by two primary chemical components: isocyanate as prepolymer and polyol. This research attempts to incorporate bio fillers such as luffa fibres in various percentages in short fibres. Further, this material is experimentally tested for acoustical properties to obtain a sound absorption coefficient (α), and as follows:

1. To determine the transfer of function method of two microphones which purely measures the sound pressure in an impedance tube at two flushed mounted microphones positions. This determines the acoustic parameters, such as the degree of absorption with a frequency range up to 6.3 kHz.
2. To determine sound absorption coefficient using numerical and analytical approach using finite element analysis using COMSOL Multiphysics software.

16.5 METHODOLOGY

Steps followed for the fabrication of bio-based PU foam incorporated with luffa fibre. The flowchart represented in Figure 16.2 gives a brief procedure involved in the fabrication process.

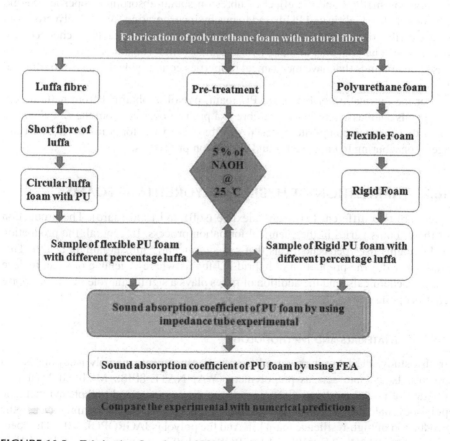

FIGURE 16.2 Fabrication flowchart of PU foam.

16.6 LITERATURE SURVEY

Ju Hyuk Park et al. [2] have investigated the optimisation and modelling of low-frequency sound absorption by controlling the cell size of the porous material. He also explained the modelling of acoustic and manipulating microstructural irradiating ultrasonic waves by mixing resin during the foaming process. This research also resulted in the numerical simulation to predict the effect of cell size on the performance of acoustic properties using the multi-scale porous acoustical method. Jung Hyeun Kim et al. [3] investigated the sound absorption behaviour of flexible polyurethane foams (FPFs) with various cell structures. This research established the relationship between the cell structure and acoustic properties. The results revealed that voids and pores were controlled by two types of gelling catalysts, namely DBTDL, and DBTDL and water, which are directly influenced by the sound absorption property of the flexible foams. Moises Pinto et al. [4] described the formulation, production and characterisation of PU foams. The PU foam produced was explained using ratios and mixtures of chemicals. And the calculation of the bulk density of foams has been carried out. Bulent EKICI et al. [5] studied the sound absorption property of PU foams by adding tea-leaf fibres. This study explains the two-microphone transfer function method and the effect of fibres on sound-absorbing properties. Sergio Mendez et al. [6] fabricated PU-based foams by incorporating "Waste" glycerol from transesterification synthesis into "Waste" Agricultural Residues (e.g., chopped rice hull fibres). The goal was to make hybrid foams with a substantial portion of plant-based compounds that have mechanical properties comparable to those of traditional PU foam [7].

For the fabrication of bio-based PU foam, polyol is obtained from natural vegetable oils such as castor oil. Luffa fibre is a product derived from the *L. cylindrica* plant that forms a composite material with bio-based PU foam in varying percentages, contributing to enhanced sound absorption properties.

16.7 FABRICATION OF HYBRID POLYURETHANE FOAMS

There are two different PU foams, flexible (soft) and rigid (hard). The production of these foams varies in the chemical formation process. In general, the production of PU foams uses petrochemicals that can significantly impact human health. This study provides the production of hybrid PU foams where vegetable oils can replace the petrochemicals, and the addition of fibres plays a significant role in various noise control applications.

16.7.1 MATERIALS AND METHODOLOGY

In this study, PU foam is produced from two constituents, namely isocyanates and polyols. Isocyanate acts as prepolymers (WANNATE-8018). Modified MDI is a mixture of polyol-modified diphenylmethane diisocyanate and polyphenyl methane polyisocyanate. It is a brown liquid at room temperature. It can be mainly used in the production of high-resilience foam [8]. And the polyol is JAGROPOL-400. The specification of WANNATE-8018 and JAGROPOL-400 is given in Tables 16.1 and 16.2.

TABLE 16.1
Properties of WANNATE-8018

Property	Values of WANNATE-8018
Appearance	Brown liquid
Isocyanate equivalent weight	142.4
Viscosity at 25°C, MPa s	70
Specific gravity at 25°C	1.20
Acidity (as HCL), %	<0.05

TABLE 16.2
Properties of JAGROPOL-400

Property	Values of JAGROPOL-400
Appearance	Amber viscous liquid
Sp.gr. at 25°C	0.950–0.975
Hydroxyl value, hydroxyl equivalent, wt.	70–100, 625
Viscosity, 25°C	40–60 Poise
Acid value	3 Max

Isocyanates are considered the essential component in the synthesis of PU foam. Generally, isocyanates are di- or polyfunctional monomers containing two or more – NCO groups per molecule [9–10]. TDI, MDI, xylene diisocyanate, 1,6-hexamethylene diisocyanate, 2,2,4-trimethyl hexamethylene diisocyanate, isophorone diisocyanate, 4,4'-di-cyclohexyl methane diisocyanate and norbornane diisocyanate are 4,4'-dibenzyl diisocyanate are classified under aromatic isocyanates [11–13]. Figures 16.3 and 16.4 show few examples of some typically used isocyanates in research and development.

Another component used for the synthesis of PU foams is polyols. Polyols play an essential role in the synthesis of PU foam industries as they are traditionally derived from petrochemicals. Concerning environmental aspects and sustainability, it has paved the way for producing polyols derived from vegetables such as soybeans, castor and palm oil [14–15]. This aspect grabbed a lot of interest and scope by various researchers as they are economically affordable [16–18]. They can also satisfy multiple features like renewability, flexibility to the process, biodegradability with cost-effectiveness, and comparable performance with petrol-based polyols [19–20].

The Luffa fibre (pictured) is used to improve the acoustic properties of PU foam obtained from the *L. cylindrica* plant. These fibres are hollow and punctured and have a low specific gravity of 1.48 g/cm^3. Luffa Loaf PU fruits are commercially available in the market for various purposes. Due to natural porous structures, these can be used as a suitable engineered material for sound absorption applications [21].

The pretreatment of the luffa fibres is done before the sample preparation as it is crucial to reduce the moisture content in the luffa fruit and improve the pore size.

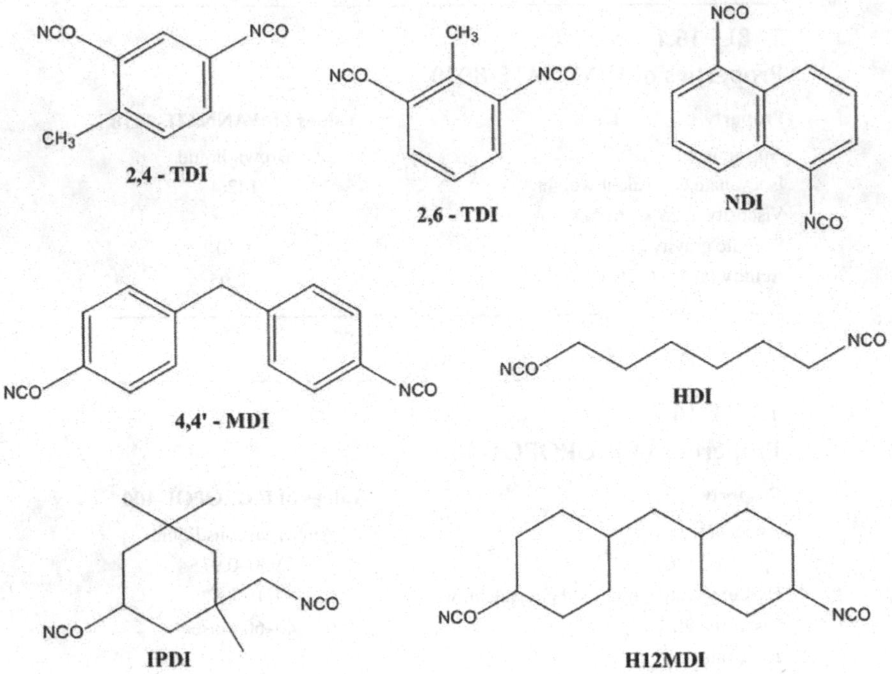

FIGURE 16.3 Examples of isocyanates.

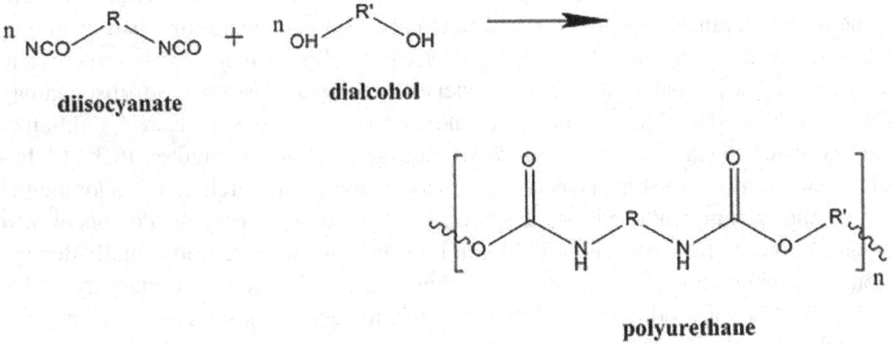

FIGURE 16.4 PU foam formation of polyol and isocyanate.

Raw luffa is moist, and the pores are uneven. To overcome all these issues, luffa fibres are treated with a 5% NaOH solution at room temperature for 1 day, as shown in Figure 16.5. After the treatment, these fibres are washed with ordinary water and dried at room temperature for a day to improve the material's properties. The effect of alkali treatment with 5% of NaOH improves the surface interactions and gets apparent homogenous voids, which is vital in the sound absorption process. The naked eye can observe the appearance change, as the colour of the luffa has changed

FIGURE 16.5   Pretreatment of luffa fibre with 5% of NaOH.

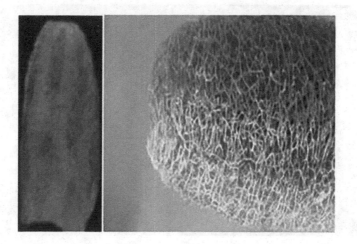

FIGURE 16.6   Luffa fibre after alkali treatment.

from dusky brown to pale yellow colour as shown in Figure 16.6. Luffa fibre consists of the inner core and outer shell, and once the pretreatment is done, the inner core is removed and the outer shell is made into mat forms, as shown in Figure 16.7. This study's main interests are in short fibres and circular mats. The short fibres are of length 2–4mm, and the circular mat of a diameter of 33 mm is made from the outer shell shown in Figure 16.8.

In this study, PU foam is produced from isocyanate (WANNATE-8018). Modified MDI is a mixture of polyol-modified diphenylmethane diisocyanate and polyphenyl methane polyisocyanate. It is a brown liquid at room temperature. It can be mainly used in the production of high-resilience foam. And the polyol is JAGROPOL-400. The specification of WANNATE-8018 and JAGROPOL-400 is given in Tables 16.1 and 16.2, respectively.

FIGURE 16.7 Separation of inner core from luffa fibre.

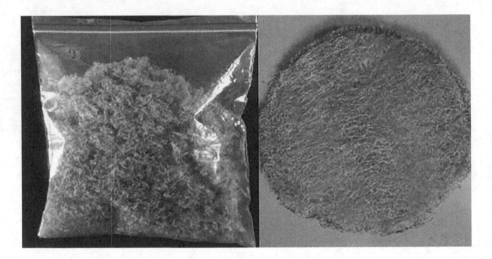

FIGURE 16.8 Short form of luffa fibre and circular shape of luffa.

16.7.2 Sample Preparation

The reaction between polyol (A) and isocyanate (B). The mixing ratio of A:B leads to the formation of PU foam. The mixing ratio of polyol and isocyanate varies by the foam phase. If the polyol content is higher than the isocyanate content, the result is a flexible foam. When the isocyanate weight ratio is higher than the polyol, the foam produced becomes very rigid, known as rigid polyurethane foam (RPF). A rectangular mould is used to manufacture PU foam. Polyol and isocyanate with different wt% fibres are mixed in a beaker. A mixing tool at up to 2000 rpm was used to mix the composition to obtain a homogeneous composition. After mixing, the foam was allowed to rise freely at room temperature for 1 hour. The circular samples with a diameter of 33 and a height of 40 mm are made for impedance tube tests.

FIGURE 16.9 Open-cell PU foam incorporated with luffa fibre.

Flexible Foam: If polyol (A) is higher than isocyanate (B), the foam formed is soft and dimensionally unstable, as shown in Figure 16.9.

Polyol	JAGROPOL	400
Polyol (A)	EMPEYOL WSF	100/S
Isocyanate (B)	WANNATE	8018
Mixing ratio	A:B	100:65

Rigid Foam: If polyol (A) is less than isocyanate (B), then the foam formed was rigid, as shown in Figure 16.10 (Table 16.3).

Polyol (A)	JAGROPOL	400
Isocyanate (B)	EMPEYONATE	
Mixing ratio	A:B	100:115

Both open-cell and closed-cell foams are incorporated with fibres to maintain bio-based and improve noise reduction properties. Short fibres of 2 mm are made from luffa fibre mat and combined with flexible and RPFs with three different percentages like 5%, 10% and 15%. A total of eight samples of flexible and rigid foams will be prepared for testing experimental and numerical predictions, as shown in Figures 16.11 and 16.12.

FIGURE 16.10   Closed-cell PU foam incorporated with luffa fibre.

TABLE 16.3
Flexible and Rigid PU Foam with Different Luffa Fibre Percentages

Samples	Rigid Polyurethane Foam (wt.%)	Soft Polyurethane Foam (wt.%)	Luffa Fibre (wt.%)	Sample Thickness (mm)
100	100	100	0	40
95	95	95	5	40
90	90	90	10	40
85	85	85	15	40

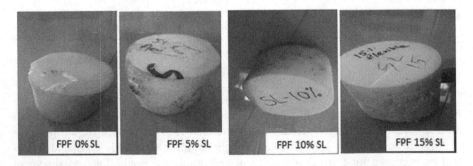

FIGURE 16.11   Flexible samples of PU foam with different luffa fibre percentages.

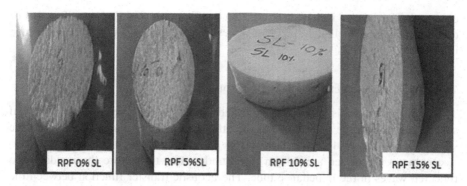

FIGURE 16.12 Rigid samples of PU foam with different luffa fibre percentages.

16.8 EXPERIMENTAL INVESTIGATION

16.8.1 IMPEDANCE TUBE THEORY

There are two methods of sound absorption properties: the reverberation room test and the impedance tube test. Through these two methods, engineers and scientists can determine the absorption properties of the material in the early stages of the acoustic environment. These two methods can evaluate absorption properties such as sound absorbance, reflectance and surface impedance. The main disadvantage of the reverberation room test is that it requires a lot of space, an experienced acoustician and large samples to derive properties [22–23].

Therefore, due to all these problems, the reverberation room test was not considered and mainly, the impedance tube test was used for evaluation. The impedance tube is a circular, hollow, rigid tube of varying dimensions concerning frequency. The low-frequency impedance tube had a diameter of 96 mm and a frequency of up to 4 kHz. The high-frequency impedance tube had a diameter of 34 mm and a frequency range of up to 6.3 kHz.

Two methods for acoustic impedance measurement with an impedance tube are presented. The first uses continuous white noise with two microphones, and the other uses transient acoustic excitation to excite a single microphone. Acoustic theory can be used to derive equations for the two methods [24]. The first method, i.e., the two-microphone method, is the most commonly used in this work. The acoustic sample is placed at one end of a tube, and a loudspeaker is mounted at the other end. The speaker creates sound, resulting in a forward-propagating sound wave. Some of the sounds are reflected, creating a backward-travelling sound wave [25]. The reflection coefficient is determined by measuring the impinging pressure on the reflected pressure. It can be interesting to find the absorption coefficient (fraction of energy dissipated by the porous material at total incidence) and the reflection coefficient R during the impedance measurement.

There are three types of sound field incidence angles: normal incidence, random incidence and oblique incidence. The absorption coefficient of the material increases up to a certain angle; beyond the rise, the drop in the absorption coefficient can be observed. The average incidence absorption coefficient is slightly lower than that of random incidence. Obliquely incident waves are used to measure average surface impedance.

16.8.2 IMPEDANCE EXPERIMENTAL PROCEDURE

The impedance tube is a straight, rigid and smooth speaker known as the sound source that creates the plane wave at the other end. At the other end is placed the sample of the required diameter for which the properties will be measured. Two microphones are placed away from the sound source to capture the incident and reflected waves in the impedance tube. The acoustic transfer function between the microphones is used to calculate the absorption coefficient, the reflection coefficient R and the test samples.

The frequency in impedance tube measurement depends on the diameter of the tube and the distance between the microphones' impedance ratio. To increase the frequency, change the diameter and spacing. The experimental procedure at normal incidence depends entirely on the reflection factor r. The reflection factor r is determined using the transfer function H12 between two microphone distances on the impedance tube near the sample.

The incident pressure P_i and reflected pressure P_r produced from the sound source are given below:

$$P_i = P_I\, e^{jk_0 x} \quad \text{incident plane wave,} \tag{16.1}$$

And

$$P_r = P_R\, e^{-jk_0 x} \quad \text{reflected plane wave,} \tag{16.2}$$

where P_I and P_R are the magnitudes of the P_i and P_r at the reference plane $(x=0)$, respectively, and $k_0 = k_0 - jk_0$ is a complex wavenumber.

The pressure at two microphones, P_1 and P_2, can be now defined by

$$P_1 = P_I\, e^{jk_0 x_1} + P_R\, e^{-jk_0 x_1} \tag{16.3}$$

and

$$P_2 = P_I\, e^{jk_0 x_2} + P_R\, e^{-jk_0 x_2}. \tag{16.4}$$

The transfer function at incident wave H_I is defined by

$$H_I = \frac{P2I}{P1I} = e^{-jk_0(x_1 - x_2)} = e^{-jk_0 S}, \tag{16.5}$$

where S is the separation between the two microphones $S = (x_1 - x_2)$.

Likewise, the transfer function at reflected wave H_R is defined by

$$H_R = \frac{P2R}{P1R} = e^{-jk_0(x_1 - x_2)} = e^{-jk_0 S}.$$ (16.6)

Therefore,

$$H_{12} = \frac{P2I}{P1I} = \frac{e^{\wedge}jk0x2 + re^{\wedge}-jk0x2}{e^{\wedge}jk0x1 + re^{\wedge}-jk0x1},$$ (16.7)

where $P_R = r*P_I$.

From H_{12} on rearranging, we get the reflection coefficient

$$r = \frac{H12 - HI}{HR - H12} e^{2jk_0 x_1}.$$ (16.8)

From reflection coefficients, the absorption coefficient is derived by the following relation:

$$\alpha = 1 - |r|^{\wedge}2.$$ (16.9)

16.8.3 EXPERIMENTAL PROCEDURE

The impedance tube is a straight hollow tube with a similar cross-section throughout the tube. The diameters of 35 and 96 mm are chosen according to the frequency-dependent diameter of the tube. The 35 mm diameter is used to calibrate for the high-frequency range up to 6.3 kHz, and the 96 mm diameter is used to calibrate for the low-frequency range up to 4 kHz. As a rule, the length of the tube should be three times greater than the diameter of the tube. The thickness of the hose is recommended as 5% of the diameter of the hose (Figure 16.13).

According to ISO standard, the material recommended for the tube is aluminium, which is used to avoid the vibrations caused due to frequency as well as to avoid the

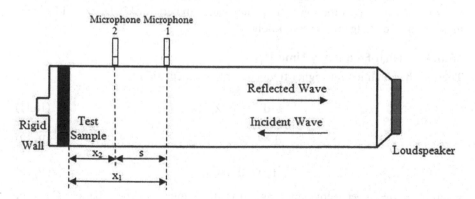

FIGURE 16.13 Schematic representation of impedance tube setup.

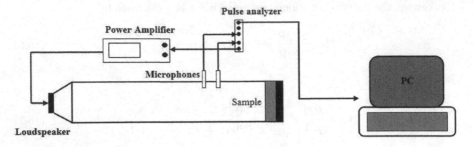

FIGURE 16.14 Schematic representation of the experimental setup.

cracks; there are no quench voids present on the aluminium tube due to its smooth surface. The microphones used in this project are selected according to the ISO standard 10534, and they are placed on the wall tightly to avoid the slightest escaping of pressure from the microphone walls. As per the literature reviews, microphones of length ½ or ¼ are generally referred in inches (Figure 16.14).

16.8.4 WORKING FREQUENCY RANGE

As mentioned above, the working frequency of the impedance tube is determined using the diameter of the tube. The lower frequency depends on the microphone spacing, like 150–300 Hz. The upper frequency range depends on the diameter of the tube sequence.

$$F_u < \frac{kc}{d},\tag{16.10}$$

where c is the speed of sound, d is the diameter of the tube and k is constant (0.586).

The lower frequency is considered because of better accuracy in signal processing equipment. The upper frequency is specified to maintain better plane wave generation without any occurrences of non-plane wave generation.

The working frequency range is $F_l < F < F_u$, where F is the operating frequency in hertz or kilohertz, F_l is the lower frequency range in hertz or kilohertz and F_u is the upper frequency range in hertz or kilohertz.

16.8.4.1 High-Frequency Limit [F_u]

There is the condition for higher frequency range calculation:

$$d < 0.58\, \lambda_u\tag{16.11}$$

$$d < 0.58\, C_0 / F_u\tag{16.12}$$

$$F_u d < 0.58 C_0,\tag{16.13}$$

where F_u is the upper frequency range, d is the diameter of the tube and C_0 is the speed of sound.

16.8.4.2   Low-Frequency Limit [F$_l$]

The condition for lower frequency range calculation [F$_l$]:

$$F_l > \frac{0.75(343)}{(1-d)}, \qquad (16.14)$$

where F$_l$ is the lower frequency range, l is the length of the tube, d is the diameter of the tube and 343 m/s is the speed of the sound.

The cut-off frequency of the impedance tube is calculated as follows:

$$F_l > \frac{0.75(343)}{(1-d)} \qquad (16.15)$$

$$F_l > \frac{0.75(343)}{(0.272-0.03499)}. \qquad (16.16)$$

Similarly

$$F_l > 796.5 \text{ Hz}. \qquad (16.17)$$

For condition F$_u$,

$$F_u < \frac{kc}{d} \qquad (16.18)$$

$$F_u < \frac{0.58*343}{d} \qquad (16.19)$$

$$F_u < \frac{0.58*343}{0.03499} \qquad (16.20)$$

$$F_u < 6.3 \text{ kHz}. \qquad (16.21)$$

Therefore, the frequency range of the impedance tube is calculated as

$$96.5 \text{ Hz} < F < 6.3 \text{ kHz}. \qquad (16.22)$$

The mic spacing S can also be calculated as follows:

$$S < 0.45 \, \lambda_0 \qquad (16.23)$$

$$S \ll \frac{Co}{2} F_u, \qquad (16.24)$$

where S is the microphone spacing, Co is the speed of sound and F$_u$ is the upper frequency.

16.9   DYNAMIC SIGNAL ANALYSIS

M+P analyser is an integrated method for finding noise and vibrations using dynamic signal measurement. It is used for analysis and advanced reporting of noise and vibration, acoustics and general dynamic signalling applications. Comprehensive time and frequency analyses are available with online and offline data processing. With the advanced application, M+P's dynamic signal analyser makes it easy to collect data, view results, perform analysis and create customised reports, socialised all within one user interface. The M+P analyser is designed for noise and vibration applications in the field and in the laboratory.

M+P dynamic analyser is used to examine acoustic properties with the 4/8 channel. The M+P analyser collects time history data for fast Fourier transform (FFT) spectra. Real-time and triggered data acquisition modes with a 50 Hz to 10,000 Hz sampling frequency are recorded. Further FFT block sizes of up to 256k in the higher frequency range were noted. The data collected by the M+Panalyser is in excel format which is further converted to a CSV file analyser. Later MATLAB is used to determine the absorption coefficient and surface impedance.

16.10   MATLAB

MATLAB is a tool used in this work to plot the absorption coefficient value graphically. The algorithm developed is based on a transfer function and a few inputs as variables to define the accurate value.

Step 1: Clear all

    Clc

    freq_low_limit = 750;

    freq_high_limit = 6000;

    Dataset = xlsread ('mp.csv', 'A3:C16386');

Step 1 defines the tube's low and high-frequency range, and the data from the m+p software are represented in an mp.csv file, which needs to be read by MATLAB.

Step 2: %Transfer function

    duration = 60; % duration of recording

    temp = 21; % room temperature

    ap = 101.325; % atmospheric pressure, kPa

    F_s = 12,600; % sampling rate

    rho = 1.21; % density of air (kg/m^3) normally 1.21

c = 343; % speed of sound in air (m/s) normal

Zair = rho*c; % characteristic impedance of air (kg/m²/s)

s = 0.0248; % microphone spacing (m)

Step 2: This steps defines the variables and all parameters which is required to run the procedure for finding absorption coefficient.

Step 3: % Coefficients

k = (2*pi*freq)/c;   (16.25)

R = (H12-exp(-1i.*k.*s)). /(exp(1i.*k.*s) -H12).
    *exp (2. *1i.*k.*x1)   (16.26)

Alpha = 1-abs(R). ^2; % Absorption coefficient   (16.27)

Zs = Zair. *((1+R). /(1-R)); % Surface impedance   (16.28)

Zs_n = ((1+R). /(1-R)); % Normalised Surface Impedance   (16.29)

Step 3 defines the required output from the data given. The output is given as absorption coefficient, surface impedance and normalised surface impedance.

16.10.1 EXPERIMENTAL RESULTS

The absorption coefficient was determined for the two different types of flexible and rigid foams. PU foam is a bio-based material with three different percentages of Luffa fibres incorporated into it. The sample thickness is 40 mm, and the diameter is 33 mm. The measurements are carried out in a frequency range from 500 Hz to 6.3 kHz. The plot of all samples is shown to understand the effects of different fibre percentages on the absorption coefficient.

16.10.2 OPEN-CELL POLYURETHANE FOAM

The open-cell PU foam is also known as a flexible, lower-density foam. In general, open-cell foams are used in many applications because of their absorbent properties. Due to its overall structure, the observed percentage of porosity is higher than the standard material. To improve the porosity of the structure in this study, the short fibres are used to penetrate the structure to enlarge the pores further. The sample size for the experimental research is 33 mm in diameter.

The absorption coefficient of the three samples of three different percentages of luffa with open-cell PU foam is represented in Figure 16.15. The percentage of luffa plays an essential role in improving the quality of absorption properties. Each sample has been calibrated a minimum of three times, and the average is considered as the final result. From Figure 16.16, it is clear that luffa fibre of 15% has a better absorption coefficient. The pore size will improve by adding the fibres, subsequently improving the alpha value with porous structure improvement.

FIGURE 16.15 M+P dynamic analyser.

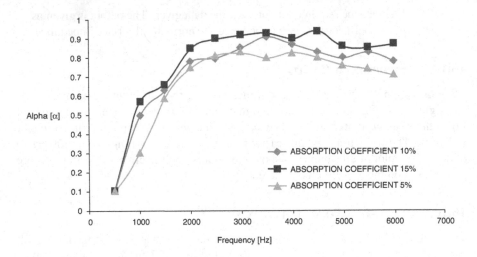

FIGURE 16.16 Absorption coefficient of flexible foam.

16.10.3 CLOSED-CELL POLYURETHANE FOAM

Closed-cell PU foam is used for applications where the structures require thermal insulation. This work paper on closed-cell foam aims to predict the effect of fibres in closed-cell foam compared to open-cell foam. The density of the closed-cell foam is greater than that of the open-cell foam. The change in properties with fibre is investigated experimentally.

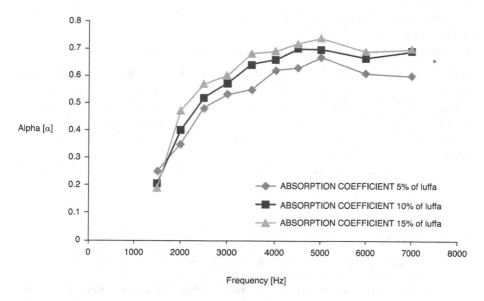

FIGURE 16.17   Absorption coefficient of rigid foam.

The absorption coefficient of the three samples with three different percentages of luffa with closed-cell PU foam is shown in Figure 16.17. The level of luffa plays a vital role in improving the quality of the absorption properties. Each sample is calibrated at least three times, and the average is considered the final result. It is clear from Figure 16.17 that luffa fibre of 15% has a better absorption coefficient. The effect of porous size will improve by adding fibres, which will improve the alpha value with porous structure improvement.

When comparing open-cell and closed-cell PU foams, it was observed that open-cell foam has the highest absorption coefficient than closed-cell foam, which is due to the physical properties of foams.

16.11   SUMMARY

16.11.1   EXPERIMENTAL OBSERVATIONS

16.12   CONCLUSION

From the experimental investigations, it is clear that the overall comparison between FPF with 15% of luffa fibres has an alpha value of 0.93. And the lowest value of alpha is 0.61 for the RPF having 5% of luffa fibre. The varying importance of porosity, tortuosity and flow resistivity also changes the value of the absorption coefficient value as well (Table 16.4).

TABLE 16.4
Experimental Results of the Sound Absorption Coefficient

Samples	FPF 15% Fibre	FPS 5% Fibre	RPF 15% Fibre	RPF 5% Fibre
Porosity (ε)	0.995	0.93	0.84	0.75
Tortuosity (τ)	1.009	1.151	1.258	1.342
Flow resistivity	9800	17,800	26,000	32,000
Frequency	4500	4500	4500	4500
Alpha	0.93	0.9	0.70	0.61

REFERENCES

[1] N. Atalla, R. Panneton. F.C. SGard, X. Olny. Acoustics absorption of macro perforated porous materials. *J. Sound Vib.* 243(4) (2001) 659–678.

[2] J.H. Park, S.H. Yang, H.R. Lee, C.B. Yu, S.Y. Pak, C.S. Oh, Y.J. Kang, J.R. Youn. Optimization of low-frequency sound absorption by cell size control and multiscale poroacoustics modelling. *J. Sound Vib.* 397 (2017) 17–30.

[3] J.G. Gwon, S.K. Kim, J.H. Kim. Sound absorption behaviour of flexible polyurethane foams with distinct cellular structures. *Mater. Des.* 89 (2016) 448–454.

[4] M.L. Pinto. Formulation, preparation, and characterization of polyurethane foams. *J. Chem. Educ.* 87 (2010) 212–215.

[5] B. Ekici, A. Kentli, H. Kucuk. Improving sound absorption property of polyurethane foams by adding tea-leaf fibers. *Arch. Acoust.* 37 (2012) 515–520.

[6] M.A. Ibrahim, R.W. Melik, Optimized sound absorption of rigid polyurethane foam, *Arch. Acoust.* 28 (2003) 305–312.

[7] A. Nilesson, Wave propagation in and sound transmission through sandwich plates, *J. Sound Vib.* 138 (1990) 73–94.

[8] J. Zaarek, Sound absorption in porous materials, *J. Sound Vib.* 61 (1978) 205–234.

[9] J. Lee, G.H. Kem, C.S. Hae, Sound absorption properties of polyurethane/nano-silica nanocomposite foams, *J. Appl. Polym. Sci.* 123 (2012) 2384–2390.

[10] T.J. Lu, F. Chen, D. He, Sound absorption of cellular metals with semi-open cells, *J. Acoust. Soc. Am.* 108 (2000) 1697–1709.

[11] T.J. Lu, A. Hes, M. Ashby, Sound absorption in metallic foams, *J. Appl. Phys.* 85 (1999) 7528–7539.

[12] X. Sagartzazu, L. HervelleaNieto, J.M. Pagalldday, Review in sound-absorbing materials, *Arch. Comput. Methods Eng.* 15 (2008) 311–342.

[13] M.T. Hoang, G. Bonnet, H. Tuan Luu, C. Perrot, Linear elastic properties derivation from micro structures representative of transport parameters, *J. Acoust. Soc. Am.* 135 (2014) 3172–3185.

[14] S. Fatima, A.R. Mohanty, Acoustical and fire-retardant properties of jute composite materials. *Appl. Acoust.* 72 (2011) 108–114.

[15] I. Curtu, M.D. Stanciu, C. Cosereanu, O. Vasile, Assessment of acoustic properties of biodegradable composite materials with textile inserts. *Mater. Plast.* 49 (2012) 68–72.

[16] H. Binici, M. Eken, M. Dolaz, O. Aksogan, M. Kara, An environmentally friendly thermal insulation material from sunflower stalk, textile waste and stubble fibres. *Constr. Build Mater.* 51 (2014) 24–33.

[17] C.W. Shan, M.I. Idris, M.I. Ghazali, Study of flexible polyurethane foams reinforced with coir fibres and tyre particles. *Int. J. Appl. Phys. Math.* 2(2) (2012) 123–130.
[18] Y. Li, H.F. Ren, A.J. Ragauskas, Rigid polyurethane foam reinforced with cellulose whiskers: Synthesis and characterisation. *Nano-Micro Lett.* 2(2) (2010) 89–94.
[19] B. Ekici, A. Kentli, H. Kucuk, Improving sound absorption property of polyurethane foams by adding tea-leaf fibers. *Arch. Acoust.* 37(4) (2012) 515–520.
[20] M. Nar, C. Webber III, N.A. D'Souza, Rigid polyurethane and kenaf core composite foams. *Polym. Eng. Sci.* 55 (2015) 132–144.
[21] J. McIntyre, *Synthetic Fibres: Nylon, Polyurethane, Acrylic and Polyolefin.* Cambridge, UK: Woodhead Publishing Limited; 2004.
[22] R. Sinclair, *Textiles and Fashion. Material Design and Technology.* Wood head publishing services in Textiles, London; 2015, 3–27.
[23] M.N.N. Hassan, A.Z.M. Rus, The acoustic performance of green polymer foam from renewable resources after UV exposure. *Int. J. Automot. Mech. Eng.* 9 (2014) 1639–1648.
[24] Standard SR EN ISO 10534-2, Determination of sound absorption coefficient and acoustic impedance with the interferometer. Part 2. Transfer function method, 2002.
[25] A. Borlea (Tiuc), T. Rusu, S. Ionescu, Research on obtaining soundproof materials from wastes. *Acta Tech. Napoc., Ser. Environ. Eng. Sustain. Dev. Entrepreneurship* 1(3) (2012) 13–20.

Index

Note: **Bold** page numbers refer to tables and *italic* page numbers refer to figures.

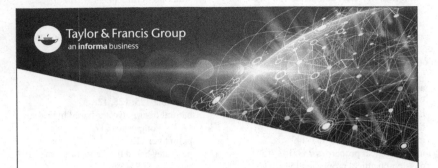

Printed in the United States
by Baker & Taylor Publisher Services